Membrane Technology for Water and Wastewater Treatment in Rural Regions

Rosalam Sarbatly
Universiti Malaysia Sabah, Malaysia

IGI Global
DISSEMINATOR OF KNOWLEDGE

A volume in the Practice, Progress, and Proficiency in Sustainability (PPPS) Book Series

Published in the United States of America by
IGI Global
Engineering Science Reference (an imprint of IGI Global)
701 E. Chocolate Avenue
Hershey PA, USA 17033
Tel: 717-533-8845
Fax: 717-533-8661
E-mail: cust@igi-global.com
Web site: http://www.igi-global.com

Library of Congress Cataloging-in-Publication Data

Names: Sarbatly, Rosalam Hj., author.
Title: Membrane technology for water and wastewater treatment in rural regions / by Rosalam Sarbatly.
Description: Hershey, PA : Engineering Science Reference, 2020. | Includes bibliographical references and index. | Summary: "This book explores the fundamentals of membrane process, membrane fabrication and characterization, and membrane processes and operation. It also examines the application of membrane technologies in water and wastewater treatment in rural regions"-- Provided by publisher.
Identifiers: LCCN 2019042063 (print) | LCCN 2019042064 (ebook) | ISBN 9781799826453 (h/c) | ISBN 9781799826460 (s/c) | ISBN 9781799826477 (eISBN)
Subjects: LCSH: Water--Purification--Membrane filtration--Malaysia--Sabah Region. | Sewage--Purification--Reverse osmosis process--Malaysia--Sabah Region. | Water-supply, Rural--Malaysia--Sabah Region.
Classification: LCC TD313.M32 S37 2020 (print) | LCC TD313.M32 (ebook) | DDC 628.1/64--dc23
LC record available at https://lccn.loc.gov/2019042063
LC ebook record available at https://lccn.loc.gov/2019042064

This book is published in the IGI Global book series Practice, Progress, and Proficiency in Sustainability (PPPS) (ISSN: 2330-3271; eISSN: 2330-328X)

British Cataloguing in Publication Data
A Cataloguing in Publication record for this book is available from the British Library.

Practice, Progress, and Proficiency in Sustainability (PPPS) Book Series

Ayman Batisha
International Sustainability Institute, Egypt

ISSN:2330-3271
EISSN:2330-328X

Mission

In a world where traditional business practices are reconsidered and economic activity is performed in a global context, new areas of economic developments are recognized as the key enablers of wealth and income production. This knowledge of information technologies provides infrastructures, systems, and services towards sustainable development.

The **Practices, Progress, and Proficiency in Sustainability (PPPS) Book Series** focuses on the local and global challenges, business opportunities, and societal needs surrounding international collaboration and sustainable development of technology. This series brings together academics, researchers, entrepreneurs, policy makers and government officers aiming to contribute to the progress and proficiency in sustainability.

Coverage

- ICT and knowledge for development
- Strategic Management of IT
- Eco-Innovation
- Knowledge clusters
- Sustainable Development
- Innovation Networks
- Global Content and Knowledge Repositories
- Green Technology
- Outsourcing
- Global Business

IGI Global is currently accepting manuscripts for publication within this series. To submit a proposal for a volume in this series, please contact our Acquisition Editors at acquisitions@igi-global.com or visit: https://www.igi-global.com/publish/.

The Practice, Progress, and Proficiency in Sustainability (PPPS) Book Series (ISSN 2330-3271) is published by IGI Global, 701 E. Chocolate Avenue, Hershey, PA 17033-1240, USA, www.igi-global.com. This series is composed of titles available for purchase individually; each title is edited to be contextually exclusive from any other title within the series. For pricing and ordering information please visit https://www.igi-global.com/book-series/practice-progress-proficiency-sustainability/73810. Postmaster: Send all address changes to above address.

Titles in this Series

For a list of additional titles in this series, please visit: www.igi-global.com/book-series

Smart Materials for Environmental and Industrial Aplications
Khaled Salah Abou-El-Sherbini (National Research Centre, Egypt) Nasrallah Mohamed Deraz (National Research Centre, Egypt) and Alia Hosni Salama (National Research Center, Egypt)
Engineering Science Reference • © 2020 • 300pp • H/C (ISBN: 9781799829317) • US $215.00

Handbook of Research on Health and Environmental Benefits of Came Products
Omar Amin Alhaj (University of Petra, Jordan) Bernard Faye (Independent Researcher, France) and Rajendra Prasad Agrawal (Independent Researcher, India)
Medical Information Science Reference • © 2020 • 480pp • H/C (ISBN: 9781799816041) • US $285.00

Interdisciplinary Approaches to Public Policy and Sustainability
Rituparna Das (Adamas University, India) and Nivedita Mandal (Adamas University, India)
Information Science Reference • © 2020 • 320pp • H/C (ISBN: 9781799803157) • US $195.00

African Perspectives on Reshaping Rural Development
Mavhungu Abel Mafukata (Department of Development Studies, University of South Africa, South Africa) and Khathutshelo Alfred Tshikolomo (Limpopo Department of Agriculture and Rural Development, South Africa)
Information Science Reference • © 2020 • 300pp • H/C (ISBN: 9781799823063) • US $195.00

Implications of Mobility as a Service (MaaS) in Urban and Rural Environments Emerging Research and Opportunities
António Manuel Amaral (University of Minho, Portugal) Luís Barreto (Applied Research Center for Digital Transformation (ARC4DigiT), Escola Superior de Ciências Empresariais, Instituto Politécnico Viana do Castelo, Portugal & Instituto de Telecomunicações (IT), Universidade de Aveiro, Portugal) Sara Baltazar (Escola Superior de Ciências Empresariais (ESCE), Instituto Politécnico Viana do Castelo (IPVC), Portugal & Research Center for Systems and Technologies (SYSTEC), University of Porto, Portugal & Institute of Systems and Robotics (ISR), University of Coimbra, Portugal) João Pedro Silva (Escola Superior de Tecnologia e Gestão, Politécnico de Leiria, Portugal) and Luísa Gonçalves (Câmara Municipal de Leiria, Portugal & Escola Superior de Tecnologia e Gestão, Instituto Politécnico de Leiria, Portugal & INESCC, NOVA IMS, Universidade NOVA de Lisboa, Portugal)
Engineering Science Reference • © 2020 • 293pp • H/C (ISBN: 9781799816140) • US $150.00

Social, Economic, and Environmental Impacts Between Sustainable Financial Systems and Financial Markets
Magdalena Ziolo (University of Szczecin, Poland)
Business Science Reference • © 2020 • 383pp • H/C (ISBN: 9781799810339) • US $245.00

IGI Global
DISSEMINATOR OF KNOWLEDGE

701 East Chocolate Avenue, Hershey, PA 17033, USA
Tel: 717-533-8845 x100 • Fax: 717-533-8661
E-Mail: cust@igi-global.com • www.igi-global.com

Table of Contents

Section 3

Preface

Clean water scarcity and water pollution are the two prevalent issues that commonly occur in many developing countries. Most municipal water and wastewater industries in third world countries are unable to provide sufficient and clean water for drinking and sanitation to all of its residents, especially those who live in rural areas. Correspondingly, they experience poor water quality due to inherent pollutions and inadequate access to these treatment facilities, leading to many water-borne diseases that affected both the public health and environment. To tackle these perpetual problems, there is a need to extend the discussion on the prominent solutions that can be effectively applied to the shortcomings experienced in the rural regions. In particular, this book is written to deal with the issues of water and wastewater treatment by focusing on the utilization of membrane technology.

This book mainly discusses the application of membrane technology for water and wastewater treatment in rural regions. Special interests are given to the rural regions situated in the tropical zone, as there are many developing countries in the tropics experiencing similar issues of water and wastewater. Sabah, Malaysia is chosen as the main example for case studies discussed in this book. Sabah is a state in Malaysia that has a population of 3.9 million censuses in 2018 (Department of Statistic Malaysia, 2019). Being the poorest and least developed state in Malaysia, Sabah has a lot of rural districts that experience the lowest water supply coverage and sanitation. The prospect of the environmental future of Malaysia, particularly in Sabah is highly dependant on the waste management and treatment applied, along with the public and authority awareness. Severe consequences on environmental quality deterioration are highly potential if the current trend of waste management is used without any improvement in the technologies utilized.

With the accelerated increase in the number of population, growth in demand for life essentials such as food and energy will follow. Rapid development of the industrial sector will occur in order to cope with the demand. Consequently, a larger amount of wastewater will be produced. Accordingly, rapid developments of other sectors will lead to smaller land available to be utilized for wastewater treatment facilities development. Therefore, technologies with higher efficiency and smaller physical and environmental footprint are required to ensure Sabah's environmental quality is not compromised.

Over the past decades, treatment of wastewater has been conducted by using conventional technologies such as biological treatment and physical/chemical treatment. These technologies are proven to have some complication and low treatment efficiency. Optimistic discoveries have been made on new technologies with higher efficiency, low cost and lower footprint. Many of these technologies can be applied in a stand-alone system or coupled together with the conventional or as a new wastewater treatment system.

One of the technologies that have recently gained interests in the water and wastewater treatment field is membrane technology. Membrane technology has been around for many years and has been utilized in dairy production, juice production, drinking water production, and biotechnology. Application of membrane technology for water and wastewater treatment is also highly potential. Combining the membrane technology with the conventional biological treatment, membrane bioreactor has the ability to produce effluent that is high in quality with shorter time requirement. Modifications and improvements have been made to this technology such as by using the anaerobic membrane bioreactor. This system requires less energy due to the absence of aeration to operate and can compensate its energy usage by producing methane during its operation.

There are many industrial applications of water and wastewater treatments that suit the operation of membrane technology. However, the utilization of membrane technologies in rural regions are exceptionally low, as most industrial processes are still using traditional methods due to non-readiness to migrate into newer or more advanced technologies. This brings the discussion up to a higher level as there is a critical need of introducing the superiority of membrane technologies over the conventional treatments, owing to their excellent properties for various applications.

Several imperative applications are chosen in the discussion, whereby emphasize are given on clean water production from the surface and underground water in rural areas, brackish water desalination to produce drinking water, palm oil mill industry wastewater treatment, sewage treatment, and microalgae application for wastewater treatment. The processes and technologies discussed are selected by taking into account their suitability based on the demographic and geographic considerations of the rural regions.

In this book, the author's discussion on membrane technology for water and wastewater treatment in rural regions span over several purposeful objectives which are:

1. To introduce the fundamentals of the membrane process, mass transfer phenomenon, membrane fabrication and characterization, and membrane processes and operation;
2. To describe the main sources of water and wastewater, their conventional physical treatment, physicochemical treatment, and biological treatment processes.
3. To discuss a few selected, promising aspects of membrane usage in water treatment for the rural regions, brackish water treatment for isolated areas, wastewater from palm oil mills, sewage management and treatment, and microalgae and membrane treatment system.

This book will provide essential teaching materials for courses related to the subject areas (e.g. membrane technology, water treatment, wastewater treatment). It covers extensive learning aspects on both fundamental knowledge and rural application so that the readers can easily relate to the discussions. Prior discussing these specific applications, the readers are equipped with the core theoretical concepts pertaining the core subject areas which include the basic fundamentals, membrane fabrication, membrane characterization, membrane processes, source of water and wastewater in Sabah, conventional treatments of water and wastewater as well as the potential membrane technological application in these treatment processes. The readers would also be introduced with the current scenarios of water and wastewater, important policies and standards governing the two subjects, and the traditional treatment used along with proposed treatment incorporating membrane technology.

This book will be an excellent guide not only for graduate students and professors who are engaged in this particular field but also for anyone who is interested in the subjects concerned. By being an excellent reference for membrane technology as an emerging application for water and wastewater treatment in Sabah, it can be used by engineers, water and wastewater companies, and also to policymakers who are interested in this topic. Hence, it is hoped that the book will provide the path needed for the expansion of knowledge and establishment of local and global alliances in this field, from classrooms to industries that could benefit the society as a whole.

OVERVIEW OF THE BOOK

The book is divided into three main sections which discuss the key ideas of this book. Section 1 comprised of Chapter 1 to Chapter 5 that describes the central concepts in membrane technology. The discussions span from the fundamentals, mass transfer phenomenon, membrane fabrication, membrane characterization, and finally the membrane processes. In Section 2, the application of membrane technology for water treatment in rural regions is focused in Chapter 6 until Chapter 8. The conventional water treatment systems are initially examined before incorporating membrane technology into the systems. Technically, the systems are planned to tackle the constraints faced in rural applications. Each rural regions are unique with its intrinsic conditions, essentially in terms of populations, geographical settings, and availability of water resources, which are the important aspects to be considered for the design of water treatment systems. Subsequently, nanofibre is assessed as a promising membrane material for membrane distillation, especially for desalination of brackish water. Section 3 discusses on the topic of wastewater treatment specifically on the conventional wastewater technologies, membrane technology applications to treat sewage and palm oil mill effluent (POME), and the application of microalgae membrane bioreactor for wastewater treatment which are reviewed in Chapter 9 to 12. Wastewater treatments are prevalent in developing countries as it involves the sanitary and well-being of the people. Focus is also given to the industrial wastewater such as POME, as oil palm is a commodity plant that is commercially grown and produces a high volume of wastewater in many tropical countries in the developing region of Asia.

The first chapter of the book describes the critical aspects of membrane and membrane processes with a focus on the fundamental concepts of the membrane as a separation technology. The chapter covers the key examples of the membrane process parameters, membrane structure, and the membrane structural parameters. On the whole, the opening chapter begins as an introductory discussion to present the readers with the basics of membrane and membrane processes.

The book then presents the phenomenon of mass transfer in membrane processes. Membrane processes involved in water and wastewater treatments are generally driven by mechanical pressure, partial pressure, osmotic pressure, concentration, or thermal gradients. Chapter 2 also reviews the transport resistances resulting from the nature of the process and external constraints such as the concentration polarization and membrane fouling. The heat transfer phenomenon is analysed particularly on the membrane process driven by thermal gradient, specifically on the membrane distillation.

Chapter 3 studies the membrane fabrication techniques. The fabrication techniques typically differ depending on the material used to construct the membrane, such as a polymer, ceramic, and metallic. Several fabrication methods are mentioned for each class of material. The chapter also introduces the different types of modules for which the membranes can be assembled. This includes the two main categories which are flat membrane module and tubular membrane module.

In Chapter 4, the methods of membrane characterization are explored briefly on the structure, morphology, chemical and physical properties, and transport properties of the membrane. The characterization techniques are generally categorized into transport/flow, electron microscopy, scattering, and spectroscopy methods.

Chapter 5 gives an account on the important membrane processes used in pressure-driven membrane processes such as the microfiltration, ultrafiltration, and reverse osmosis. A brief introduction is also provided on some advanced membrane processes like the pressure-retarded osmosis, membrane distillation, osmotic membrane distillation, and pervaporation. This chapter also presents a detailed account of the membrane bioreactor as a promising technology for water and wastewater treatment. Important aspects of membrane bioreactor are the membrane material and module, system configuration, the principle of operation, critical flux operation, and fouling control and prevention.

The next chapter reviews the conventional water treatment systems which are widely used in water treatment plants in most regions. This system uses the physicochemical treatment process which combines the physical and chemical treatments to produce potable water. Example of several water treatment plants in Sabah is provided as well as reviewing the issues commonly encountered in the conventional water treatment plants.

Chapter 7 continues the discussion on small scale water treatment that is suitable for rural regions application especially in the isolated areas with geographical constraints and scatters of population. Different design of the water treatment system is presented for each type of water resources discussed, which includes the surface water, groundwater, rainwater, and seawater. The small scale water treatment systems are suitable for isolated areas due to its simplicity and modularity which makes construction, operation, and maintenance to be less challenging.

Membrane distillation is a promising technology for rural regions with limited freshwater availability but abundant supply to seawater. This technology is widely studied for the desalination process to produce potable water. Chapter 8 looks at this technology's usability especially in small islands and offshore that have little to no access to freshwater. Membrane distillation can efficiently produce clean water from seawater, provided that a special membrane that exhibits distinct characteristics to separate salt from water is used. The nanofibre membrane is highlighted as one of the excellent options of the membrane as it can be fabricated to exhibit hydrophobic, thin, and open pores characteristics.

Chapter 9 deals with conventional wastewater treatments that are extensively applied in most wastewater treatment plants. Many types of wastewater are discussed on the characteristics, treatment, and management status, for instances the oil and gas wastewater, urban wastewater, nonpoint source pollution, wastewater derived from floating residential etc. The technologies vary from chemical, biological, physicochemical, and mechanical processes to remove organic loading, solids and nutrient contents from wastewater.

Chapter 10 specifically concentrate on sewage management and treatment. Accumulation of sewage is a global issue that occurs primarily due to the rising rate of population growth. Sewage management is crucial to prevent threats to human health and the environment that can arise from improperly managed sewage. Some technologies used in the sewage treatment is reviewed with membrane technology highlighted as a potential to be used in rural regions.

The eleventh chapter talks about the treatment of Palm Oil Mill Effluent (POME) that is currently produced in large amount by the palm oil mills. In Sabah, the POME is still treated using traditional technologies that are outdated and have several drawbacks making it a less sustainable approach. Most palm oil mill in Sabah is using open ponding system to treat POME, even though it requires large land area, long treatment duration, and low efficiency. Discharge of poorly treated POME directly to the surrounding leads to detrimental impacts on the environment, as it contains high contents of oil and BOD. This chapter discusses the treatment method of POME from a newer perspective of membrane technology integrated into the biological treatment. Various methods such as clarification system with nanofibre unit, cooling system, activated sludge system, and anaerobic membrane bioreactor is reviewed on their principles of operation.

In the final chapter, the application of microalgae-based wastewater treatment is introduced to the potential of microalgae membrane bioreactors for wastewater treatment, microalgae cultivation, and harvesting. Microalgae membrane bioreactors combine the biological treatment of microalgae with the conventional membrane bioreactor. Furthermore, several other technologies of immobilized microalgae are introduced which can potentially reduce the membrane fouling occurrence and concurrently remove the need for microalgae harvesting process.

Acknowledgment

It is my great honor to work for over 22 years as an academician at the Universiti Malaysia Sabah, a public-funded university that provides me supports to complete my textbook writing of Membrane Technology for Water and Wastewater Treatment in Rural Regions. The honor also go to the State Government of Sabah for providing me access to some information required to complete this book. It has indeed been a wonderful experience to work with the world-leading book publisher, the IGI Global. The heartiest gratitude to my wife, Maslipah Idris and three of my children Fatin Arisa, Fadlin Zaydan, and Fahim Wafi for their great support. I extend the gratitude to Jamilah Sariau, a hard-working research assistant, for the technical support in preparing contents, figures, tables and schemes of this book. Furthermore, I would like to acknowledge the following persons for contributing in different ways; ideas, comments, supports, and inputs: Professor Dr Duduku Krishnaiah, Professor Dr Awang Bono (UMS), Mr Mitsuhiro Takahashi (Yokohama-shi, Japan), Professor Dato' Dr Ahmad Fauzi Ismail (UTM), Professor Dr Abdul Latiff Ahmad (USM), Professor Dato' Ir Dr Abdul Wahab Muhammad (UKM), Professor Ir Dr Mohd Sobri Takriff (UKM), Professor Dr Pogaku Ravindra (University of Tennessee, USA), Professor Datin Dr Azlina Bt Harum Kamarudin (USM), Professor Ir. Dr. Jailani Salihon (UiTM), Associate Professor Dr Abu Zahrim Yasser (UMS), Dr Chiam Chel Ken (UMS), Dr Emma Suali (UMS), Dr Suffian Misaran Misran (UMS), Associate Professor Dr. Nurmin Bolong (UMS), Dr. Maizurah Ismail (UMS), Farhana Abdul Lahin (UMS), Zykamilia Kamin (UMS) and not limited to others. And to Emeritus Professor John Howell and Dr Richard England from the University of Bath UK, thank you for allowing me to pursue a PhD in membrane technology which allows me to become the expert in this field.

Section 1

Chapter 1
Fundamentals of Membranes and Membrane Processes

ABSTRACT

Research into the wide possibilities of membrane-based applications is an interesting subject for the modern study of membrane science and technology. Membrane processes have been established as viable and recognized separation techniques in water and wastewater treatment processes. Membranes can be prepared into many forms, each with its intrinsic properties which ultimately determine its suitability for specific applications as well as the overall performance of the process. Thus, this chapter highlights the fundamental concepts of membranes and membrane processes. The critical parameters in membrane processes, and membranes' structural characteristics and parameters are reviewed.

INTRODUCTION

Separation, purification, and concentration are some of the major unit operations that are highly important in many industrial applications. One of the major challenges in the separation of molecular mixtures is to obtain products with high recovery and high quality. Over the last couple of centuries, the development in these operations has been snowballing with multi-scale techniques being established such as the distillation, centrifugation, adsorption, extraction, crystallization, etc. which is continuously progressed to overcome this challenge. The more recent development in these processes has been in the membrane which acts as a semipermeable barrier that allows the separation of components.

Since membranes have variation in their properties, structures, and usages, a comprehensive definition of the membrane which relates all of its aspects is challenging. In the general context of this book, the membrane is a synthetic film which acts as a selective barrier between two phases and controls the transport of various chemical components through it in a specific manner. A membrane may be porous or nonporous, symmetric or asymmetric, thin or thick, homogeneous or heterogeneous, solid or liquid, or neutral or carries positive or negative charges. The mass transport across a membrane may be due

DOI: 10.4018/978-1-7998-2645-3.ch001

Figure 1. The basic of membrane and membrane process

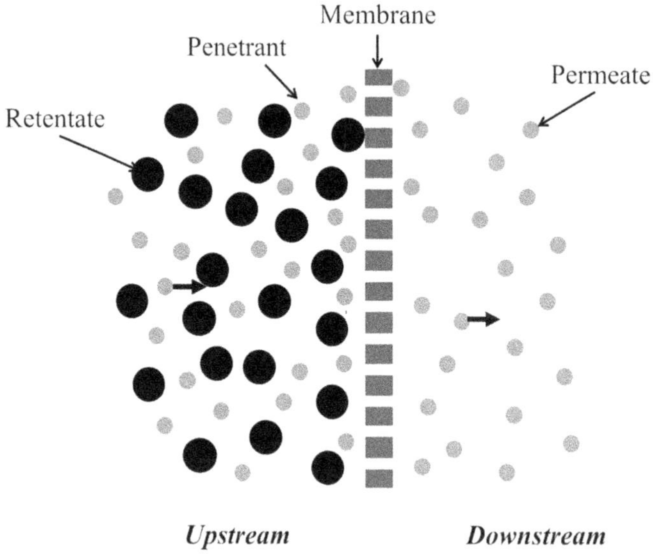

to convective flow or diffusion of individual molecules. Membrane processes comprised of different techniques which are mainly classified by the membrane pore sizes and the separation driving force. The membrane thickness can range from less than 100 nm to more than a centimetre, and the membrane pore diameter can be less than 0.1 nm to more than several micrometres. The membrane processes can be driven by the force gradients in terms of applied pressure, osmotic pressure, concentration, electrical, or thermal, or a combination of the driving forces. Although membrane can be described in a wide range of ways, a membrane can fit into more than one of the above-mentioned classes. For example, a membrane can be porous, asymmetric, and carries electrical charges.

The selection of suitable membrane or membrane process depends on some factors such as the characteristics of the mixture feed, degree of separation intended, and the volume of feed needs to be handled. Each membrane is unique, as there are no two membranes that are completely identical; each one will have differences to an extent in terms of the membrane structure and basic characteristics such as the porosity, pore sizes and pore size distributions, surface roughness, tortuosity, and thickness, even if they are made from the same type of material and fabrication method. These characteristics are important to classify the different types of membranes which are suitable for different applications. Furthermore, the characteristics of a membrane influence the parameters obtained such as the membrane flux, the membrane permeability, the achievable retention and recovery rate, and consequently the overall performance.

Membranes and membrane processes were first introduced as analytical tools in laboratory scale and have since been developed rapidly into industrial-scale utilization which shows major technical and profitable advancements. In essence, Figure 1 introduces the basic of membrane and membrane process, showing the components and separation mechanism involved. Many attentions have been shifted towards membrane with extensive studies being performed on membrane and membrane processes as promising tools due to its many advantages over the other conventional techniques. The main advantage of membrane processes that distinguish them from other conventional separation processes is the ability to perform separation via physical means at ambient temperature without any alteration on the components, physically and chemically. These attributes are especially important to the food, pharmaceutical and bio-industries. Moreover, the extent of separation up to the molecular scale attainable by membrane implies that a very high degree of separation can be achieved by membrane processes. Most membrane processes do not involve the need of phase change of the fluid feed (with the exception of the pervaporation process), and thus low energy requirements are common. However, in some membrane processes, high energy is required to provide high-pressure force in order to drive the permeating species through the membrane such in the case of the reverse osmosis process. Membrane processes also require no moving parts and consist of simple configurations that are easy to assemble, easy to operate, and require less maintenance. The membrane can be fabricated from a wide range of polymer and inorganic materials to possess high selectivity of components, requires a small footprint, and are potentially better for the environment since no harmful materials are used to produce membranes.

Nonetheless, membrane and membrane processes do have some limitations in several areas. Membrane processes hardly produce two pure products streams, as one of the 2 streams (i.e. the downstream and upstream) will still contain contaminants of the other components especially in the upstream. This shows that the separation between the mixture of two components is not 100% achievable. In some cases, the permeate obtained is not 100% pure, as the membrane selectivity is also not absolute leaving some materials to still be able to infiltrate. Besides, membrane processes are not easily staged to obtain a product that is almost pure as in the case of membrane distillation. Membrane processes typically consist of only one and sometimes two to three stages. This can be compensated by utilizing a membrane with high selectivity in order to obtain a product with high purity. The trade-off comparison between membrane processes and other conventional separation is usually high selectivity/low number of stages against low selectivity/a large number of stages respectively. The membrane also has limitations in some chemical processes and high temperature. Since most synthetic membranes are made from polymer-based, they are not suitable in extreme chemical conditions which contain a high concentration of organic compounds that can interact with the membrane (dissolving, swelling, shortening lifespan, lowering selectivity etc.). Also, the high temperature will affect the physical integrity of polymeric membranes. Membrane processes also experience fouling phenomenon if the optimum conditions are not met which can greatly lower the permeation flux and thus lowers the recovery rate. Fouling is usually encountered when the feed quality is too polluted with a large number of contaminants or when the applied pressure exceeds the allowable operating conditions.

The study of membranes and membrane processes are very extensive and it is not without any challenges. This is especially true for membrane design and manufacturing in order to perfect the application in industries. Some challenges that still remain in membrane field are producing a membrane that is thin but able to withstand high-pressure differences in the case of reverse osmosis. This is because the thin membrane is desirable, as the membrane flux increases when its thickness decreases. Moreover, the main challenges in the fabrication sector lie on producing a membrane with exceptional selectivity, thermally stable, and uniform pore sizes. In membrane operations, membrane fouling and cleaning are major constraints that need to be resolved.

The context of this book mainly covers the membrane technology for water and wastewater treatment. Membrane technology in the perspective of this book refers to the key technologies that utilize membrane as the central core of the processes. Water and wastewater are two areas of concern globally which are highlighted in this book. Treatment of water is important especially to produce drinking water, while wastewater needs to be treated as they present hazards to the environment if left untreated. Besides, treated wastewater can be reused for many other purposes. The rural regions become the subjects of interest because these regions are far away from urban environments which benefit from relatively high water quality as a result of advanced technologies. Thus, in particular, this book addresses several applications of membrane technologies for water and wastewater treatments that are suited for rural environments. In this introductory chapter, the fundamentals of membranes and membrane processes are covered on the basic terminologies, the critical parameters, and characteristics of membrane structures.

TERMINOLOGIES AND CONCEPTS IN MEMBRANE PROCESS

The International Union of Pure and Applied Chemistry (IUPAC) on 1996 has assembled a basic set of terms which include the key vocabulary used in the literature concerned with scientific, technical and commercial aspects of the membrane field. Some selected general terms and definitions of membranes and membrane processes recommended by the IUPAC are listed as follows:

1. **Asymmetric Membrane:** Membrane constituted of two or more structural planes of non-identical morphologies.
2. **Cake Layer:** Layer comprised of rejected particulate materials residing on the upstream face of a membrane.
3. **Co-current Flow:** Flow pattern through a membrane module in which the fluids on the upstream and the downstream sides of the membrane move parallel to the membrane surface and in the same directions.
4. **Completely-mixed (Perfectly-mixed) Flow:** Flow through a membrane module in which fluids on both the upstream and downstream sides of the membrane are individually well-mixed.
5. **Composite Membrane:** Membrane having chemically or structurally distinct layers.
6. **Counter-current Flow:** Flow through a membrane module in which the fluids on the upstream and downstream sides of the membrane move parallel to the membrane surface but in opposite directions.
7. **Cross Flow:** Flow through a membrane module in which the fluid on the upstream side of the membrane moves parallel to the membrane surface and the fluid on the downstream side of the membrane moves away from the membrane in the direction normal to the membrane surface.
8. **Dead-end Flow:** Flow through a membrane module in which the only outlet for upstream fluid is through the membrane.
9. **Dense (Non-porous) Membrane:** Membrane with no detectable pores.
10. **Downstream:** Side of a membrane from which permeate emerges.
11. **Fouling:** Process resulting in loss of performance of a membrane due to the deposition of suspended or dissolved substances on its external surfaces, at its pore openings, or within its pores.
12. **Homogeneous Membrane:** Membrane with essentially the same structural and transport properties throughout its thickness.
13. **Membrane Distillation:** Distillation process in which the liquid and gas phases are separated by a porous membrane, the pores of which are not wetted by the liquid phase.
14. **Membrane Module (Cell):** Manifold assembly containing a membrane or membranes to separate the streams of feed, permeate, and the retentate.
15. **Membrane:** Structure, having lateral dimensions much greater than its thickness, through which mass transfer may occur under a variety of driving forces.

16. **Microfiltration:** Pressure-driven membrane-based separation process in which particles and dissolved macromolecules larger than 0.1 μm are rejected.
17. **Molecular-weight Cut-off:** Molecular weight of a solute corresponding to a 90% rejection coefficient for a given membrane.
18. **Nanofiltration:** Pressure-driven membrane-based separation process in which particles and dissolved molecules smaller than about 2 nm are rejected.
19. **Penetrant (Permeant):** Entity from a phase in contact with one of the membrane surfaces that pass through the membrane.
20. **Permeate:** Stream containing penetrants that leave a membrane module.
21. **Pervaporation:** Membrane-based process in which the feed and retentate streams are both liquid phases while permeant emerges at the downstream face of the membrane as a vapour.
22. **Potable Water:** Term used to indicate water having a total dissolved solids content of less than 500 ppm with a sufficiently low level of biological agents, suspended solids, organic odour and colour-generating components to be safe and palatable for drinking.
23. **Retentate (Raffinate):** Stream that has been depleted of penetrants that leaves the membrane modules without passing through the membrane to the downstream.
24. **Reverse Osmosis:** Liquid-phase pressure-driven separation process in which applied transmembrane pressure causes selective movement of solvent against its osmotic pressure difference.
25. **Synthetic (Artificial) Membrane:** Membrane formed by a process not occurring in nature.
26. **Ultrafiltration:** Pressure-driven membrane-based separation process in which particles and dissolved macromolecules smaller than 0.1 μm and larger than about 2 nm are rejected.
27. **Upstream:** Side of a membrane into which penetrants enter from the feed stream.

DIMENSIONS AND UNITS

A dimension is a measurable extent or quantity of a particular kind such as length, depth, height, mass, weight, volume or time. The dimensions are primitives and not definable in any simpler forms. Every dimension has a specific unit of size. The International System of Units (abbreviated SI, for Système International) has codified the primary units. Multiples and decimal fractions of SI units are designated by prefixes. Those in common use are listed in Table 1.

MEMBRANE PROCESS PARAMETERS

Flux

Flux can be calculated in many ways depending on the membrane process and its driving force. In general, flux is defined as the number of moles, volume or mass of a specified component *i* passing per unit time through a unit of membrane surface area normal to the thickness direction.

$$J_i = \frac{m_i}{A_m t} \tag{1}$$

Table 1. Prefixes for SI units

Multiple	Prefix	Symbol	Multiple	Prefix	Symbol
10^{-24}	yocto	y	10^{1}	deca	da
10^{-21}	zepto	z	10^{2}	hecto	n
10^{-18}	atto	a	10^{3}	kilo	k
10^{-15}	femto	f	10^{6}	mega	M
10^{-12}	pico	p	10^{9}	giga	G
10^{-9}	nano	n	10^{12}	tera	T
10^{-6}	micro	μ	10^{15}	peta	P
10^{-3}	milli	m	10^{18}	exa	E
10^{-2}	centi	c	10^{21}	zetta	Z
10^{-1}	deci	d	10^{24}	yotta	Y

where J_i and m_i are the flux and the mass of component *i* respectively. A_{m} is the membrane surface area and t is the time. The units of flux commonly used are [kmol/m^2 s], [kg/m^2 s], [m^3/m^2 s] and [m^3 (measured at standard temperature and pressure)/m^2 s].

For pressure-driven operations, the pressure differences between the two sides of the membrane are the driving force for the solvent to flow across the membrane, and this is what is called the transmembrane pressure. Since most membrane processes are pressure-driven, the flux can also be defined as:

$$J_T = \frac{\Delta P}{\mu_T R_{\mathrm{T}}} \tag{2}$$

where J_T is the flux at operating temperature, ΔP is the transmembrane pressure, μ_T is the viscosity at operating temperature, and $R_{M,T}$ is the total resistance in the membrane system.

Equation 2 shows that the flux increases when the transmembrane pressure increases but decreases if the viscosity and membrane resistances increase. Similarly, a membrane with a significant amount of concentration polarization and fouling will also experience reduced in flux. Thus, for fluid with significant temperature effect on viscosity will highly impact the flux obtained. For such cases, the membrane flux is typically normalized as flux at 20 °C, and thus flux for membrane operation operated at any other temperature can be calculated as:

$$J_{20} = \frac{\mu_T}{\mu_{20}} J_T = TCF.J_T \tag{3}$$

where J_{20} and μ_{20} are the flux and viscosity at 20 °C respectively, J_T and μ_T are the flux and viscosity at certain operating temperature respectively, and TCF is the temperature correction factor.

From Equation 2, the relationship between transmembrane pressure and viscosity at different temperatures is derived which can be used to predict the pressure that should be applied when changes in temperature are experienced (i.e. changes in fluid viscosity) to maintain a constant flux.

$$\frac{\Delta P_{T1}}{\mu_{T1}} = \frac{\Delta P_{T2}}{\mu_{T2}} \tag{4}$$

The linear relationship between flux and transmembrane pressure can be expressed using specific flux. The specific flux describes the membrane flux per unit of transmembrane pressure of the membrane itself at 20 °C. The higher the specific flux value, the pressure loss is low through the membrane system. This parameter can be used to measure the performance of the system with respect to energy consumption, i.e. the operating pressure and costs.

$$\left(Specific\ flux\right)_{20} = \frac{J_{20}}{\Delta P} \tag{5}$$

EXAMPLE 1: Flux

Problem. Calculate the flux of a membrane filter if it contains 7 m^2 of membrane area and operates at a flow rate of 1.15 m^3/h for 24 h at a temperature of 20 °C. Express the answer in the unit of LMH.

Solution. The flux can be calculated directly using Equation 1. The unit LMH is expressed from L/m^2.h. Thus, the flux of the membrane filter,

$$J = \frac{1.15\ \mathrm{m}^3\ /\ \mathrm{h}}{7m^2} \times \frac{1000L}{1\mathrm{m}^3} = 164L\ /\ \mathrm{m}^2.\mathrm{h}\ (\mathrm{LMH}).$$

EXAMPLE 2: Specific Flux

Problem. Calculate the specific flux of a membrane filter if it contains 7 m^2 of membrane area and operates at a flow rate of 1.15 m^3/h for 24 h at a temperature of 20 °C. The transmembrane pressure of the system is 1.5 atm.

Solution. The specific flux can be calculated directly using Equation 5 by simply dividing the flux value obtained in EXAMPLE 1 by the transmembrane pressure which gives,

$$\left(Specific\ flux\right)_{20} = \frac{164L\ /\ \mathrm{m}^2.\mathrm{h}}{1.5atm} = 110164L\ /\ \mathrm{m}^2.\mathrm{h.atm}.$$

Rejection Factor

A rejection factor is a parameter equals to one minus the ratio the concentration of component i on the downstream and upstream sides of the membrane.

$$R_i = 1 - \frac{C_{i_{downstream}}}{C_{i_{upstream}}} \tag{6}$$

where R_i is the rejection factor of component i. $C_{i_{upstream}}$ and $C_{i_{downstream}}$ are the concentration of component i on the upstream and downstream sides of the membrane respectively.

'Apparent rejection factor' is called when the concentrations of component i in the bulk phases are used. As the concentrations of component i at the membrane surface are considered, a term called 'intrinsic rejection factor' is expressed. The concentrations of the component i dissolved in the liquids which reside in the membrane module. The rejection factor is dimensionless.

Retention Factor

A retention factor (r_i) is a parameter equals to one minus the ratio of permeate concentration of component i ($C_{i_{permeate}}$) to the retentate concentration of component i ($C_{i_{retentate}}$).

$$r_i = 1 - \frac{C_{i_{permeate}}}{C_{i_{retentate}}} \tag{7}$$

The concentrations of the component i dissolved in are measured when the feed or retentate and permeate leaving the membrane modules.

Relative Recovery

The relative recovery also called substance efficiency is defined as an amount of substance of the component i collected in a useful product divided by the amount of substance of that component entering the process.

$$\eta_i = \frac{n_{i_{out}}}{n_{i_{in}}} \tag{8}$$

where η_i is the relative recovery, $n_{i_{in}}$ and $n_{i_{out}}$ are the amount of substance of component i in the entering stream and the useful product, respectively. The useful product may be either the retentate or the permeate. The relative recovery is dimensionless.

Separation Coefficient

Separation coefficient (S_c) is defined as the ratio of the composition of component A and B in the downstream relative to the ratio of compositions of these components in the upstream.

$$S_c = \frac{\left[n_A / n_B\right]_{downstream}}{\left[n_A / n_B\right]_{upstream}} = \frac{\left[C_A / C_B\right]_{downstream}}{\left[C_A / C_B\right]_{upstream}} \tag{9}$$

where n and C are the mole and concentration for components A and B respectively.

Separation Factor

Separation factor (S_F) is the ratio of the compositions of components A and B in the permeate relative to the composition ratio of these components in the retentate.

$$S_F = \frac{\left[n_A / n_B\right]_{permeate}}{\left[n_A / n_B\right]_{retentate}} = \frac{\left[C_A / C_B\right]_{permeate}}{\left[C_A / C_B\right]_{retentate}} \tag{10}$$

EXAMPLE 3: Separation Factor

Problem. An experiment was conducted for a reverse osmosis membrane system at 20 °C. The feed is brackish water with a flow rate of 1000 L/h containing 1.6 wt% sodium chloride (NaCl). The retentate contains 4.4 wt% NaCl while the permeate 0.01 wt%. The transmembrane pressure was set at 60 atm.

1. What are the flow rates of the permeate and the retentate?
2. Calculate the relative recovery of permeate.
3. Calculate the separation factor of the membrane system.

Solution. Assume that the density is uniform at 1000 g/L.

1. 1. The concentration for both permeate and retentate is calculated as:

$$C_P = \frac{0.01g}{\left(100 - 0.01\right)g} \times 1000g / L = 0.10g / L$$

$$C_R = \frac{4.4g}{\left(100 - 4.4\right)g} \times 1000g / L = 46.03g / L$$

Similarly, the feed concentration, $C_F = 16.26\, g / L$.
From the volumetric flow rate balance,

$$Q_P = Q_F - Q_R = 1000 - Q_R \tag{1}$$

Then, from the mass balance of solute,

$$C_P Q_P = C_F Q_F - C_R Q_R$$

$$0.10 Q_P = (16.26)(1000) - 46.03 Q_R . \tag{2}$$

Solving (1) and (2) simultaneously gives $Q_P = 648\, L / h$ and $Q_R = 352\, L / h$.

2. The relative recovery, $\eta = \frac{Q_P}{Q_F} = \frac{648}{1000} = 0.648$.
3. 3. The separation factor,

$$S_F = \frac{[C_W / C_{NaCl}]_{permeate}}{[C_W / C_{NaCl}]_{retentate}} = \frac{[x_W / x_{NaCl}]_{permeate}}{[x_W / x_{NaCl}]_{retentate}} = \frac{(100 - 0.01) / 0.01}{(100 - 4.4) / 4.4} = 460.20 .$$

Removal Efficiency

Removal efficiency (R_v) is a parameter defined as one minus the ratio of a target component final concentration to the initial concentration in the retentate.

$$R_v = 1 - \frac{C_f}{C_0} \tag{11}$$

where C_0 and C_f are respectively the initial and final concentration of the target component in the retentate.

Concentration Factor

For a component *i* is the target product (non-volatile) in the retentate, the concentration factor is defined as the ratio of the concentration of component *i* in the retentate to its concentration in the feed.

$$c_F = \frac{C_{i_{retentate}}}{C_{i_{feed}}} \tag{12}$$

For a component *j* is the target product (volatile) in the permeate, the concentration factor is defined as the ratio of the concentration of the component *j* in the permeate to its concentration in the feed.

$$c_F = \frac{C_{i_{permeate}}}{C_{i_{feed}}} \tag{13}$$

Permeability Coefficient

The permeability coefficient is defined as a transport flux per unit transmembrane driving force per unit membrane thickness.

$$P_i = J_i \frac{\delta}{\Delta P} \tag{14}$$

where P_i, J_i, δ, and ΔP are the permeability coefficient, flux, membrane thickness and driving force of pressure difference, respectively. The common units used for the permeability coefficient includes [kmol m m^{-2} s^{-1} kPa^{-1}], [m^3 m m^{-2} s^{-1} kPa^{-1}], [m^3 (measured at standard temperature and pressure) m m^{-2} s^{-1} kPa^{-1}] or [kg m m^{-2} s^{-1} kPa^{-1}].

Permeance

Permeance, also known as the pressure normalised flux is defined as the transport flux per unit transmembrane driving force.

$$\mathcal{P}_i = \frac{J_i}{\Delta P} = \frac{P_i}{\delta} \tag{15}$$

The common units used are [kmol m^{-2} s^{-1} kPa^{-1}], [m^3 m^{-2} s^{-1} kPa^{-1}], [m^3 (measured at standard temperature and pressure) m^{-2} s^{-1} kPa^{-1}] or [kg m^{-2} s^{-1} kPa^{-1}].

EXAMPLE 4: Permeance

Problem. In an experiment, a saline solution having 15 kg NaCl/m^3 solution is fed to a reverse osmosis membrane system at 25 °C. The membrane area is given as 2.2 x 10^{-3} m^2. When a constant transmembrane pressure of 58.48 atm is applied, the permeate is measured to be 0.42 kg NaCl/m^3 solution at the rate of 6.95 x 10^{-5} m^3/h.

Note: In any desalination application, it is important to take into account the osmotic pressure difference which is the pressure that needs to be overcome in order to allow water transport against the concentration gradient. The difference between the applied pressure and the osmotic pressure is the net driving force of the process. The osmotic pressure difference can be approximated from the permeate and retentate concentrations, using van't Hoff's correlation for osmotic pressure, $\pi = nMRT$, where n is the number of ions dissociation, M is the molecular concentration, R is the universal gas constant, and T is the temperature in Kelvin. However, in reality, the osmotic pressure difference should be calculated by integrating the concentrations of feed and retentate along the membrane area. In this example, the osmotic pressure difference is approximated from the known concentration of feed and permeate.

1. Calculate the permeance of water.
2. What is the permeance of the salt?
3. Find the rejection factor.

Solution. Normally, the permeance can be directly calculated or given once the flux and transmembrane pressure are provided. This example shows how the permeance of solvent and solute can be calculated experimentally.

1. 1. For a pressure driven operation, the solvent flux of the solvent is given by,

$$J_W = \mathcal{P}_W\left(\Delta P - \Delta\pi\right) = \frac{Q_{W,permeate}}{A_m} = \frac{6.95\times10^{-5}m^3/h}{2.2\times10^{-3}m^2} = 0.032m^3/m^2.h\ .$$

$$C_F = C_R = \frac{15000g/m^3}{58.5g/mol} = 256.41mol/m^3$$

$$C_P = \frac{420g/m^3}{58.5g/mol} = 7.18mol/m^3\ .$$

Since both streams are the same mixture of the solution with different concentrations, the osmotic pressure difference can be found below. The number of ion dissociation for a saline solution is 2 for Na^+ and Cl^- ions.

$$\Delta\pi = n\left(\Delta M\right)RT$$

$$= 2\times\left(256.41-7.18\right)mol/m^3\times8.206\times10^{-5}m^3.atm/mol.K\times298\,K$$

$$= 12.19\,atm\ .$$

Thus, solving for permeance,

$$\mathcal{P}_W = \frac{J_W}{\left(\Delta P - \Delta \pi\right)} = \frac{0.032 m^3 / m^2.h \left(\frac{h}{3600 s}\right)}{\left(58.48 - 12.19\right) atm} = 1.92 \times 10^{-7} m^3 / m^2.s \,.$$

2. 2. The permeance for salt can be calculated through the change in a concentration gradient,

$$J_{NaCl} = \mathcal{P}_{NaCl}\left(C_R - C_P\right)$$

$$J_{NaCl} = J_W C_P \,.$$

Thus,

$$\mathcal{P}_{NaCl} = \frac{J_W C_P}{\left(C_R - C_P\right)} = \frac{0.032 m^3 / m^2.h \left(\frac{h}{3600 s}\right) \times 7.18 mol / m^3}{\left(256.41 - 7.18\right) mol / m^3} = 2.56 \times 10^{-7} m / s \,.$$

3. 3. The rejection factor,

$$R_{NaCl} = 1 - \frac{C_P}{C_F} = 1 - \frac{7.18}{256.41} = 0.97 \; or \; 97\% \,.$$

MEMBRANE STRUCTURE

Generally, the structure of the membranes consists of two types: symmetric and asymmetric structures. These structures affect the transport properties as well as the separation performance of the membranes. Figure 2 shows the illustration of the symmetric and asymmetric structure of the membranes.

Symmetric Structure

Symmetric membranes are tailored from a single material to form a homogeneous structure. The membrane structure is uniform in the density of pore structure throughout the cross-section.

Symmetric membranes have been developed for commercial distribution by several corporations specializing in membrane fabrications. For example, SepraTek has commercially produced polypropyl-

Figure 2. Illustrative representation of various symmetric and asymmetric membranes

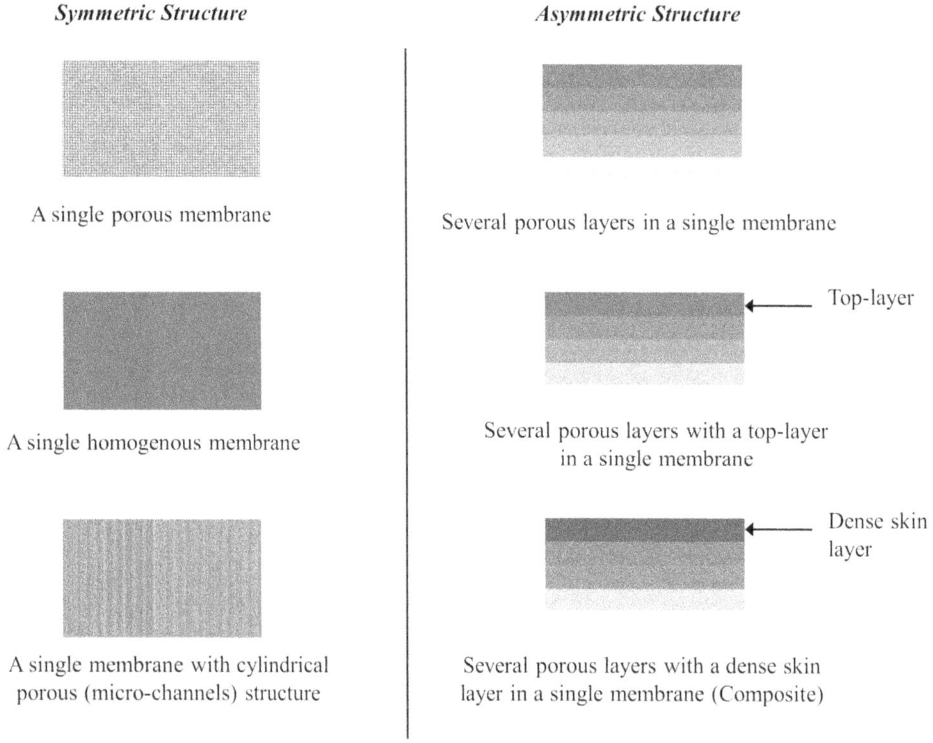

ene (PP) hollow fibre membrane which has a symmetric porous structure by using Thermally Induced Phase Separation (TIPS) technique. The porous structure is also known as 'sponge-like' structure. The PP hollow fibre membrane is specially fabricated for the application in membrane contactor and membrane distillation.

Superstring MBR Technology Corporation has successfully developed a series of SuperUF micropore membranes using TIPS method commercially. One of the membranes is symmetric and 'sponge-like' porous PP hollow fibre membrane with a nominal pore size of 1 μm. The symmetric PP membranes have two types; they are hydrophobic and hydrophilic. These membranes are specially produced for membrane bioreactor (MBR) process. The pure water flux that can be attained by the symmetric membrane is approximately 3600 L/m^2 h atm. The membrane can reject most of the protozoon and fungi as well as bacteria.

Membranes made from other polymeric materials also shows symmetric structure depending on the fabrication methods. A study by Santos, Borges, and Fonseca (2015) involves six membranes made from different materials characterized by the change in membrane properties due to ozone attack. Two membranes namely the polyvinylidenefluoride (PVDF) and polytetrafluoroethylene (PTFE) membranes show 'sponge-like' with interconnected pores properties. The Scanning Electron Microscopy (SEM) images of the cross-sectional structure of the membrane are shown in Figure 3.

The symmetric structure also includes cylindrical, conical, cigar-shaped and rhombohedral structures, as shown in Figure 4. These membranes are fabricated by using the track-etching technique. The

Figure 3. SEM image of the cross-section of polymeric membranes before the contact with ozone (a) PVDF, and (b) PTFE
Source: Santos, Borges, & Fonseca, 2015 (Licensed under CC BY 4.0)

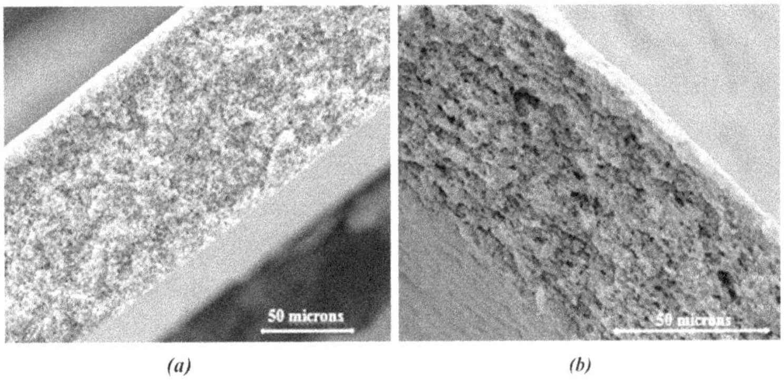

polymers that can be used include polyimide (PI), polyethene terephthalate (PET), polycarbonate and inorganic materials such as mica and glass.

The symmetric structure of glass membranes with integrated arrays of nanospiked microchannels and a narrow pore size distribution is investigated by Ma et al. (2009). The peak of the pore diameter is 3.4 µm with 90% of the pore diameters falling in the range of 3 to 4 µm. These membranes are fabricated for the application of desalination through membrane distillation. The flux ranges from 10 to 12 kg/m^2 h when the membrane thickness is 500 µm.

Asymmetric Structure

Asymmetric membrane constitutes of two or more structural planes of non-identical morphologies. There is a change in the density of the membrane material across the cross-sectional area of the membrane. Asymmetric membranes consist of a skin layer or top layer with a thickness range of 0.1 to 0.5 µm. The skin layer or top layer determines the resistance to mass transfer significantly. The asymmetric membrane is a composite membrane when the top layer and sublayers in a single membrane are tailored from different materials whereby each layer can be optimized independently. Generally, a thin and dense layer

Figure 4. Symmetric membranes with (a) cylindrical channels in PI, (b) conical channels in PI, (c) cigar-shaped channels in PET and (d) cylindrical channels with a rhombohedral cross-section in mica
Source: Toimil-Molares, 2012

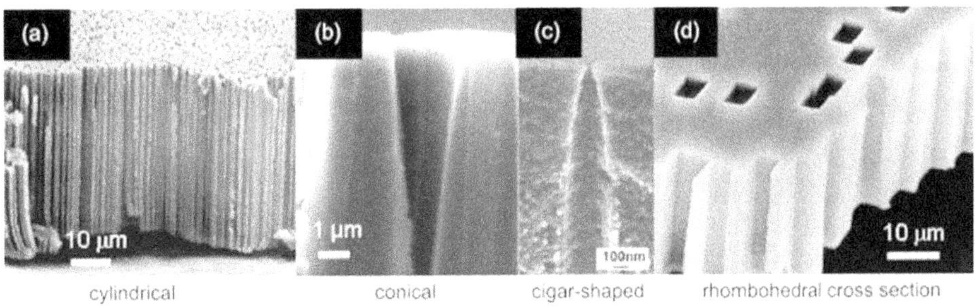

is deposited on an asymmetric sublayer and the sublayer is the support of the dense layer. The structure of the sublayers may comprise of sponge-like, finger-like, cellular-like, macrovoids and cavities. The overall mechanical strength of the membrane is determined by the structure of the sublayers. Membrane composition and formulation condition are the factors to determine the asymmetric membrane structure.

For example, TFC polyamide membranes prepared by Joshi, Singh, and Bhattacharya (2011) indicates a polymer-rich phase at the top whereas channels and macrovoids are beneath the top layer in the polyethersulfone (PES) membrane. The SEM image indicates the asymmetric structure in the PES membrane when the membrane is coagulated in a water bath. When the PES membrane is coagulated in the 0.1% 1,3-phenylene diamine aqueous solution, the macrovoids become more and finger-like structures appeared.

Danish Separation Systems AS commercially manufactures polysulfone (PSf) ultrafiltration membrane with 10,000 Dalton cut-off. The PSf membrane is an asymmetric membrane. The membrane consists of a very thin skin layer with very small porosity. Membrane structure with 'short finger-like' is formed at the middle section of the membrane. The bottom part of the membrane mainly consists of a porous structure (Castro et al., 2002).

In a research conducted by Morihama and Mierzwa (2014), a composite membrane made from polyvinylidene difluoride (PVDF) and clay exhibits asymmetric structure. The membranes comprised of a superficial layer with macropore as the top layer. With the addition of 4% clay, the membrane structure has a bottom layer which consists of a sponge-like pore structure with micro-pores and an intermediate layer that has finger-like pores. The sponge-like structure at the bottom layer disappears when 1% polyvinylpyrrolidone (PVP) and 4% clay were added.

The asymmetric membrane is also developed by embedding cellulose acetate (CA) into a polyester mesh. The mesh acts as the mechanical support of the membrane. This membrane has been developed by Osmotek Inc. (Albany, Oregon) which is currently known as Hydration Technologies Inc. (HTI). This membrane is actively tested for forwarding osmosis. The HTI–CA membrane is further researched. Thin-film composite (TFC) membranes are favourably employed for engineered osmosis. The TFC membrane comprises an electrospun polymeric nanofibre support layer and a polyamide skin layer formed by *in situ* polymerization. The skin layer made from polyamide in TFC is approximately 0.1 μm which is thinner than the skin layer in the commercial membrane. The porous support for the polyamide skin layer can be tailored in the form of an electrospun nanofibrous porous layer or asymmetric porous layer which comprises sponge-like, finger-like and macro-voids structures. The common materials of these porous support layers are polyethersulfone and polysulfone (Bui et al., 2011).

Terra et al. (2016) fabricated several types of asymmetric alumina hollow fibre membranes using the phase inversion/sintering method. Four different hollow fibre membranes (HF1-1400, HF2-1300, HF3-1300, HF3-1400) are fabricated and characterised to study the hydrogen permeation in composite membranes. Characterization using SEM reveals that the asymmetric membranes possess 'sponge-like' property on the outer layer, while the inner layer contains microchannels covering a large area of the fibre walls. The 'sponge-like' outer layer is denser than the microchannels layer, thus providing mechanical support of the hollow fibres. Figure 5 shows the SEM image of the hollow fibres, showing the cross-section, inner surface, and the fibre wall.

MEMBRANE STRUCTURAL PARAMETER

Flux through a membrane is proportional to the membrane porosity and pore size. The flux decreases when the membrane thickness and tortuosity increase. The relationship between the flux and the membrane structural parameters is mathematically represented as follows:

$$J \propto \frac{\varepsilon \Delta d_p^{\varphi}}{\tau \delta}. \tag{16}$$

where J is the flux, ε is the membrane porosity, d_p is the mean pore size, τ is the membrane tortuosity, δ is the membrane thickness and φ is a factor equal to 1 for Knudsen diffusion and equal to 2 for viscous flow.

Thickness

The membrane thickness gives information on both the mechanical strength and the fluxes through the membrane. The strength of the membrane is weaker for a thinner membrane. The mechanical strength of the membrane increases when the membrane thickness increases. However, the membrane resistance increases with the membrane thickness which in turn reduces the flux through the membrane. In general, the membrane thickness can be measured by using a micrometre and scanning electron microscopy (SEM). Other Techniques of profilometry, ellipsometry, atomic force microscopy (AFM), Rutherford backscattering spectrometry (RBS), transmission electron microscopy (TEM), and quartz crystal microbalance (QCM) (Lin et al., 2016).

In the first method, a micrometre gauge is a simple tool that can be employed to measure the membrane thickness up to the accuracy of 0.01 mm. In this method, a systematic error can be encountered due to membrane compression by the micrometre gauge while obtaining the reading. For very thin membranes, several membrane sheets of similar thickness are stacked together before the micrometre is used to measure the thickness. Then, the average thickness for each membrane is determined by dividing the measurement obtained with the number of membrane sheets.

The SEM is a powerful analytical tool that is widely used to characterize the membrane characteristics such as its thickness, porosity, pore size, and surface roughness. To measure the membrane thickness using SEM, the membrane sample is first prepared and analysed under the system. The images captured in magnifications of up to 40,000x – 80,000x can be analysed by software images to obtain the membrane thickness.

Porosity

The membrane porosity is defined as the volume of the pores divided by the total volume of the membrane. The porosity of the membrane is calculated as follows:

$$\varepsilon = 1 - \frac{\rho_m}{\rho_p} \tag{17}$$

Figure 5. (a) Cross-section, (b) inner surface and (c) fibre wall SEM images of the hollow fibres (1) HF1-1400, (2) HF2-1300, (3) HF3-1300, (4) HF3-1400
Source: Terra et al., 2016 (Licensed under CC BY 4.0)

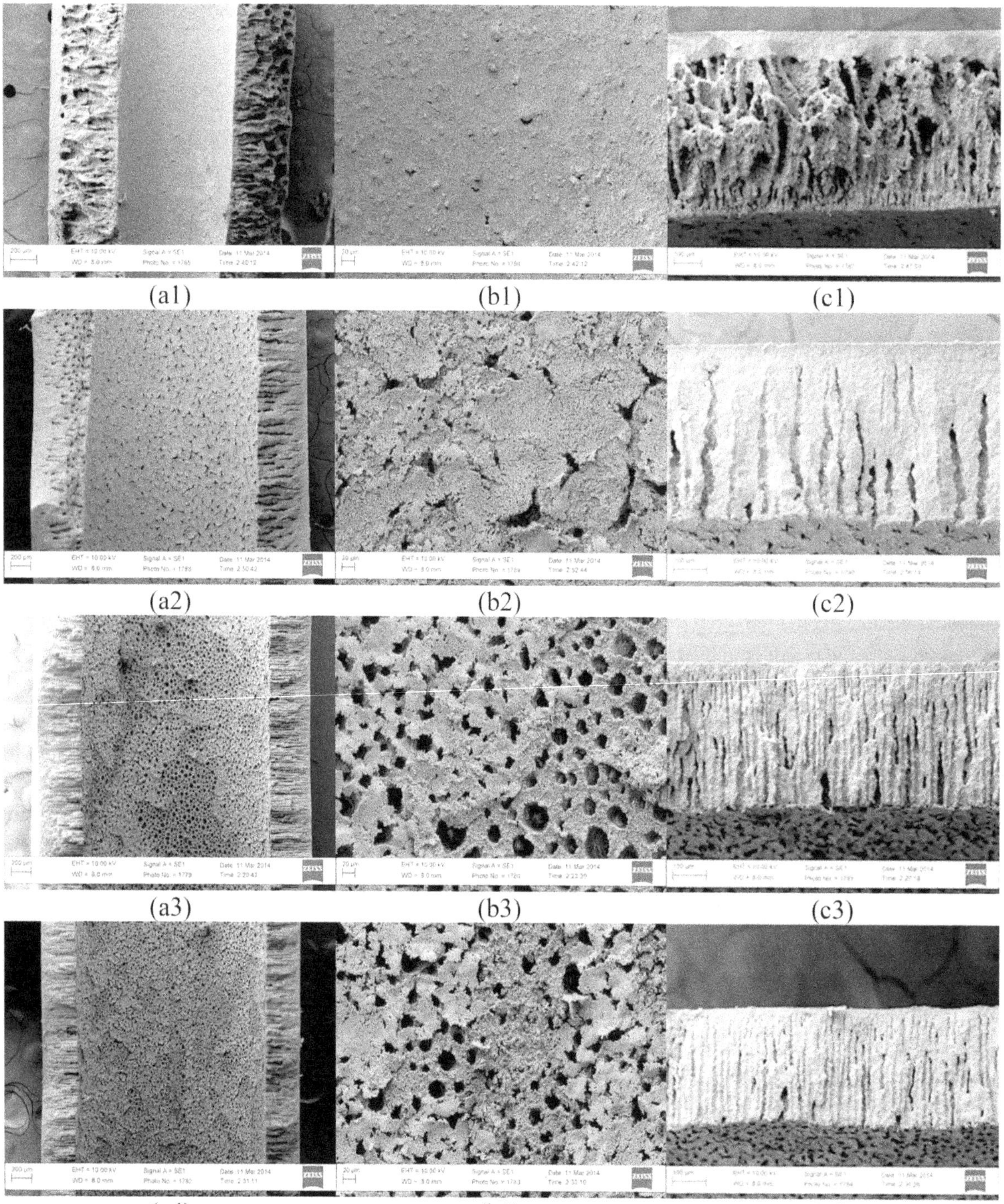

(a1) (b1) (c1)
(a2) (b2) (c2)
(a3) (b3) (c3)
(a4) (b4) (c4)

Figure 6. Loosely packed spheres

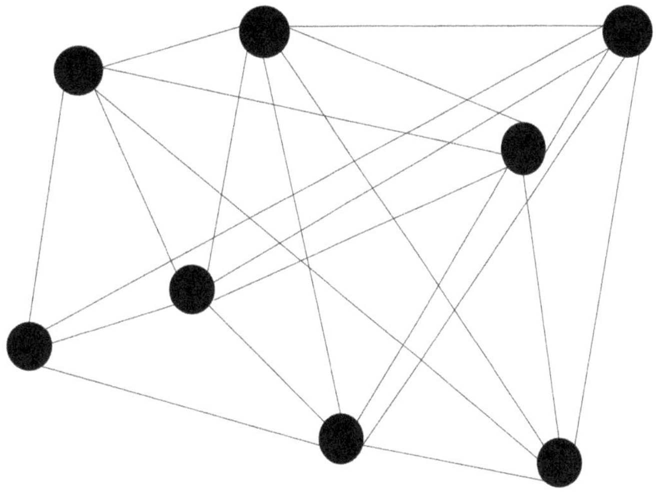

where ε is the membrane porosity, ρ_m is the density of the membrane and ρ_p is the density of the polymer.

Gravimetric method is employed to determine the membrane porosity experimentally. The gravimetric equation is derived as follows:

$$\varepsilon = \frac{\left(m_w - m_d\right) / \rho_w}{\left(m_w - m_d\right) / \rho_w + m_d / \rho_p} \quad (18)$$

where m_w is the weight of the wetted membrane, m_d is the weight of the dried membrane and ρ_w is the density of the wetted membrane. Isopropyl alcohol is a common wetting agent.

Tortuosity

The membrane tortuosity is defined as the effective pore length divided by the membrane thickness. The flux decreases when the tortuosity increases. The geometry of the membrane structure determines the tortuosity factor. The tortuosity factor can be estimated based on two models: (1) loosely packed spheres (Figure 6), and (2) interstices between closed packed spheres (Figure 7).

$$\tau = \frac{1}{\varepsilon} \quad (19)$$

$$\tau = \frac{\left(2 - \varepsilon\right)^2}{\varepsilon} \quad (20)$$

Figure 7. Interstices between closed packed spheres

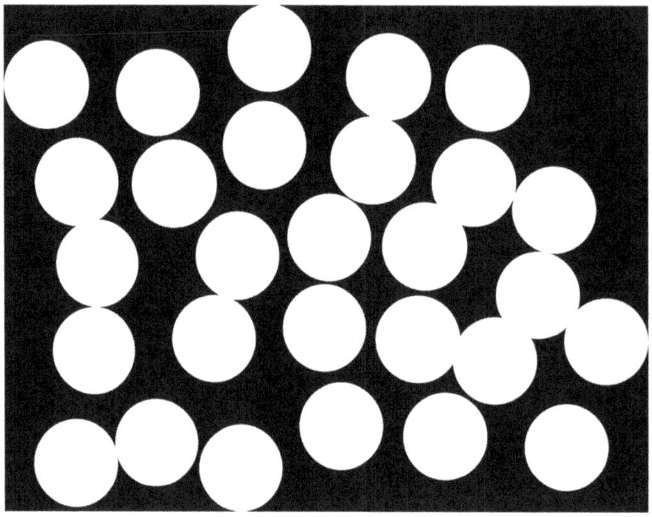

Pore Size

Nominal pore size is essential to predict the fluxes. The membrane pore size which includes maximum pore size, average pore size and pore size distribution can be measured by using bubble-point and gas permeation tests. The bubble-point and gas permeation tests are also known as wet and dry flow method. The apparatus of the bubble-point and gas permeation tests is shown in Figure 8. In the bubble-point test, isopropyl alcohol (IPA) is used as the wetting agent. The procedures of the IPA bubble-point are the same as described in ASTM F-316 (Smolders & Franken, 1989).

First, the gas permeation velocity (F_{dry}) is measured through a dried membrane and plotted as a function of the transmembrane pressure difference (ΔP). A straight line is generally obtained. Next, the membrane is wetted by IPA. Then, the gas permeation velocity (F_{wet}) is measured through the wetted membrane at different values of ΔP . The test can be conducted at room temperature and the downstream side of the membrane is maintained at atmospheric pressure.

The maximum pore size ($d_{p,max}$) can be measured by using the ΔP as in the equation as follows:

$$d_{p,max} = 2r_{max} = \frac{4B\gamma}{\Delta P} \quad (21)$$

where B is the pore size morphology constant and γ is the surface tension of IPA. The pore size morphology constant B is 1 for a circular pore and less than 1 for an elliptical or irregularly shaped pore.

Various pore sizes at different ΔP are calculated according to the Equation 21. A graph of the pore size distribution that is the ratios of F_{wet} / F_{dry} as a function of the pore sizes is plotted. The average pore size is determined as the pore size at the corresponding to F_{wet} / F_{dry} equal to 0.5. Figure 9 illustrates the graphical relationship between F_{wet} / F_{dry} and the radius.

Figure 8. Bubble-point and gas permeation tests

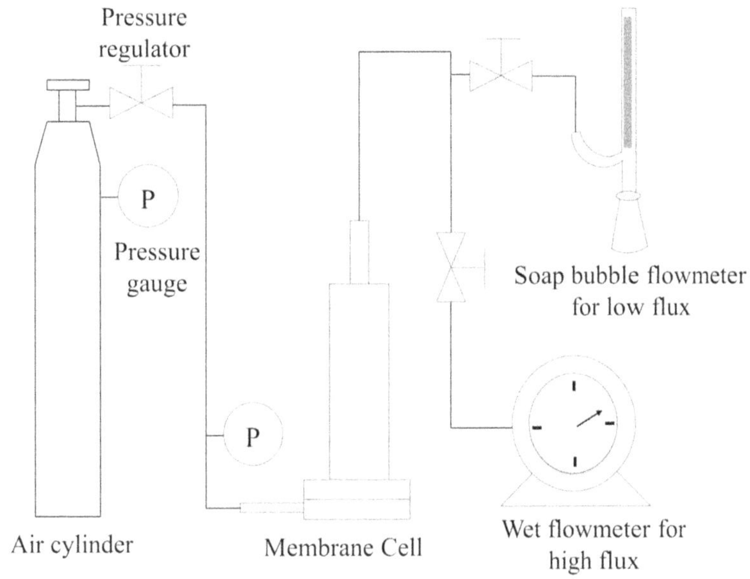

Figure 9. Relationship between the ratio of dried membrane gas permeation velocity to wetted membrane gas permeation velocity (F_{wet}/F_{dry}) versus pore radius (r)

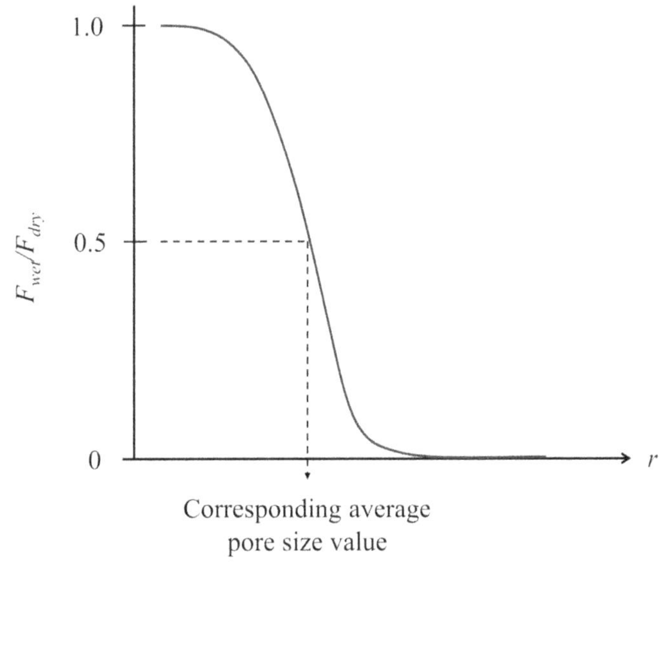

REFERENCES

Bui, N. N., Lind, M. L., Hoek, E. M. V., & McCutcheon, J. R. (2011). Electrospun nanofiber supported thin-film composite membranes for engineered osmosis. *Journal of Membrane Science, 385–396*, 10–19.

Castro Vidaurre, E. F., Achete, C. A., Gallo, F., Garcia, D., Simão, R., & Habert, A. C. (2002). Surface modification of polymeric materials by plasma treatment. *Materials Research*, *5*(1), 37–41. doi:10.1590/S1516-14392002000100006

Joshi, R. N., Singh, K., & Bhattacharya, A. (2011). Approaches to prepare TFC polyamide membranes by coating diamine during, and/or post formation of asymmetric membranes and their performances. *Brazilian Journal of Chemical Engineering*, *28*(3), 457–465. doi:10.1590/S0104-66322011000300011

Lin, L., Feng, C. C., Lopez, R., & Coronell, O. (2016). Identifying facile and accurate methods to measure the thickness of the active layers of thin-film composite membranes – A comparison of seven characterization techniques. *Journal of Membrane Science*, *498*, 167–179. doi:10.1016/j.memsci.2015.09.059

Ma, Z., Hong, Y., Ma, L., & Su, M. (2009). Superhydrophobic membranes with ordered arrays of nanospiked microchannels for water desalination. *Langmuir*, *25*(10), 5446–5450. doi:10.1021/la900494u PMID:19341278

Morihama, A. C. D., & Mierzwa, J. C. (2014). Clay nanoparticles effects on performance and morphology of poly(vinylidene fluoride) membranes. *Brazilian Journal of Chemical Engineering*, *31*(1), 79–93. doi:10.1590/S0104-66322014000100009

Santos, F. R. A., Borges, C. P., & Fonseca, F. V. (2015). Polymeric Materials for Membrane Contactor Devices Applied to Water Treatment by Ozonation. *Materials Research*, *18*(5), 1015–1022. doi:10.1590/1516-1439.016715

Smolders, K., & Franken, A. C. M. (1989). Terminology for Membrane Distillation. *Desalination*, *72*(3), 249–262. doi:10.1016/0011-9164(89)80010-4

Terra, N. M., Lemos, C. O. T., da Silva, F. B., Cardoso, V. L., & Reis, M. H. M. (2016). Characterisation of Asymmetric Alumina Hollow Fibres: Application for Hydrogen Permeation in Composite Membranes. *Brazilian Journal of Chemical Engineering*, *33*(3), 567–576. doi:10.1590/0104-6632.20160333s20150074

Toimil-Molares, M. E. (2012). Characterization and properties of micro- and nanowires of controlled size, composition, and geometry fabricated by electrodeposition and ion-track technology. *Beilstein Journal of Nanotechnology*, *3*(2012), 860–883. Adapted from https://www.ncbi.nlm.nih.gov/pmc/articles/PMC3556775/figure/F3

ADDITIONAL READING

Baker, R. W. (2012). *Membrane Technology and Applications*. United Kingdom: Wiley. doi:10.1002/9781118359686

Calabro, V., & Basile, A. (2011). Fundamental membrane processes, science and engineering. In A. Basile & S. P. Nunes (Eds.), Advanced Membrane Science and Technology for Sustainable Energy and Environmental Applications (pp. 3–21). Sawston, Cambridge: Woodhead Publishing. doi:10.1533/9780857093790.1.3

Chen, J. P., Mou, H., Wang, L. K., & Matsuura, T. (2006). Membrane filtration. In L. K. Wang, Y. T. Hung, & N. K. Shammas (Eds.), *Advanced Physicochemical Treatment Processes, Handbook of Environmental Engineering* (Vol. 4). New York: Humana Press. doi:10.1007/978-1-59745-029-4_7

Drioli, E., & Giorno, L. (Eds.). (2016). *Encyclopedia of Membranes*. Berlin: Springer. doi:10.1007/978-3-662-44324-8

Hwang, S. T. (2011). Fundamentals of membrane transport. *Korean Journal of Chemical Engineering*, *28*(1), 1–15. doi:10.100711814-010-0493-z

Magara, Y., Kunikane, S., & Itoh, M. (1998). Advanced membrane technology for application to water treatment. *Water Science and Technology*, *37*(10), 91–99. doi:10.2166/wst.1998.0383

Peinemann, K. V., & Nunes, S. P. (Eds.). (2010). *Membranes for Water Treatment* (Vol. 4). Britain: Wiley-VCH. doi:10.1002/9783527631407

Strathmann, H. (1981). Membrane separation processes. *Journal of Membrane Science*, *9*(1–2), 121–189. doi:10.1016/S0376-7388(00)85121-2

KEY TERMS AND DEFINITIONS

Concentrate: The stream leaving the membrane module on the same side as the feed. The concentrate is one of the two outlets leaving the membrane system aside from the permeate stream.

Dense Membrane: Membrane with dense structure showing no visible pore in the range of electron microscopy.

Feed: The stream flowing into the membrane system which generally separated into permeate and retentate streams.

Flux: The amount (in mass, volume, or mole) of the permeating species passing through a unit of membrane area per unit of time.

Permeability: The amount of permeating species across a defined membrane area per unit of driving force per unit of membrane thickness.

Permeance: The membrane characteristics which display the ability of a species to permeate across a membrane with a certain thickness.

Permeate: The stream whole mass permeated through the membrane which contains one or more species on the opposite feed side.

Pore Size: The average or range of sizes of holes or pores on the membrane surface.

Porous Membrane: Membrane consisting of a solid matrix with defined pores on the membrane surface.

Rejection Factor: The ratio or percentage of solutes retained by the membrane.

Retentate: *See Concentrate.*

Separation Factor: Often called selectivity, it shows the ratio of components in the permeate side to the ratio of components in the retentate side.

Chapter 2
Mass Transfer Phenomenon and Transport Resistances in Membrane Separation

ABSTRACT

The inevitable decline of membrane performance in membrane separation processes can be optimized through a good understanding of the mass transfer phenomenon and the transport resistances involved in the operation. Thus, this chapter focused on the discussions of mass transfer mechanisms and models in membrane separation based on several types of driving forces. This includes the pressure from a mechanical operation, partial pressure, osmotic pressure, concentration, and also thermal gradients. The chapter elaborates on the transport resistances in membrane resulting from membrane fouling and concentration polarization. The author hopes that readers, especially engineers and technical operators, gain a deep understanding and comprehensive knowledge regarding the theories and are able to utilize the knowledge to optimize the membrane operation.

INTRODUCTION

A membrane may be defined as a thin sheet of material that regulates the passing of certain components through it, enabling the separation of components to occur. It acts as a permselective interface between phases (such as gas-liquid, liquid-liquid and liquid-solid interfaces). The potentials of the membrane as a separation tool spans over a broad spectrum of applications, from water and wastewater treatment, water desalination, juice concentration etc. Membrane possesses many advantages such as simplicity in operation, uses relatively low energy, and works without the needs of chemicals. The core properties of the membrane that tailors particularly to the separation process requirements established it as a more superior tool over the conventional technologies.

DOI: 10.4018/978-1-7998-2645-3.ch002

The transport of masses passing through the membrane governs the mass transfer phenomenon in the membrane process. The concept of mass transfer in this scope of study focuses on the movement of a particular component (permeate) across the membrane interface and the factors affecting the phenomenon. Several major factors affecting mass transfer concentrate particularly on the interface of membrane-components, the structure of the membrane, and the operating parameters used in membrane operations. The identification of mass transport resistances is crucial when handling any membrane operations. This is because mass transport resistances affect the membrane flux, and consequently, the yield obtained and performance of the membrane separation process. Membrane flux is one of the most critical parameters in membrane separation, as it determines the amount of permeate obtained per unit area of membrane surface per unit time. However, one of the major hindrance in membrane operation is that its performance will decrease with operation time, primarily due to the phenomena known as concentration polarization and membrane fouling.

In this chapter, the discussion will focus on the fundamentals of mass transfer and transport resistances in membrane separation. An overview of the driving forces, transport resistances, operation configuration and types of membrane and resistances in pressure-driven operation and mass transfer modelling is reviewed. Different models are discussed by categorizing them in terms of driving force used (i.e. pressure and partial pressure) by considering the membrane structure either porous or dense membrane, and centralizing the modelling discussion on the beginning of the operation, at steady state, and when membrane fouling starts to occur.

THE DRIVING FORCES IN MEMBRANE SEPARATION PROCESSES FOR WATER AND WASTEWATER

In general, Figure 1 illustrates the membrane separation whereby the membrane separation process takes place as a result of the specific driving force that transport component across the membrane from one phase to the other. In membrane separation, a feed mixture comprising of two or more components is separated to an extent through a semipermeable membrane which acts as an interphase between two phases. The membrane separation process can be classified based on several factors including the driving force, operation mechanism, membrane structure, and phases involved. In essence, it focuses on a certain component that can be readily transferred across the membrane from other components (selectivity) and the transfer rate of that particular component (flux). These parameters, selectivity and flux largely determine the efficiency and performance of the membrane separation process respectively.

The operation of membrane separation is mainly governed by the specific driving force mechanism of mass transport used. The driving forces in membrane separation can be classified into pressure-driven (in reverse osmosis, ultrafiltration, and microfiltration processes), concentration gradient-driven (in forward osmosis, and dialysis), partial pressure-driven or thermal-driven (in pervaporation, and membrane distillation processes), and electrical potential-driven (in electrodialysis). More than one driving force may be involved in some membrane operations, such as concentration and pressure in gas separation.

This section will focus on the operations by different driving forces that are typically used in water and wastewater treatment, which are the pressure, and partial pressure (thermal) gradients.

Figure 1. Schematic of the membrane separation process showing different types of driving forces

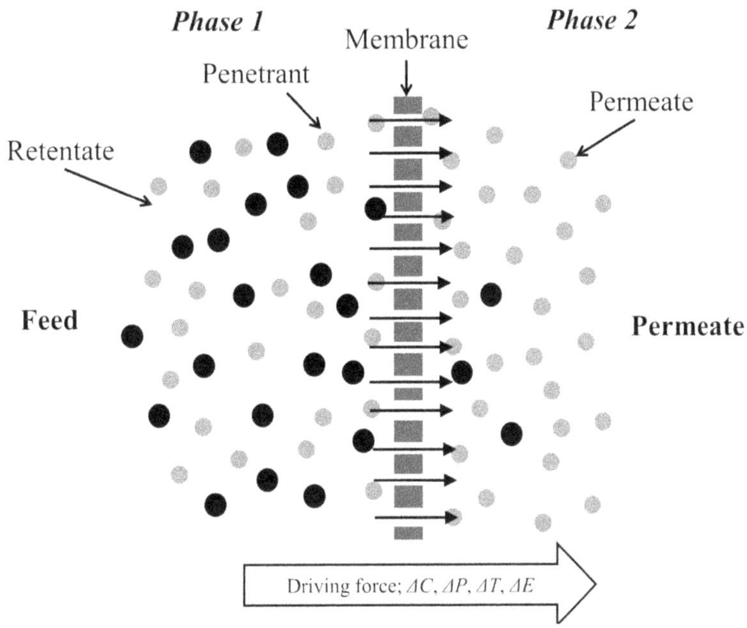

Pressure-Driven Operations

Pressure-driven membrane operations such as microfiltration, ultrafiltration, and reverse osmosis are the most commonly used industrial processes (Zhu & Bai, 2017) which are widely used in water and wastewater treatment. These three pressure-driven membrane processes (Table 1) are classified based on their membrane pore sizes range, and also on the transmembrane pressure during operation: microfiltration (0.1 – 10.0 mm, 0.1 – 3 bar), ultrafiltration (1 – 100 nm, 0.5 – 10 bar), and reverse osmosis (0.1 – 1 nm, 20 – 35 bar). While reverse osmosis is widely used for water purification and desalination, microfiltration and ultrafiltration are generally employed in the water and wastewater treatments.

Pressure-driven operation is the opposite of the natural phenomenon of passive transport, whereby molecules flow from high concentration region to low concentration region. In pressure-driven membrane operations, external pressure is applied on the feed side to enable the smaller molecules to pass through a semipermeable membrane from the feed stream to the permeate side, retaining the larger molecules. The pressure difference between the feed and permeate sides (transmembrane pressure) permits the permeate passing through the membrane at steady state, thus allowing more concentrated or purer permeate (or retentate) as the final product.

Osmotic Pressure-driven Operations

Mixing of solute and solvent creates a solution which has different properties than its individual component. Depending on the ratio of the number of solute to solvent molecules (i.e. concentration of the solution), the properties of the solution varies which are collectively known as the colligative properties. Colligative properties refer to the changes in physical properties of a solution as a result of solutes addition to the solvent, which is dependent on the concentration. Some examples of colligative properties

Table 1. Classification of pressure-driven membrane processes

Membrane Process	Pore Size	Operating Pressure (bar)	Driving Force	Membrane Type	Transport Regime	Separation Mechanism
Microfiltration	0.1 – 10.0 mm	0.1 – 3	Pressure	Porous	Macroporous	Size exclusion
Ultrafiltration	1 – 100 nm	0.5 – 10	Pressure	Porous	Microporous	Size exclusion
Reverse osmosis	0.1 – 1 nm	20 – 35	Pressure	Dense	Molecular	Solution diffusion

are the boiling point elevation, vapour pressure lowering, freezing point depression, and osmotic pressure. The latter property is the main focus in this section since it is one of the important driving force in membrane processes which are gaining more interest in this field.

In a membrane system with two solutions of different concentrations being separated by a semipermeable membrane, the solvent (in most cases being water) tends to move across the membrane from the low concentration solution into the higher concentration solution. This is because, water moves from a region with high water activity to region with low water activity, in order to equalize the water activity on both sides of the membrane. On the other hand, the solutes are retained in its original compartment, prohibited from passing through the membrane. The amount of pressure needed to stop water diffusion (i.e. no net flow of water into and out of the membrane) via osmosis is known as the osmotic pressure difference between the two solutions. The osmotic pressure of a solution is proportional to the molar concentration of the dissolved solutes.

Experimental Design of Osmotic Pressure Calculation

The phenomena of osmosis and osmotic pressure can be further explained by experimental set-up in laboratory, which consists of a container with two compartments, each filled with pure water solvent (or a low solute concentration solution) and a solution (or a high solute concentration solution) such as sucrose or salt solution is separated by a semipermeable membrane. Assuming that the membrane is ideal, only water molecules are allowed to pass across it, while the solutes are completely retained. This means that water molecules can pass through the membrane from both sides, but the net water flow is what is mainly considered in this context. Also, both sides exhibit the same temperature. The schematic of the system is illustrated in Figure 2.

In the beginning of the osmosis (or forward osmosis) process, the pure water and the solution have the same liquid level and volume. Since the liquid levels are equal, no hydrostatic pressure acts as opposing of the osmotic pressure. Nevertheless, the water activity of the pure water solution is the highest compared to the solution (Schultz, 2016). In other words, it can be said that pure water has higher chemical potential than the solution, since chemical potential increase with respect to the chemical activity. The water activity indicates the "free" water molecules that are available to interact with solutes or surfaces. It tends to reduce when more water molecules interact with other components through bonding with ions and other molecules. At this stage of the process, the water molecules from the pure water will permeate across the membrane to the solution. This causes an increase in pressure due to the rise in the liquid level of the solution.

As the process progressed, the water activity is gradually decreasing on the pure water side due to the loss of water molecules (volume). On the opposite side, the water activity is gradually increasing on

Figure 2. Experiment set-up to study the osmosis and osmotic pressure phenomena

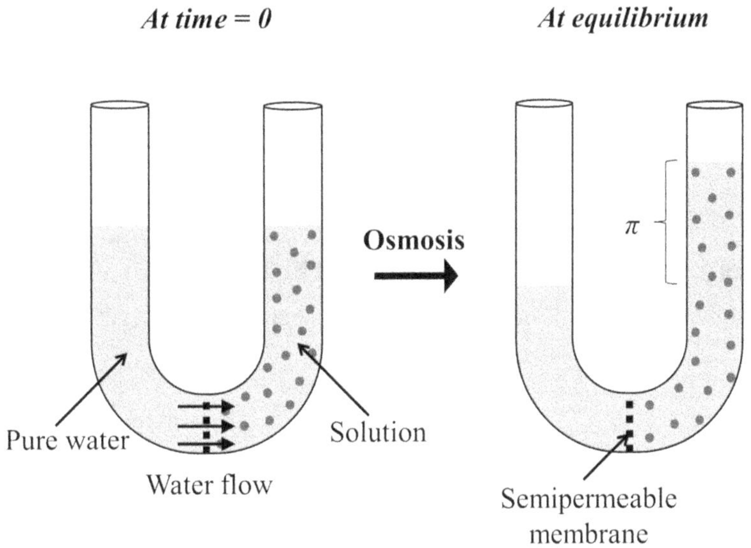

the higher concentration solution due to the gain of water molecules. Furthermore, the increase in the liquid level of the solution results in the increase of opposing hydrostatic force. Therefore, the overall driving force has been reduced and consequently decreasing the water flow through the membrane.

At the final stage, the system is said to attain osmotic equilibrium. During this time, the magnitude of water activity on both sides of the membrane are equal. This is achieved when the osmotic pressure difference becomes equivalent to the opposing hydrostatic pressure. Thus, the net driving force is now zero and completely stop the net water flow through the membrane.

Osmotic Pressure Derivation

In thermodynamics, the chemical potential is also known as the partial molar free energy. Chemical potential is a type of potential energy that is released or absorbed during chemical reaction or phase change. In a mixture or solution, the chemical potential of a component is defined by the rate of changes of the Gibbs free energy to the change in the number of particles of the component in the system, provided that the temperature, pressure and number of particles of other components are constant. Mathematically, the chemical potential of a component is represented as:

$$\mu_i = \left(\frac{\partial G}{\partial N_i}\right)_{T,P,N_j} \tag{1}$$

where μ_i is the chemical potential of component i, G is Gibb's free energy of the solution, N_i is the number of molecules of component i, T is the temperature, P is the pressure, and N_j is the number of molecules of component j.

Also, Gibb's free energy can be expressed in the thermodynamics derivative term with respect to pressure under the condition of constant temperature and composition, which gives the volume of the solution (V).

$$\left(\frac{\partial G}{\partial P}\right)_{T,x_j} = V \tag{2}$$

Differentiation of Equations 1 with respect to P and Equation 2 with respect to n_i while other parameters are kept constant provide Equation 3 and 4 respectively. The term $\bar{V}_i$ is the molar volume of component i.

$$\left(\frac{\partial \mu_i}{\partial P}\right)_{T,x_j} = \frac{\partial^2 G}{\partial P \partial N_i} \tag{3}$$

$$\frac{\partial^2 G}{\partial N_i \partial P} = \left(\frac{\partial V}{\partial N_i}\right)_{T,P,N_j} = \bar{V}_i \tag{4}$$

Equating both Equations 3 and 4, the following expression is obtained:

$$\left(\frac{\partial \mu_i}{\partial P}\right)_{T,x_j} = \bar{V}_i \tag{5}$$

The ideal expression of the chemical potential can be obtained as:

$$\mu_i = \mu_i^\circ + RT \ln a_i \tag{6}$$

where μ_i° is a constant expressing the chemical potential of the pure compound, a_i is the compound activity, $a_i = \gamma_i x_i$ and determined by the activity coefficient (γ_i) and mole fraction (x_i).

If the solvent vapour obeys the ideal gas law, Equation 6 can be expressed as:

$$\mu_i = \mu_i^\circ + RT \ln \frac{p_i}{p_i^\circ} \tag{7}$$

where p_i is the vapour pressure of solution i at temperature, T, and p_i° is the vapour pressure of pure solvent i at temperature, T.

Thus, the magnitude of activity can be calculated using the formula:

$$a_i = \frac{p_i}{p_i^\circ} \tag{8}$$

Equation 8 resembles Raoult's law which states that the vapour pressure of a solution is equal to the sum of molar fraction times the vapour pressure of pure component for each volatile species. Mathematically, Raoult's law can be defined as:

$$p_i = x_i p_i^\circ \tag{9}$$

For a solution that is sufficiently dilute, Raoult's law can be applied to the system which gives:

$$a_i = x_i = \frac{p_i}{p_i^\circ} \tag{10}$$

Therefore, by substituting Equation 10 into Equation 6, we obtain a new expression to calculate the chemical potential of the solvent.

$$\mu_i = \mu_i^\circ + RT \ln x_i \tag{11}$$

For the isothermal process such as the osmosis process (Figure 2), the chemical potential of the solvent is a function of both the external pressure and concentration of solute as shown in Equation 12. Note that from the above expressions, the component i is represented by the solvent (s-subscript) while the component j is denoted by the solute (sol-subscript).

$$\mu_s = f\left(P, x_{sol}\right) \tag{12}$$

where x_{sol} is the solute mole fraction. By differentiating Equation 12 at a constant temperature, we obtained the expression:

$$d\mu_s = \left(\frac{\partial \mu_s}{\partial P}\right)_{T,x_{sol}} dP + \left(\frac{\partial \mu_s}{\partial x_{sol}}\right)_{T,P} dx_{sol} \tag{13}$$

From Figure 2 (left), the necessary pressure needed to stop the osmosis flow of pure water solvent and keep the liquid level on both compartments at the same level is what is known as the osmotic pressure. The magnitude can be measured from the difference in liquid levels once the process has reached equilibrium. When there is no solvent flow occurs, the chemical potential of solvent, μ_s in the solution is kept constant at μ_s°. Thus, the differential term $d\mu_s$ under this condition is equal to zero. Substituting $d\mu_s = 0$ into Equation 13 and by rearranging it, the following expression is obtained:

$$\left(\frac{\partial \mu_s}{\partial P}\right)_{T,x_{sol}} dP = -\left(\frac{\partial \mu_s}{\partial x_{sol}}\right)_{T,P} dx_{sol} \tag{14}$$

From Equation 11, rewriting in terms of solute and solvent of a solution gives:

$$\mu_s = \mu_s^{\circ} + RT\ln x_s = \mu_s^{\circ} + RT\ln\left(1 - x_{sol}\right) \tag{15}$$

Differentiating the above equation with respect to x_{sol} at constant T and P provides:

$$\left(\frac{\partial \mu_s}{\partial x_{sol}}\right)_{T,P} = -\frac{RT}{1 - x_{sol}} \tag{16}$$

Substituting Equations 5 and 16 into Equation 14 and writing it in integral term leads to:

$$\int_{P}^{P+\pi} \bar{V}_s\, dP = \int_{0}^{x_{sol}} \frac{RT}{1 - x_{sol}} dx_{sol} \tag{17}$$

Based on the system in Figure 2, we can assume that the value of $\bar{V}_s$ is independent of pressure over the range of P to $P+\pi$, i.e. constant, as a liquid is generally incompressible. By integrating Equation 17, we get:

$$\bar{V}_s \pi = -RT\ln\left(1 - x_{sol}\right) \tag{18}$$

For a sufficiently dilute solution, the mole fraction of solute is very small compared to the mole fraction of the solvent. This leads to the approximation that is assumed to be valid for a dilute solution which obeys the Raoult's law:

$$\ln\left(1 - x_{sol}\right) = -x_{sol} - \frac{x_{sol}^2}{2} - \frac{x_{sol}^3}{3} - \ldots - \frac{x_{sol}^{\mathrm{n}}}{\mathrm{n}} \approx -x_{sol} \tag{19}$$

This approximation also allows us to express x_{sol} in terms n_{sol} and n_s which are the number of moles of solute and solvent respectively.

$$x_{sol} = \frac{n_{sol}}{n_s + n_{sol}} \approx \frac{n_{sol}}{n_s} \tag{20}$$

By substituting Equation 19 and Equation 20 into Equation 18, we obtain:

Figure 3. Set-up of juice concentrating process using brine solution by forward osmosis process

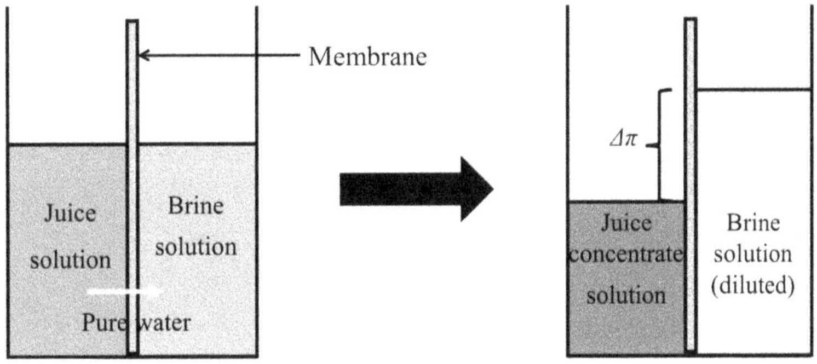

$$\pi = \frac{RT}{\bar{V}_s}\frac{n_{sol}}{n_s} \tag{21}$$

Simplifying the above equation with $V = n_s\bar{V}_s$ and $\frac{C}{MW} = \frac{n_{sol}}{V}$ yields the infamous correlation of van't Hoff to calculate the osmotic pressure of the solution as follows:

$$\pi = \frac{C}{MW}RT \tag{22}$$

where V is the volume of solution, C is the concentration of the solution, and MW is the average molecular weight.

EXAMPLE 1: Concentrating juice by forward osmosis using a brine solution

Problem. Figure 3 shows the set-up of the apparatus for a forward osmosis process operated in batch mode. The objective of the process is to concentrate a juice solution (J) using brine solution (B) with a semipermeable membrane separating the two solutions. The forward osmosis process is driven by the osmotic pressure gradient between the two solutions. An equal amount of volume is used for both solutions.

1. A fruit juice containing 5 wt% of dissolved sucrose ($C_{12}H_{22}O_{11}$) (MW = 342) is to be concentrated at 20 °C by forward osmosis using a brine solution containing 15 g of sodium chloride (NaCl) (MW = 58.5) per 100 g of water. The membrane is permeable to water but not to sucrose and salt. Calculate the osmotic pressure of juice solution and brine solution in bar. Then, find the value of $\Delta\pi$.
2. Calculate the membrane resistance assuming a clean membrane with no cake layer formed or fouling.
3. Assuming that the water extracted from the juice solution is 50% by volume, calculate the final concentration of the juice concentrate.

Solution.

1. Using van't Hoff's correlation, $\pi = nMRT$, where π is the osmotic pressure, n is the van't Hoff's dissociation factor ($n_J = 1, n_B = 2$), M is the molar concentration of the solution, R is the ideal gas constant ($0.0821\,L.atm / mol.K$), and T is the temperature in K. Density of juice solution and brine solution are 1.25 g/mL and 1.15 g/mL respectively.

The osmotic pressure of juice solution:
The concentration of juice,

$$C_J = wt\% \times \rho_J = \frac{5\,g\,sucrose}{100\,g\,solution} \times 1250g / L = 62.5g / L$$

$$\pi_J = n_J M_J RT = n_J \frac{C_J}{MW_J} RT$$

$$\pi_J = (1)\left(\frac{62.5\,g / L}{342\,g / mol}\right)(0.0821\,L.atm / mol.K)(293K) = 4.40\,atm = 4.46\,bar$$

The osmotic pressure of the brine solution:
The concentration of brine solution,

$$C_B = \frac{m_B}{m_{B,soln}} \times \rho_B = \frac{15\,g\,NaCl}{15\,g\,Nacl + 100\,g\,water} \times 1150g / L = 150\,g / L$$

$$\pi_B = n_B M_B RT = n_B \frac{C_B}{MW_B} RT$$

$$\pi_B = (2)\left(\frac{150\,g / L}{58.5\,g / mol}\right)(0.0821\,L.atm / mol.K)(293K) = 123.36\,atm = 125.00\,bar$$

The osmotic pressure difference between juice and brine solutions:

$$\Delta\pi = \pi_B - \pi_J = 125.00\,bar - 4.46\,bar = 120.54\,bar$$

2. Membrane resistance is typically found through the experimental procedure. Thus, the membrane resistance can be calculated once the water flux is known from the solution diffusion equation of membrane,

$$J_w = A_w\left(\Delta\pi - \Delta P\right) = \frac{\Delta P\left(Driving\ force\right)}{\mu R_m},$$

where A_w is the membrane permeability constant (L/m².h.bar), $\Delta\pi$ is the osmotic pressure gradient, ΔP is the pressure gradient, μ is the water viscosity, and R_m is the membrane pressure.

The membrane permeability constant of water is given by 1.776 L/m².h.bar and the viscosity of water at 20 °C is 1.002(10^{-3}) Pa.s (2.783(10^{-12}) bar.h)

Since no pressure is applied ($\Delta P = 0$) on forward osmosis and the driving force is by osmotic pressure, the water flux equation is reduced to:

$$J_w = A_w\Delta\pi = \frac{\Delta\pi}{\mu R_m}$$

$$R_m = \frac{1}{\mu A_w} = \frac{1}{\left(2.783\times10^{-12}\,bar.h\right)\left(1.776\,\frac{\mathrm{L}}{\mathrm{m}^2}.\mathrm{h.bar}\right)}$$

$$= 2.023\times10^{11}\ \mathrm{m}^2 / L = 2.023\times10^{14}\ \mathrm{m}^{-1}$$

3. Using the basis of 1 L for both juice and brine solutions, the solutes present in juice and brine solutions are assumed only $C_{12}H_{22}O_{11}$ and NaCl.

The amount of water drawn from juice solution, $V_{w,o}$:

$V_{w,o} = 0.5\times0.95V_J$, where V_J is the initial volume of juice solution.

$$V_{w,o} = 0.5\times0.95\left(1L\right) = 0.475\,L$$

The amount of sucrose solutes in the 1 L juice solution, $m_{J,s}$:

$$m_{J,s} = C_J\times V_J = 62.5g / L\times1L = 62.5\,g$$

The final concentration of the juice solution, $C_{J,f}$:

$C_{J,f} = \frac{m_{J,s}}{V_{J,f}}$, where $V_{J,f}$ is the final volume of juice concentrate.

$$C_{J,f} = \frac{62.5\,g}{1L - 0.475\,L} = 119.05\,g / L$$

Partial Pressure-driven (Thermal) Operations

Membrane processes using partial pressure or thermal as the driving force has gained considerable attention in membrane separation technology. Processes with partial pressure as the driving force involve gaseous phase(s) in the separation operation. In membrane separation for water and wastewater treatment, the most prevalent membrane process that is driven by partial pressure (or thermal), is membrane distillation. Membrane distillation is an emerging membrane process which uses thermal energy to assist mass transport across a hydrophobic microporous membrane. The vapour pressure gradient between the hot feed stream and cold permeate side acts as the driving force to facilitate the process. There are many attractive benefits associated with membrane distillation, for instance, it requires less energy, low operating costs, and can produce ultra-pure water from brackish or saline water. Despite the many appealing advantages, membrane distillation has yet been implemented in large scale industrial level. It is regarded as a very promising separation method by many researchers, but extensive researches need to be fully executed, especially in understanding the fundamentals of heat and mass transport of the membrane as well as the mechanism of operation, in order to make it as a practical industrial unit.

TYPE OF MEMBRANES AND TRANSPORT RESISTANCES IN PRESSURE DRIVEN PROCESS

General Membrane Classifications

The membrane can be organic or inorganic, natural or synthetic, symmetric (isotropic) or asymmetric (anisotropic), porous or dense (non-porous), thick or thin, or electrically neutral or charged. That being said, the membrane may be classified differently according to one's viewpoint. The membrane is generally classified as organic (i.e. polymeric) or inorganic (e.g. ceramic, metal) membranes, or porous or dense membranes and symmetric or asymmetric membranes which are attributed by the type of material used or morphologies of the membrane. The selection of membrane is an important process as its properties will determine the efficiency of the membrane to execute a specific separation process. Membrane selectivity is dependent on its charge and porosity. In addition, the range of pore sizes of the membrane will determine its structure uniformity, whereby symmetric pores are uniform sizes, while asymmetric pores have varying pore sizes.

Organic vs. Inorganic

The selection of membrane material relies on various factors such as the operating conditions, nature of the application, working feed composition, and separation objectives. Though each organic or inorganic membrane has its own advantages and disadvantages, it is vital to determine what type of membrane is the best fit for each application. The selection of membrane material should be based on the material properties that suit best to the intended application. Some essential membrane material properties include high porosity, high strength, narrow pore size distribution (sharp molecular weight cut-off (MWCO)), good flexibility, broad pH range stability, good tolerance to chlorine, and low costs. Also, the transport resistance in membrane separation may differ in different types of membrane, attributed by the properties of the membrane being used.

Most of the industrial membranes are made from natural or synthetic polymers, whereby both are categorized as an organic membrane. Some examples of naturally-occurred polymers are cellulose, rubber, and wool, while the man-made synthetic polymers include polypropylene (PP), polysulfone (PSf), and polyacrylonitrile (PAN). Synthetic polymers are fabricated through the polymerization process of a monomer or copolymerization of more than one monomers. Polymerization may give rise to three major polymeric structure, which is a linear chain such as polyvinyl chloride (PVC), branched-chain such as polyethene, and cross-linked such as melamine-formaldehyde. Linear and branched polymers have a thermoplastic characteristic, whereby they are mouldable when heated and are readily soluble in organic solvents. Cross-linked polymer, on the other hand, is known as thermosetting polymers, in which it cannot be remoulded upon heating following its initial heat-forming. The cross-linked polymer is also nearly insoluble in organic solvents.

For lower pressure membrane filtration such as microfiltration and ultrafiltration, the commercial polymeric membranes are usually made from synthetic polymers such as cellulose acetate (CA), polypropylene (PP), polysulfone (PSf), polyacrylonitrile (PAN), polyethersulfone (PES), polyvinyl chloride (PVC), polyvinylidene fluoride (PVDF), or polyethene (PE).

Inorganic membranes are membranes fabricated from metals, ceramics, zeolites, or other inorganic materials. These can be fabricated as only sheets or tubes if they are satisfactorily permeable, or as thin films mounted on strong, permeable supporting layers (Verweij, 2012). Moreover, the inorganic membrane is preferred in some application that cannot be satisfied by a conventional polymeric membrane, especially in processes requiring high thermal and chemical stability, high resistance to pressure drop and harsh environment, endurance for microbial degradation etc. However, inorganic membranes require huge capital expenses due to the needs to fulfil the thickness requirement in order to control the pressure drop. They also generally have lower permeability at moderate temperature, brittle, and low membrane surface per module volume.

The most common metallic membrane application is the gas separation between hydrogen (H_2) and other gas mixtures. Palladium (Pd) and its alloy based membrane are the leading material used in this application due to its high permeability and solubility for H_2. Due to its expensive feature, Pd membrane is currently undergoing development through process intensification to produce a thin metallic film on a support material, with possible advantages of less material cost, higher flux., and better mechanical strength. However, a major issue linked to the metallic membrane is the surface poisoning which is more extensive in thin films. The ceramic membrane is made from a variety of metal oxides and is used in the separation process for fluid filtration. Aluminium and titanium oxides are the preferred materials to be used in ceramic membrane fabrication. They are in general chemically inert and can be operated at elevated temperatures. However, they are prone to crack at changing temperature gradient.

Porous vs. Dense

The porous membrane consists of a solid medium with holes or pores having diameters varying from less than 1 nm to more than 10 μm (Strathmann, 2011). Separation is achieved based on the mechanism of sieving or size exclusion, where the pore and particle sizes being the determining factors. Porous membranes are widely applied in microfiltration and ultrafiltration processes. To attain high separation efficiency, the membrane pores should possess a relatively smaller diameter than the particles in the mixture to be filtered. The major drawback of using porous membrane is the membrane fouling which causes flux to decline with time. Besides, concentration and temperature which affect the membrane selectivity and flux should also be considered by selecting material with good chemical and thermal stability.

The non-porous or dense membrane has a dense structure portraying no detectable pore at the extent of electron microscopy (Bazzarelli, Giorno, & Piacentini, 2016). Transport across the dense membrane occurs through molecules dissolving in the dense membrane and diffuse through it by the action of pressure, concentration, or electrical potential gradient. The dense membrane is mostly used in nanofiltration, reverse osmosis, and gas separation applications. One main disadvantage of the dense membrane is that it has low flux, thus most of the time dense membrane is produced as a very thin film and mounted on a support layer as an asymmetric membrane. The composite membrane composed of two separate layers, whereby the top dense membrane acts as a selective film which allows only certain material from passing through, while the thick bottom layer provides mechanical support to the top membrane layer. The selection of membrane material will affect the selectivity and permeability of the membrane, which ultimately affects the separation process which is principally driven from the differences in solubility and diffusivity. Conventional nanofiltration, reverse osmosis membranes are made from polyamide as the thin film, while the membrane support layer is made from polysulfone.

Symmetric vs. Asymmetric

Membrane structure may be symmetric (isotropic) or asymmetric (anisotropic). The symmetric membrane has a uniform cross-section structure which enables consistent flow characteristics, whereas it varies in the asymmetric membrane. Symmetric membranes are produced from the same material through processes such as track-etching. The asymmetric membrane, on the other hand, can be made from the same material in a one-step process as in the case of phase inversion which forms the skin and support layers distinctly. It can also be made from different materials in a two-step process whereby the skin layer is deposited on the support layer. Furthermore, the symmetric membrane is attributed with lower mechanical stability, hence the application is mostly in low-pressure processes such as microfiltration and dialysis where the flux is dependent on the thickness of the membrane. Due to the high mechanical stability of the asymmetric membrane, it is widely employed in a high-pressure operation like reverse osmosis and nanofiltration processes. In terms of the membrane module, symmetric and asymmetric membrane are both used in flat sheet module, but capillary and hollow fibre modules are mostly made using asymmetric membrane.

Transport Resistances in Membrane Separation

The two major phenomena leading to transport resistance in membrane separation is the concentration polarization and membrane fouling.

Concentration Polarization

Figure 4 illustrates the concentration polarization phenomenon which occurs naturally due to the build-up of concentration gradient at the membrane-solution interface due to the membrane's permselective property which allows only certain components to pass through it by the effect of transmembrane pressure. In pressure-driven membrane processes, when pressure is applied to the feed side, the membrane will partially or completely retain the solutes while allowing the solvents passing through freely. Because

Figure 4. Concentration polarization

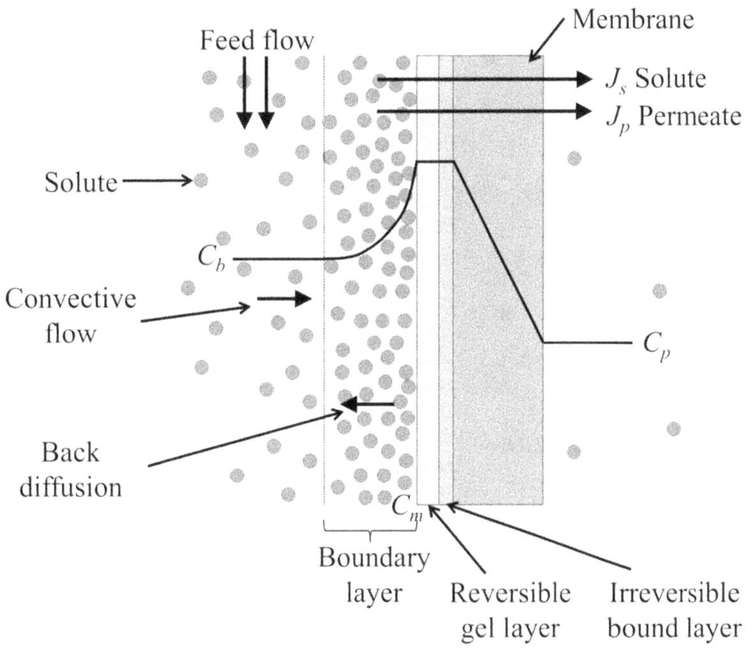

of that, the concentration of solutes in the permeate side (C_p) is lower compared to the concentration in bulk (C_b). The convective flow causes solute build-up on the membrane surface, steadily increasing the concentration of solute on the membrane surface. At steady-state, a stable concentration gradient exists on the membrane surface as balanced by the solute back diffusion from the membrane surface. Eventually, a concentration polarization profile is formed in which the concentration near the membrane, C_m is higher than C_b. The concentration polarization is a mass transport resistance that obstructs the solvent transport across the membrane as a result of reduced membrane flux.

Membrane Fouling

Membrane fouling occurs when particles or solutes accumulate on the membrane surface and/or inside the membrane pores, known as the membrane surface fouling and membrane deep fouling respectively, causing reduced in membrane flux and separation/filtration performance. There are several types of membrane fouling, including inorganic fouling (scaling), organic fouling, particulate/colloidal fouling, and biofouling. Membrane fouling happens in various mechanisms in the porous membrane, typically includes complete pore blocking, internal pore blocking, partial pore blocking, and cake formation as depicted in Figure 5. The membrane blockage caused by the foulants contributes to additional hydraulic resistance to the permeate mass transport. Membrane fouling is a serious challenge in membrane process, as it causes increase in energy requirement as higher energy is required to overcome the fouling resistance in order to maintain membrane flux, reduce in mass transfer rate due to introduction of new resistance, increase in the needs of membrane cleaning through chemical and/or physical means to remove the foulants and increase in operating expenses. Especially for the pressure-driven process during

Figure 5. Membrane fouling mechanisms in porous membrane, (a) complete pore blocking, (b) internal pore blocking, (c) partial pore blocking, (d) cake formation

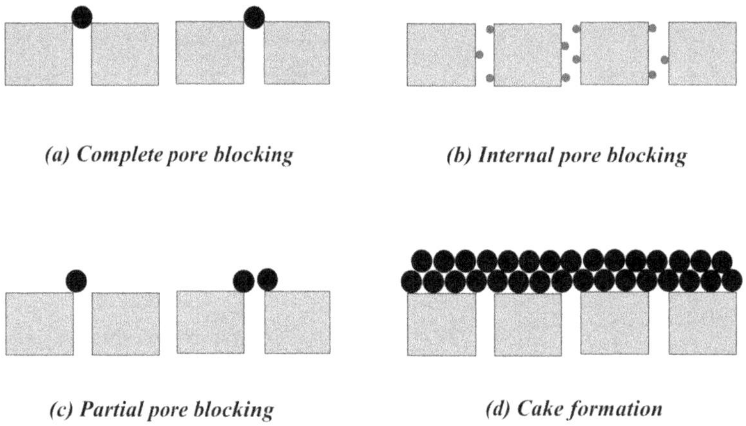

fouling, for a fixed transmembrane pressure (TMP), the flux is observed to drop, or a higher TMP is required when operated at constant flux.

THE PRESSURE DRIVEN OPERATIONAL FLOW MODES AND THE TRANSPORT RESISTANCES IN MEMBRANE SEPARATION

There are two flow modes of operation in pressure-driven membrane systems: (1) dead-end mode, and (2) cross-flow mode. Figure 6 shows a schematic of these two flow modes of operation and the effect of flux and resistance with time. Principally, the dead-end mode is more prone to fouling due to rapid concentration polarization and formation of the cake layer, while cross-flow mode experiences less membrane fouling due to the flow regime of sweeping action across the membrane surface that erode the cakes away. Several important parameters are reviewed to further understand the transport resistances in the pressure-driven operational modes.

Dead-end Mode

In dead-end operation, the feed stream flows in the membrane module perpendicular to the membrane surface which allows the permeate passing through and leave as another stream. The advantage of using dead-end is that it experiences lower energy loss than the cross-flow mode, as the pressure applied directly be used to force solvent permeate through the membrane. Over time, the retained solutes will be accumulated near the membrane surface and form a cake layer. Consequently, the solvent will experience additional resistance to cross the membrane due to the formation of the cake layer. A constant pressure operation will result in the flux to keep decreasing to an extent that the membrane will need to be cleaned.

The dead-end mode is typically only used in microfiltration for feed that is relatively clean to sterilize or filter the feed. Treating concentrated or dirty feed using dead-end mode may result in the membrane to undergo severe fouling and damage making it impractical for such feed conditions. Nonetheless, the formation of cake layer can be controlled by several means, such as (1) periodic cleaning of the mem-

Figure 6. (a) Pressure-driven operational flow modes, and (b) the effects on flux and resistance with time

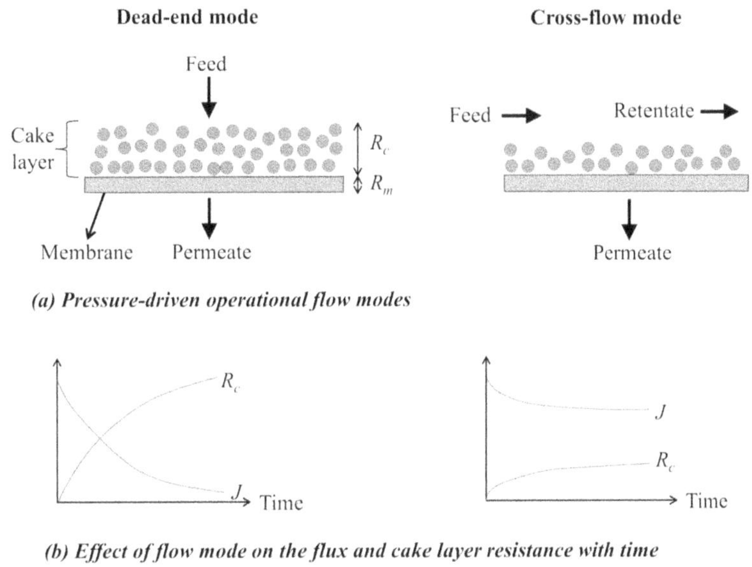

brane, (2) installing a pre-treatment as in sand filtration upstream of the membrane module, or (3) use the cross-flow mode of operation (Singh, 2005). The dead-end mode can also be operated as a batch process, where at a certain time interval, the cake formed will be removed from the filtration process. Example of a batch operated dead-end filtration is the press filter process.

Cross-flow Mode

In cross-flow mode, the feed stream flows tangentially to the membrane, whereby permeate flows across the membrane as the permeate stream while another stream leaving the module as retentate. Typically, the retentate stream in membrane operation using the cross-flow mode will be recycled back into the feed stream. Due to its flow regime, only a small amount of the water is used to produce permeate, while the rest leaves the module as retentate. This causes the cross-flow mode to consume more energy and cost for recycling and permeate recovery.

Despite that, the cross-flow mode is favoured as the flow mode can help to wash away the accumulated suspended solids near the membrane surface. The speed of feed flow in this mode is fairly high, which enables the thickness of the cake layer (resistance) to be reduced and controlled, and thus keep the permeate flux relatively stable. Regardless, periodic cleaning is still required in cross-flow mode. This flow mode operation is generally used for microfiltration, ultrafiltration, and reverse osmosis processes.

Operational Performance Indicators

In this section, several important parameters affecting transport in the membrane. There are many factors which affect the transport and contribute to the transport resistances, such as the transmembrane pressure (TMP) and retention (or rejection) factor.

Figure 7. Pressure drop in cross-flow mode

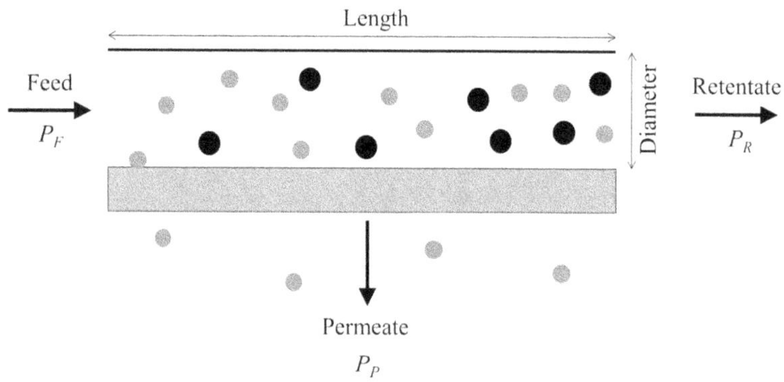

The ability of the solvent to pass through the membrane is attributed by the driving force of pressure provided that is needed to force the solvent across the membrane, which is the TMP. Pressure drop is experienced as the feed flows across the membrane module and can be calculated as shown in Equation 23.

$$\Delta P = P_F - P_R \tag{23}$$

where ΔP = pressure drop, P_F = feed pressure, and P_R = retentate pressure.

Figure 7 indicates the phenomenon of pressure drop in a cross-flow mode operation. Pressure drop is experienced in membrane system due to some factors for instances the feed flow rate, build-up of foulants decrease the flow path areas, and membrane system's spacer orientations and dimensions.

TMP is the pressure difference between two sides of the membrane and can be calculated by using Equation 24. It is very important to know the optimum TMP of the membrane used to identify how much force is required to transport the solvent through the membrane and maintain the membrane's cleanliness at the same time. In most cases, the ideal TMP varies in each membrane, which can be obtained from the manufacturer that fabricate and supply the membrane. The operating TMP can easily be found out and calculated in laboratory or industrial process to optimize the pressure-driven operation and maintain the membrane's lifespan.

First, the pressure of the feed (inlet stream) and retentate (outlet stream) on the outside of the membrane and also permeate pressure on the inside of the membrane can be identified using pressure transducer. By placing the sensor end of the transducer on the targeted areas, each of the pressures can be known. Using these pressures, the TMP of the operation can be calculated by the average of the outside of the membrane pressure minus the pressure in the permeate side.

$$\Delta P = \frac{P_F + P_R}{2} - P_P \tag{24}$$

where ΔP is the transmembrane pressure difference, P_F is the feed pressure, P_R is the retentate pressure, and P_P is the permeate pressure.

The flux of a component is affected by two major factors which are the concentration polarization and membrane fouling. In general, the degree of mass transport across the membrane for a component *i* is influenced by its concentration on the feed and permeate streams. The flux of component *i* represents the amount (in moles, volume, or mass) of component *i* passing through a unit of membrane surface area per unit of time as shown in Equation 25.

$$J_i = \frac{m_i}{A_m t} \tag{25}$$

where J_i and m_i are the flux and the mass of component *i* respectively. A_m is the membrane surface area and t is the time.

For a number of n components permeating across the membrane, the total flux, J_T is the sum of the individual fluxes, J_i of n components as follows:

$$J_T = \sum_{i=1}^{n} J_i \tag{26}$$

In most applications of pressure-driven in water and wastewater treatment, often only one species is of interest, either the solvent or solute, where the retention factor is of concern. Since the permeate flow rate is much less than the retentate flow rate, the change in retentate concentration can be neglected. The retention factor, r_i of component *i* can be defined and used as a measure of separation performance.

$$r_i = 1 - \frac{C_{i_{permeate}}}{C_{i_{retentate}}} \tag{27}$$

where r_i is the retention factor, and $C_{i_{permeate}}$ and $C_{i_{retentate}}$ are the concentration of permeate and retentate for component i respectively.

Retention of any component in the feed can be found experimentally by sampling and analysing the concentration of the feed and permeate sides simultaneously. Then, the retention factor can then be calculated using Equation 27. Retention factor can convey the data on the extent of separation achievable by the membrane, quality, and efficiency. The value of the retention factor is expressed between 0 and 1 (or 0 – 100%), based on the scenarios (Grandison, 1996). This practical method of assessing the retention of a membrane can assist in selecting the most suitable membrane for a particular use, though the retention value could also be affected by operating conditions.

- $C_{i_{permeate}} = 0$, thus $r_i = 1$; all of component *i* is retained in the feed stream;
- $C_{i_{permeate}} = C_{i_{retentate}}$, thus $r_i = 0$; component *i* is freely permeating through the membrane.

Figure 8. Mass transfer through a) pore flow model, and b) solution diffusion model

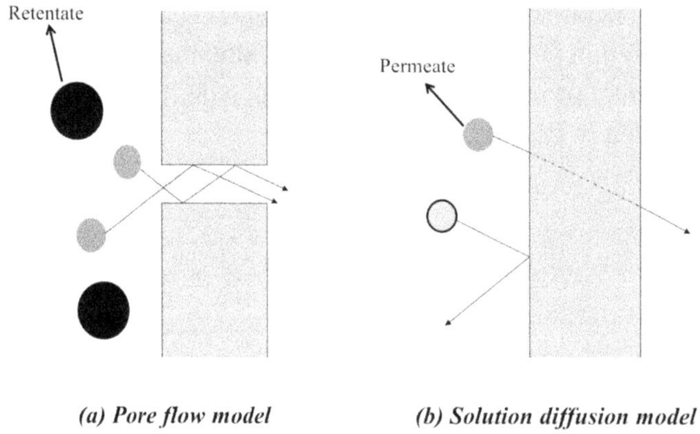

MASS TRANSFER MECHANISMS IN PRESSURE DRIVEN MEMBRANE SEPARATION PROCESS

The mass transport models in pressure-driven processes are generally divided into two which are widely used based on the permeation mechanism. The first model is based on the pore flow (sieving) model (Figure 8 (a)) used to describe the mass transfer across a porous membrane. The permeate is transported through convective flow across the membrane pores. Separation takes place by the action of the semi-permeable membrane which is selectively able to control the permeation rate of different species. Small molecules are allowed to cross the membrane, while molecules bigger than the membrane pores diameter are retained (filtered) from passing through it.

The second mass transfer model is used in dense membrane application which is the solution-diffusion model (Figure 8 (b)). In this model, permeants will dissolve into the membrane matrix and subsequently diffuse across the membrane. Concentration gradient plays a significant role in this mass transport mechanism. The separation between the permeants and other components in the feed solution are due to the differences in solubility as well as the rate of diffusivity for each component in and through the membrane respectively.

The difference in membrane porosity is the key idea behind these two mass transport models. Contrary to popular belief, the non-porous dense membrane actually does have 'pores' which provide narrow passages for the mass transport to occur. These pores are the free-volume elements on the membrane surface. These pores are the tiny gaps between polymeric chains produced from the thermal activity of the polymer molecules. In porous membrane, the pores are rather relatively large free-volume elements with fixed shapes, sizes, position, and volume.

In this section, mass transfer models are discussed by taking into considerations the membrane operation at start-up, during the operation, and after a certain time of operation. The membrane performance will inevitably reduce with time. Because of that, periodic cleaning is required to re-establish the membrane flux. Normally, membrane cleaning should be carried out when the standardized permeate flow reduces by 10 – 15%, or when deemed necessary by the discretion of membrane operators. Other parameters such as pressure drop and permeate purity should also be considered when measuring the performance of the membrane operation. For a properly operated membrane operation, the cleaning frequency is normally less than 5 times per year depending on the nature of the feed water. For a water treatment operation, the membrane can generally last for 5 years of operation, and some even up to 6 – 10 years before the membrane would need to be replaced (Bazzi et al, 2011).

In pressure-driven membrane operation, the process is operated either at constant flux or constant transmembrane pressure (TMP) operational modes. Constant flux mode is typically achieved using vacuum pressure-driven, by which the TMP variable is observed to track the condition of membrane fouling. This mode is usually used in membrane bioreactor operating with submerged membranes. In contrast, constant TMP mode is operated through positive pressure-driven, whereby the flux variable is monitored to trail the state of membrane fouling. In most membrane modules (tubular or plate and frame), the constant TMP mode is employed (Yoon, 2016). In constant TMP mode, the TMP is adjusted to obtain enough initial flux for an optimum operation requirement. The initial flux without any foulant yet is high, which causes particles in water getting pulled towards the membrane surface much faster than they are being carried away from the membrane. Thus, concentration polarization occurs at a faster rate due to the high flux at the beginning of the operation.

At the beginning of a membrane separation process, a clean membrane sheet would theoretically exhibit 100% performance or flux. Ideally, only the membrane resistance (R_M) are involved. The membrane process start-up portrays unsteady-state operation as the system is acclimatizing with the new environment. During this dynamic stage, some variations may be encountered in terms of feed and/or permeate quality, feed and/or permeate flow rate, and membrane performance. After some time, the membrane flux will inevitably decline due to the accumulation of rejected materials near the membrane surface which presents new resistance for permeation as compared to the beginning of the operation. Accumulation of the rejected materials leads to two phenomena which are the concentration polarization and membrane fouling as seen in Figure 9 and Figure 10.

During the cross-flow mode membrane operation, particles are brought close to the membrane surface by convection due to the applied pressure, where most are accumulated on the membrane surface. This condition caused concentration polarization, whereby the rejected solutes have the tendency to accumulate at the membrane-solute interface within a concentration boundary layer. The particles build up increased their concentration over the bulk feed at a rate of exponential rise with increasing flux. The thickness of the concentration boundary layer is determined entirely by the system's hydrodynamic, whereby turbulent flow may be used to decrease the thickness, whereby laminar flow may promote the increase in layer thickness. Under normal steady-state operation, the rate of water and solutes within the boundary layer towards, through and away the membrane is constant, which is controlled by the concentration polarization. Concentration polarization also increases the required operating TMP, hence promoting turbulence flow of the feed or operating at a flux below that of which concentration polarization starts to become significant. In certain cases, the presence of concentration polarization is rather significant, as the filtration would be taking place by both the concentration boundary layer and membrane itself.

The particles in the concentration boundary layer may flow across the membrane along with the permeating species if the sizes are small enough to pass through the pores. The gradual development of the concentration boundary layer can bring upon the formation of a gel or cake layer, which is a form of fouling. Over some time of operation, the occurrence of concentration polarization layer, gel layer (cake deposition), pore blockage, and adsorption on pores generate additional resistances on the feed side to the transport through the membrane. If the convective force is high, or back diffusion force is weak, the particles may be deposited and form a linked layer which is either reversible or irreversible. In the reversible type, the particles can be removed from the membrane surface back into the feed bulk

Figure 9. Formation of the concentration boundary layer and cake layer

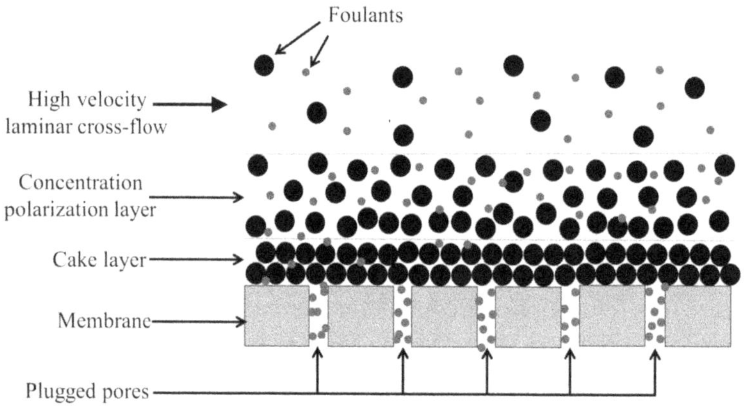

through back diffusion and other means such as of axial dynamic or Brownian diffusion. During steady-state, the convective transport of solutes to the membrane is balanced by the total of solutes flow in the permeate stream and the back diffusive transport of the solutes.

Succinctly, the membrane separation process and mass transport through membranes is a function of the membrane utilized and the components being separated, and subsequently, the theory used to define the operations and mechanisms. However, the mutually shared principle for all membrane system

Figure 10. Resistances of mass transport across the membrane due to concentration polarization and membrane fouling

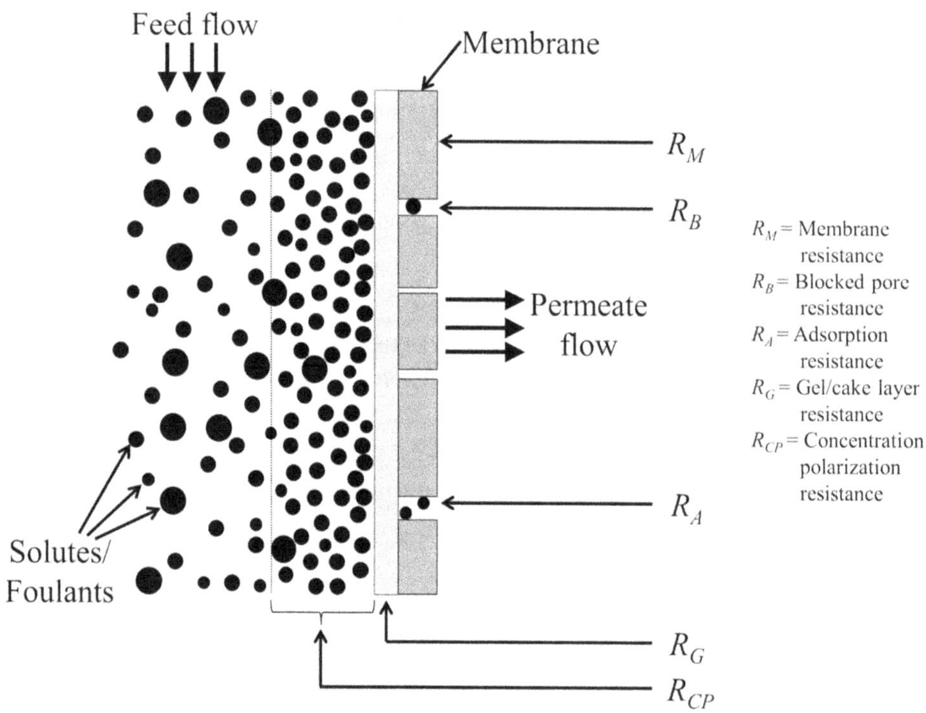

is illustrated in Figure 1 where a membrane is a thin film separating two phases or what's commonly known as the permselective barrier. Furthermore, the membrane separation process occurs due to the existence of a specific driving force which enabled the transport of a component from one phase to the other. The membrane selectivity is normally expressed as the retention factor (r_i) (Equation 27) or by separation factor (S_F).

Retention factor is usually used to define the selectivity of a dilute aqueous solution. The value derived from Equation 27 has a dimensionless unit expressed as a ratio or percentage between 0% to 100%. The value of $r_i = 0\%$ indicates that both solute and solvent are freely passing through the membrane, while of $r_i = 100\%$ means absolute retention of the solute is achieved.

Separation factor is generally used to describe the degree of separation for gas mixture or organic mixture. In a binary mixture consisting of compounds A and B, the separation factor ($S_{FA/B}$) is defined by Equation 28 with respect to the concentrations in the feed ($C_{A_{feed}}$, $C_{B_{feed}}$) and the permeate ($C_{A_{permeate}}$, $C_{B_{permeate}}$). The selectivity should be expressed using the terms by which component can pass readily across the membrane, i.e. component with a higher permeation rate. For example, the term $\alpha_{A/B}$ indicates that component A is the species with higher permeation rate. When $S_{FA/B} = S_{FB/A}$, it means that no separation is attained.

$$S_{FA/B} = \frac{C_{A_{permeate}} / C_{B_{permeate}}}{C_{A_{feed}} / C_{B_{feed}}} \tag{28}$$

A good membrane is governed by its selectivity which suits a specific separation process to ensure that the membrane can achieve the separation of molecules that is required. Once the membrane selectivity is known, the membrane flux which is the main parameter that defines the mass transport across the membrane can be determined. The magnitude of flux is largely defined by the driving force that acts on the transport mechanism. The driving force is responsible for the transport phenomena which governs the molecules or particles. The amount of the driving force will depend on the gradient of the potential, across the membrane. The main potential gradients in membrane processes are the chemical potential difference and the electrical potential difference. The chemical potential difference generally determined the driving forces of either the pressure gradient, concentration gradient, and temperature gradient between the two sides of the membrane. Besides, the electrical potential is usually only significant for certain membrane processes such as electrodialysis.

In the thermodynamic perspective, spontaneous mass transport occurs only if the chemical potential gradient of the species to be transported is negative. The chemical potential gradient across a membrane is the difference between the chemical potential of permeate and chemical potential of feed. Thus, in pressure-driven membrane processes, spontaneous mass transport occurs when the driving force is generated by increasing the feed pressure or decreasing the permeate pressure.

Figure 11. Basic pore geometries observed in a porous membrane, (a) symmetrical cylinders, (b) packed bed spheres, and (c) asymmetric/sponge-like

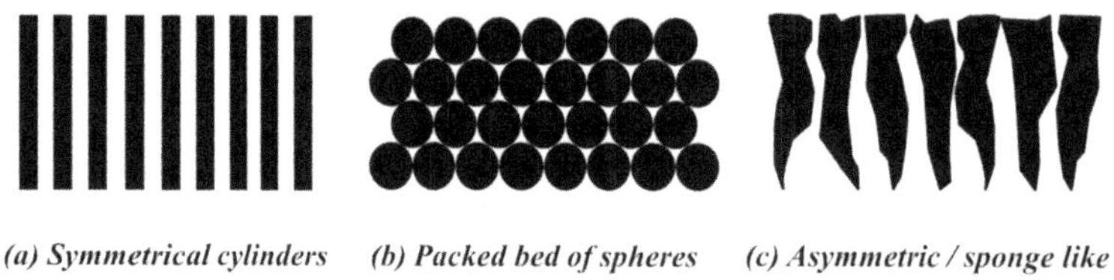

However, the transport phenomena in membrane processes are more than just understanding the fundamental of thermodynamics that is taking place, though it is significant as one of the perspective to describe the transport phenomena. There are many transport models being developed on the basis of thermodynamics principles, membrane properties, and membrane structure. In the subsequent sub-sections, the mass transport models are discussed based on the main distinction of membrane structure, i.e. porous and dense membrane. Nevertheless, some models are applicable to both types of membrane structure.

Mass Transfer Models in Porous Membrane: Convective Transport Mechanism

Porous membrane exhibits voids or pores that allow the passing of particles smaller than its sizes through the membrane by means of sieving mechanism. The conditions of membrane pores are highly varied, depending on the membrane material and fabrication method used. In a single membrane sheet, the pore sizes may vary depending on the shapes and structures. Mass transport models in the porous membrane have been developed by taking serious incorporation in terms of the membrane pore geometries and membrane characteristics that affect the transport mechanism. Parameters such as the pore sizes, porosity, pore size distribution, pore structures and geometries are important to be considered while developing the mass transport models. In general, the basic pore geometries observed in the porous membrane are represented in Figure 11.

The mass transfer across the porous membrane is best described using the capillary model which shows the flow of fluid through the microporous membrane under several assumptions. These assumptions put the membrane system under ideal conditions of uniform distribution of even membrane pore sizes, no fouling occurs on the membrane, and the concentration polarization is negligible.

The first geometry is the cylindrical pores vertical to the membrane surface, wherein an ideal condition would possess pores having identical radius size, though in reality the geometry may not be as common and well defined and have some variations on the structure. In this geometry, the flux may be expressed using the Hagen-Poiseuille equation for permeability as follows:

$$J = \frac{\varepsilon r^2}{8\mu\tau} . \frac{\Delta P}{\delta} \tag{29}$$

where J is the flux, ε is membrane porosity, r is the pore radius, μ is the viscosity, τ is the tortuosity, ΔP is the transmembrane pressure across the membrane, and δ is the membrane thickness.

For cylindrical-shaped pores, the tortuosity is 1. Different shapes of the actual membrane pores make it difficult to determine the tortuosity parameter of the membrane. Equation 29 clearly shows that the morphology and membrane structural characteristics are significant to the transport across the membrane. It gives good determination for precisely parallel pores, but this is not common in actual membranes condition which becomes a challenge for determining the flux.

The second geometry is the porous structure formed when a packed bed are made from spherical shape components to form a porous media. This type of porous structure can be found in membranes fabricated through the means of sintering and phase inversion. The mass transport through porous media of packed spheres is established by Carman-Kozeny as shown in Equation 30.

$$J = \frac{\varepsilon^3}{K\mu A_S^{\ 2}\left(1-\varepsilon\right)^2} \cdot \frac{\Delta P}{\delta} \tag{30}$$

where J is the flux, ε is membrane porosity, K is the Carman-Kozeny constant, μ is the viscosity, A_S is the internal surface area, ΔP is the transmembrane pressure across the membrane, and δ is the membrane thickness.

The parameters given by Equation 30 are dependent on the pore geometry which is difficult to calculate practically as it is the function of the medium grain size, grain shapes, and packing density. In this model, several mass transfer mechanisms can occur including convective flow, Knudsen diffusion and pore surface diffusion. However, only convective flow is prevalent for liquid transport as the other two mechanisms primarily important for gas/vapour transport through porous media as discussed in Section 2-6.

Mass Transfer Models in Dense Membrane: Solution Diffusion Mechanism

Nonporous membranes are generally fabricated from organic polymers. It can be described as a membrane with a homogenous structure which has no characterized pores. From a molecular level perspective, the morphological structure resulting from the polymer molecular strings does indicate a structure described as the molecular pores. This indictment is important when discussing transport across different types of nonporous membranes. Succinctly, transport through nonporous membranes can be explained as a solution diffusion mechanism.

Nonporous membranes are sometimes known as solution diffusion membranes since the transport takes place when molecules dissolve into and diffuse across the membrane. In theory, the mechanism of transport is similar to whether the separation occurs in the liquid or gas phase. Nevertheless, the liquid and gas have different solubility into the membrane polymer, whereby liquid possesses a higher affinity for polymers compared to gas. Thus, the transport of the liquid and gas through the membrane layer would certainly also be different.

The transport across the membrane can be explained by the membrane permeability (P_i). Permeability of membrane is largely influenced by the solubility (S_i) and diffusivity (D_i) of the material used to fabricate the membrane. Permeability of a membrane can be portrayed as:

$$P_i = S_i \times D_i \tag{31}$$

Generally, solubility is a thermodynamic parameter that refers to the ability of a solute to dissolve in a solvent. In this context, the solubility of a component dissolving into the membrane mainly depends on its state (e.g. whether it is gas or liquid). Normally, gas has low solubility in polymer and can be defined by using Henry's law. Conversely, diffusivity is a kinetic parameter that describes the speed of the component being dissolved/transported through the membrane. The diffusivity of the component is largely influenced by the properties of the molecule and geometry of the polymer membrane.

For dense membrane, the solution diffusion model (which is discussed in this section) and the free-volume diffusion model are the two most widely accepted transport model across the membrane. The theory says that the transfer of molecules within a matrix depends on the free volume and energy available to overcome the polymer-polymer interaction. The free-volume diffusion model is relatively complex to explain the theory of transport as the free-volume parameter has been rather difficult to be defined simply.

Three steps are used as the basic assumptions to describe mass transport through the membrane in the solution diffusion model, as follows:

1. Dissolution of the selective component into the feed side of the membrane;
2. Diffusion of the selective component through the membrane; and
3. Desorption from the permeate side of the membrane.

During the transport, steps 1 and 3 happen very fast relative to step 2, thus the diffusion of the component through the membrane becomes the rate-limiting step of the mass transport.

In the solution-diffusion model, the flux of a component is influenced by the driving force taking place, which is related to the potential difference. The flux of a component can be defined by using the following equation:

$$J = -L_i \frac{d\mu_i}{dx} \tag{32}$$

where the differential term $\frac{d\mu_i}{dx}$ is the chemical potential gradient of compound *i*, and L_i is a proportionality coefficient. The proportionality coefficient that connects the flux with chemical potential can be expressed in any of the driving forces used in the specific application (e.g. pressure, temperature, concentration, electrical).

Fick's first law describes the relationship between the diffusive flux and the concentration at steady state. The transport of molecules (or flux) occurs from the high concentration areas to the low concentration areas. Fick's first law can be mathematically expressed as:

$$J = -D\frac{dC}{dZ} \tag{33}$$

where J is the flux of the membrane, D is the diffusivity of diffusion coefficient, C is the concentration of diffusing substances, and Z is the space coordinate measured normal to the membrane.

Integrating this expression over the membrane thickness (δ) generates the following expression, where $C_{m,f}$ and $C_{m,p}$ are the component concentrations at the membrane interface.

$$J = D\frac{\left(C_{m,f} - C_{m,p}\right)}{\delta} \tag{34}$$

Overall, many versions of solution diffusion models have been developed by many researchers. The different models exist to take into account the limitation of nonporous membranes produced during manufacturing. Fine pores may be developed in some membrane produced, in which a solution diffusion imperfection model is used to consider the transports that may occur through the fine pores.

Resistance-in-series Model

The mass transfer models reviewed in Sections 2-5.1 – 2-5.2 were discussed on the basis of membrane structure. The convective transport and solution diffusion models are designed to describe the membrane separation processes that occur in porous and dense membranes respectively. However, these models cannot truly encompass the accurate representation of the mass transport due to their specific dependence of the structural parameters of the membrane. Membranes are manufactured for various applications that require distinct structures and compositions that may not fit the simple definition of porous/dense membrane. In many instances, a membrane may possess characteristics that are associated with both porous and dense classification. Thus, the transport for these types of membranes can be expressed using a more general transport model that can be applied for various situations.

The resistance-in-series model describes the mass transport of a component from one phase to another (e.g. liquid/liquid, liquid/gas) across the resistance. This model is analogous to the electrical circuit which states that the flow of electric current along the circuit will need to overcome a series of resistances. In the context of mass transfer across the membrane, resistance is present in every transport from one phase to another. Summation of the series of resistances will total to the effective resistance. The essence of the resistance-in-series model is to identify all the resistances to mass transfer and the definition for each resistance. Figure 12 shows the key principle of resistance-in-series model. The total resistance to transport is the sum of the series of resistances in each phase.

Based on Figure 12, the resistance in phase 1 (R_1) is usually the feed side which can be caused by the build-up of cake layer on the membrane surface, concentration polarization, or gel formation. The membrane resistance (R_M) is generally caused by the structural parameter of the membrane (e.g porous, dense) and various fouling phenomena. Furthermore, the resistance in phase 2 (R_2) on the permeate side is influenced by the efficiency of the permeate recovery from the membrane. The overall mass transfer resistance can be calculated using Equation 35, whereby the reciprocal of the overall resistance is equal to the total summation of the reciprocal resistances defined in the system. The reciprocal of the overall resistance is also the overall mass transfer coefficient of the system.

$$\frac{1}{R_{\mathrm{M,T}}} = \frac{1}{R_1} + \frac{1}{R_M} + \frac{1}{R_2} \tag{35}$$

where $R_{M,T}$ is the total resistance, R_1 is the resistance in phase 1, R_M is the resistance in the membrane, and R_2 is the resistance in phase 2.

Figure 12. The key principle of the resistance-in-series model

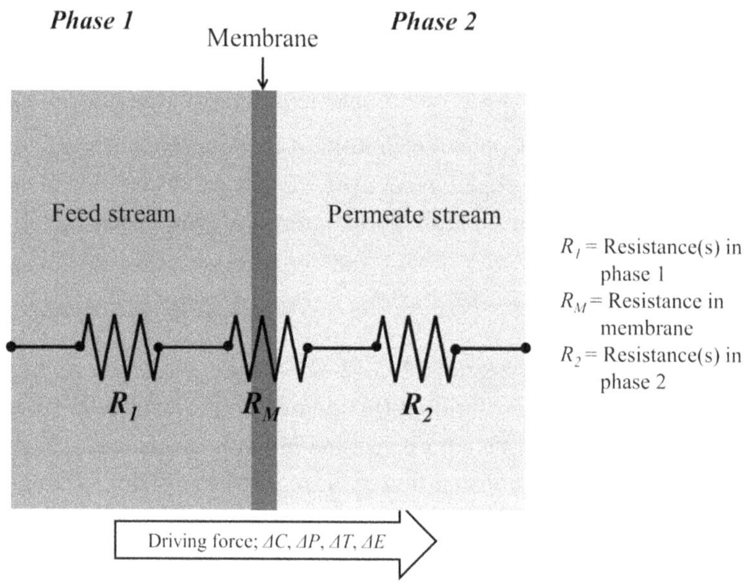

The resistance-in-series model can also be used to describe the transport in porous and dense membranes. The membrane resistance is expressed based on the membrane properties for each application. For porous membrane, the parameters of pore size and membrane geometry will determine the mode of transport being either convective diffusion, Knudsen diffusion, or molecular diffusion. In a dense membrane, the membrane resistance is expressed using the permeability and thickness of the membrane. For a composite membrane comprising both the porous support layer and active dense layer, a combination of the parameters are taken into account.

Various resistance-in-series models have been reported for the many membrane applications. In most cases, this model is used to express the mass transport across the membrane in terms of the overall resistance. This is generally true for pressure-driven membrane processes, in which only the mass transfer is particularly of significant interest. In some cases, an extension of the model to incorporate the additional resistances of the transport phenomena that significantly affect the separation process is considered. In particular, the resistance-in-series model is used to describe both the heat and mass transfers in membrane distillation, as the heat transfer is, in particular, a great influence to the extent of mass transfer. The application of resistance-in-series model for heat and mass transfer is discussed in detail in Section 2-6.2 and 2-6.3.

MASS TRANSFER MODELLING IN PARTIAL PRESSURE (THERMAL) MEMBRANE SEPARATION PROCESS

Most membrane separation processes are isothermal, operating with either pressure (e.g. microfiltration, ultrafiltration, reverse osmosis), concentration (e.g. dialysis), or electrical potential (e.g. electrodialysis) gradient as the driving force. In other cases, some membrane separation processes use partial pressure or thermal gradient as the driving force, such in the case of membrane distillation. The main focus in this section is to discuss the heat and mass transfer models in membrane distillation which is an important membrane desalination process to produce treated pure water. The mass transfer focus in membrane distillation is unique because the transport occurs in the gaseous phase, whereby volatile solvent evaporates and permeate across the membrane as vapours unlike conventional pressure-driven processes with liquid permeates. Membrane distillation is a membrane separation process based on the principle of evaporation which uses hydrophobic membrane as separating medium. Unlike the membrane processes of ultrafiltration and reverse osmosis which only focus on mass transfer, membrane distillation relies on heat transfer, as it is a key factor affecting the mass transfer efficiency. Thus, the investigation of heat and mass transfer in the membrane distillation process is the key to explore the mechanism of the membrane distillation process. Many scientists have made significant progress in the study of heat and mass transfer in the membrane distillation process.

The transport of permeate across the porous membrane in membrane distillation is driven by the partial pressure difference between the feed side and permeate side. In order to allow the permeate transfer, external energy is applied to increase the chemical potential of the permeate to enable the transfer to occur effectively. In membrane distillation, the source of energy supplied to increase the chemical potential is heat. This can be seen from the direct relationship between the chemical potential to the temperature as shown in Equation 6. Generally, the feed is brought up to the temperature of around 40 – 70 °C which is sufficient to increase the potential of permeating (the more volatile) species. At this temperature range, the heat is just adequate to allow the phase change of permeating component from liquid to vapour by evaporation, but not for the other non-volatile species. This is a significant key behind the membrane distillation process as only the vapour of the desired permeant should be transported across the membrane pores.

Another way to increase the transport of permeating species in membrane distillation is by creating a force gradient, i.e. thermal gradient or partial pressure gradient between the two sides of the membrane. Natural transport phenomena tend to happen down the force gradient, whereby in membrane distillation, the vapours move from the region of high partial pressure (feed side) to the low partial pressure region (permeate side). To maintain this condition, the feed temperature should be kept around constant to ensure steady evaporation of the permeating species. Hence, the membrane used, as well as the system's material of construction should have low thermal conductivity to prevent heat loss to the other areas. Once these are achieved, the partial pressure of the permeate side should be kept low to ensure that the vapours will migrate into there. There are several ways of keeping the partial pressure low, and it's mainly done by making sure that the permeate vapour is continuously recovered from the permeate side. If not, the accumulation of vapour molecules on the permeate side will lead to the rise of concentration and eventually disrupts the transport of vapour. Thus, there are four general configurations of membrane distillation which are designed in different ways vapour pressure difference is generated and permeate is recovered. These four configurations are the direct contact membrane distillation, air gap membrane distillation, sweeping gas membrane distillation, and vacuum membrane distillation which are discussed more in Chapter 8 of this book. Figure 13 illustrates the mass and heat transfer mechanisms for all four types of membrane distillation configurations.

The transport of vapour across the membrane distillation will only take place if the liquid feed stream is in direct contact with the membrane film as to what is called the liquid-membrane interface. On the membrane surface at the feed side, the vapour-liquid and solid-liquid interfaces emerged which enabled the transports to occur. It is important to keep the hydro pressure on the vapour-liquid interface lower than the liquid entry pressure of water (LEP_w) of the membrane. This is to prevent the undesirable effect of membrane pore wetting which disrupts the efficiency of the separation process. Hydro pressure is the total of the hydrostatic pressure (due to the fluid at rest) and hydrodynamic pressure (due to fluid in motion). During the operation of membrane distillation, the mass and heat transfers will encounter transport resistances from several sources such as the feed phase (polarization effects), membrane resistance, and permeate phase (polarization effects). These resistances are important considerations to form the transport models of the membrane operation.

The principle of operation of each membrane distillation configuration differs in the ways the permeate side is controlled to facilitate the continuous recovery of permeate. Indirect contact membrane distillation, cooling water flow is used to create the thermal gradient required and will condensate the permeating vapour. The condensation of vapour will allow the partial pressure to be maintained at the permeate side. For the air gap membrane distillation, an air gap is employed between the membrane sheet and a cooling plate which will condensate the vapour in contact. It keeps the low temperature at the permeate side and controls the partial pressure by condensing the vapour to liquid which prevents the increase of vapour concentration. In the sweeping gas membrane distillation, the inert gas is utilized to sweep the permeate vapour out of the chamber to the outside of the membrane system into a condensation unit. The sweeping gas can keep both the vapour concentration and the temperature at low, but it increases the pressure of the permeate stream. Finally, the vacuum membrane distillation uses a vacuum pump which creates suction that recovers the vapour permeate. The permeate is then transferred to a condensation unit which condenses the vapour back to liquid. The use of vacuum helps to reduce the pressure, temperature, and recover a good amount of vapour in the permeate side, but it could lead to large transmembrane pressure difference if the vacuum pressure is too low. Consequently, the transmembrane pressure may exceed the LEP_w and eventually cause the membrane to be wetted.

Characteristics of Membrane for Membrane Distillation

Some essential membrane characteristics of the vacuum membrane distillation process are the pore size, pore size distribution, membrane thickness, LEP_w, thermal conductivity, porosity, and tortuosity. These parameters are important to ensure high performance of the membrane process which are briefly discussed in this section.

Pore Size

The membranes used in membrane distillation have average pore sizes between 0.1 – 1 µm. The probability of liquid entry in vacuum membrane distillation is reported to be higher than other membrane distillation configuration, thus small pore sizes membrane should be used (Abu-Zeid et al., 2015).

Thickness

The permeate flux is inversely proportional to the membrane thickness. Thin membrane is desired to obtain higher flux, higher permeability, and ultimately higher yield. Nonetheless, the thickness of the membrane will influence the conductive heat loss through the membrane, but it can be considered negligible since the membrane has low conductivity. According to Lagana, Barbieri, and Drioli (2000), the optimum membrane thickness is between the range of 30 – 60 µm.

Figure 13. Mass and heat transfer mechanisms in (a) direct contact membrane distillation, (b) air gap membrane distillation, (c) sweeping gas membrane distillation, and (d) vacuum membrane distillation

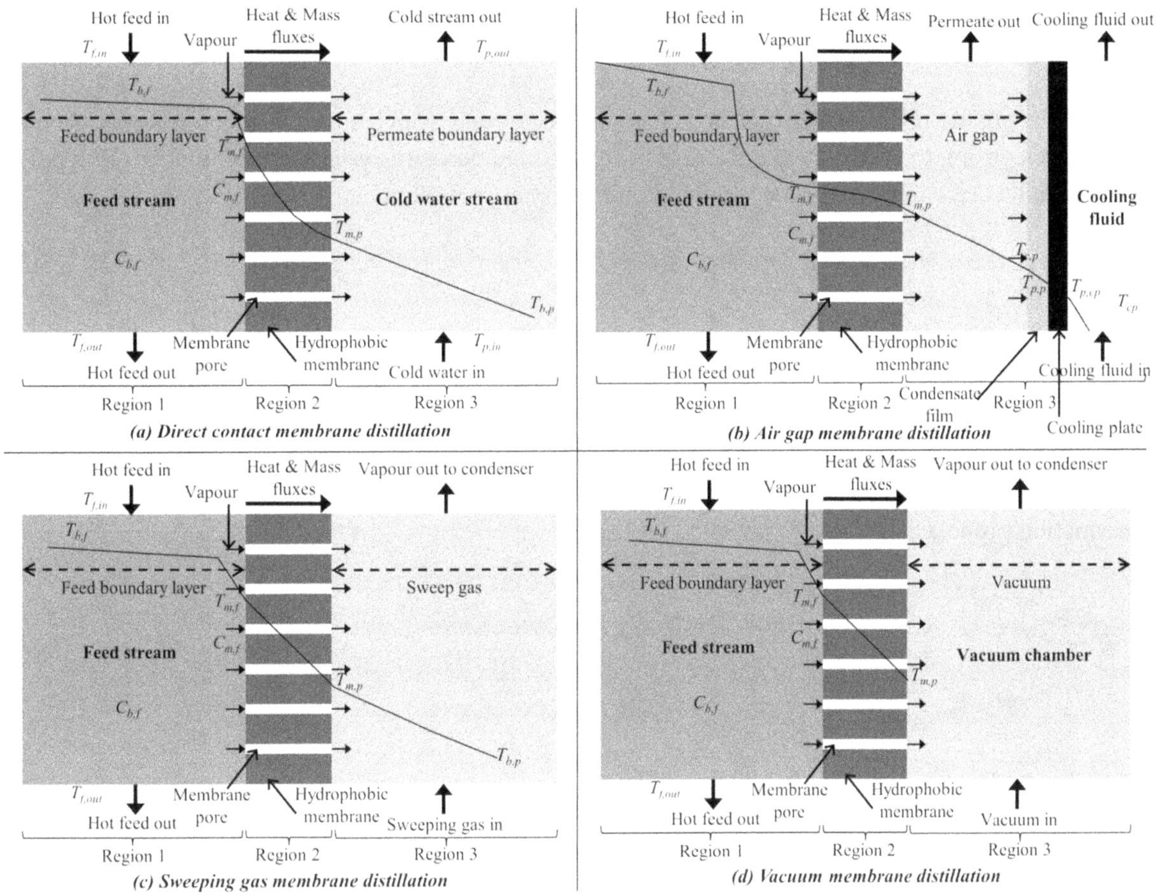

Liquid Entry Pressure (LEP)

Liquid entry pressure (LEP) is the lowest transmembrane pressure that will cause an aqueous solution to infiltrate and wet the membrane pores. This is generally known as the membrane wetting phenomenon. To prevent membrane wetting, a hydrophobic membrane with high LEP is necessary to be used in the membrane distillation application. Membranes with high LEP is generally characterized by the low surface energy material, high surface tension, and small maximum pore size (Abu-Zeid et al., 2015).

Thermal Conductivity

The thermal conductivity of polymer and gas (usually air) is used to assess the membrane thermal conductivity. Besides, temperature, degree of crystallinity and shape of the crystal have a major effect on the polymer thermal conductivity. Different hydrophobic materials, such as PVDF, PTFE and PP, have asymptotic values of coefficients of thermal conductivity. To sum up, the thermal conductivity of the membrane material should be as low as possible, to obtain higher permeation flux and further to minimize the conductive heat loss.

Porosity and Tortuosity

Membrane porosity refers to the voids present on the membrane which are significant to the rate of mass transfer. According to El-Bourawi et al. (2006), membrane porosity varies from 30 to 85%. The porosity (ε) can be determined by Smolder–Franken's equation as follows:

$$\varepsilon = 1 - \frac{\rho_m}{\rho_p} \tag{36}$$

where ρ_m and ρ_p are the densities of membrane and polymer material respectively.

Tortuosity (τ) of a membrane is the deviation of pore structure from the cylindrical shape. Accordingly, the higher the tortuosity value, the lower the permeate flux. The tortuosity can be calculated using the equation proposed by Srisurichan et al. (2006) as:

$$\tau = \frac{\left(2 - \varepsilon\right)^2}{\varepsilon} \tag{37}$$

Heat Transfer

Heat can be transferred from one location to another using three methods: (1) conduction, (2) convection, and (3), radiation. Heat is transferred through conduction when two objects with different temperature are in contact with each other. The heat will flow from the hotter object to the cooler object until a steady-state temperature is achieved. In general, solid is a better conductor than liquid and gas, while the liquid is a better conductor than gas. Convection occurs in a circulation pattern as a result of the heat transfer from the warmer liquid/gas to the cooler liquid/gas. On the other hand, the heat transfer through radiation does not require contact between the heat source and the heated object. In particular, radiation does not require a mass exchange or medium to transfer heat. Heat is transmitted in empty space by thermal radiation which is a type of electromagnetic radiation. In membrane distillation, only the heat transfer by conduction and convection are significant, whereas heat transfer by radiation is negligible and is omitted in the discussion.

Heat Transfer Resistances in Membrane Distillation

In general, the heat transfer in membrane distillation consists of three steps. Figure 14 illustrates the heat transfer resistances in membrane distillation based on the steps shown below.

Step 1: Heat transfer by convection in the feed boundary layer;

Figure 14. Heat transfer resistances in membrane distillation

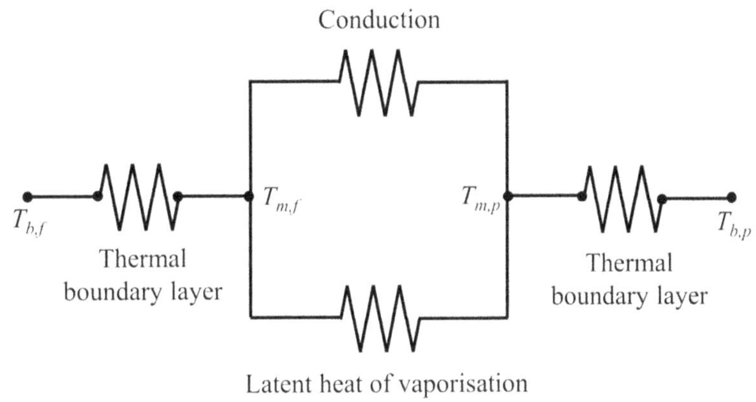

Step 2: Heat transfer by conduction through the solid membrane plus the heat flow by vapour molecules across the membrane which carries heat together (i.e. the latent heat of vaporization); and
Step 3: Heat transfer by convection in the permeate boundary layer.

For all four of the membrane distillation configurations, the heat transfer mechanisms are similar for region 1 and region 2 of Figure 13. However, differences in heat transfer mechanism are observed for region 3 for each configuration due to the different methods permeate are recovered.

Heat Transfer by Convection in the Feed Boundary Layer

On the feed side, the heat transfer by conduction in the boundary layer can be determined by the equation:

$$Q_f = h_f\left(T_{b,f} - T_{m,f}\right) \tag{38}$$

where Q_f is the heat flux through the feed boundary layer, h_f is the heat transfer coefficient of feed, $T_{b,f}$ is the bulk temperature of the feed, and $T_{m,f}$ is the membrane surface temperature on the feed boundary layer.

Heat Transfer Across the Membrane Matrix Through Conduction and Latent Heat of Vaporization

For the heat transfer across the porous membrane, heat is transferred together with the mass flux through the membrane pores. Furthermore, heat is also transferred by conduction through the membrane matrix and the vapours that are trapped inside the pores. The heat transfer across the membrane can be described as the summation of heat transfer by conduction through the membrane with the heat transfer by the latent heat of vaporization carried by the permeating vapour as shown in Equation 39 and 40.

$$Q_m = \frac{k_m}{\delta}\left(T_{m,f} - T_{m,p}\right) + J\Delta H_v \tag{39}$$

$$Q_m = h_m\left(T_{m,f} - T_{m,p}\right) + J\Delta H_v \tag{40}$$

where Q_m is the heat flux through the membrane, J is the flux, ΔH_v is the latent heat of vaporization of the feed, h_m is the heat transfer coefficient of membrane calculated by $h_m = \frac{k_m}{\delta}$, k_m is the thermal conductivity of membrane, δ is membrane thickness, $T_{m,f}$ is the membrane surface temperature on the feed boundary layer, and $T_{m,p}$ is the membrane surface temperature on the permeate boundary layer.

In cases where the thermal conductivity of the membrane is very low and negligible, the equation of heat transfer across the porous membrane from Equation 39 is reduced to:

$$Q_m = J\Delta H_v \tag{41}$$

Nonetheless, the value of k_m can be determined by a series of correlations if the mixture is complex and the value of k_m is difficult to be estimated. This is because, the thermal conductivity of membrane fill encompasses the properties of the membrane material, air and vapour which can be hard to measure. Thus, Equation 42 and 43 give the prediction of thermal conductivity of membrane (Olatunji & Camacho, 2018):

$$k_m = \left(1 - \varepsilon\right)k_{mm} + \varepsilon k_g \tag{42}$$

$$k_m = \left(\frac{1-\varepsilon}{k_{mm}} + \frac{\varepsilon}{k_g}\right)^{-1} \tag{43}$$

where k_{mm} is the thermal conductivity of the membrane material, and k_g is the thermal conductivity of gas in the pores of the membrane.

Thermal conductivity of gas is considered the same for both air and water vapour and can be defined by Equation 43. Similarly, the thermal conductivity of the hydrophobic membrane material can be predicted at various temperatures as shown in Equation 45.

$$k_g = 1.36 \times 10^{-3} + 3.885 \times 10^{-5} T_m + 1.66 \times 10^{-3}\sqrt{T_m} \tag{44}$$

$$k_{mm} = \alpha \times 10^{-4} T_m + \beta \times 10^{-2} \tag{45}$$

where T_m is the membrane temperature calculated from the average of membrane surface temperature at feed and permeate sides, and α and β are the membrane material constants.

Heat Transfer by Convection in the Permeate Boundary Layer

For the convective heat transfer in the permeate boundary layer, the heat transfer equation is generally given by:

$$Q_p = h_p\left(T_{b,p} - T_{m,p}\right) \tag{46}$$

where Q_p is the heat flux through the permeate boundary layer, h_p is the heat transfer coefficient of permeate, $T_{b,p}$ is the bulk temperature of the permeate, and $T_{m,p}$ is the membrane surface temperature on the permeate boundary layer.

In the permeate side region, the different types of membrane distillation configurations will affect the heat transfer mechanism that occurs. Equation 46 is applicable for direct contact membrane distillation and sweeping gas membrane distillation, but air gap membrane distillation and vacuum membrane distillation have different mechanisms due to the presence of air gap separating the membrane surface and the cooling plate and the vacuum presence respectively.

For the air gap membrane distillation, the heat transfer occurs across the air gap, the condensate layer, the cooling plate, and the cooling fluid on the other side of the plate. The heat transfer equations for the permeate side are given by:

$$Q_{p,ag} = \frac{k_g}{\delta_g}\left(T_{m,p} - T_{c,p}\right) + J\Delta H_v \tag{47}$$

$$Q_{p,c} = h_c\left(T_{c,p} - T_{p,p}\right) \tag{48}$$

$$Q_{p,p} = \frac{k_p}{\delta_p}\left(T_{p,p} - T_{p,cf}\right) \tag{49}$$

$$Q_{cf} = h_{cf}\left(T_{p,cf} - T_{cf}\right) \tag{50}$$

where $Q_{p,ag}$, $Q_{p,c}$, $Q_{p,p}$, and Q_{cf} are heat transfer from the membrane to condensate layer across the air gap, heat transfer from condensate layer to the cooling plate, heat transfer from cooling plate to the cooling fluid, and heat transfer at the cooling fluid side respectively. k_g and k_p are the thermal conductivity of air and cooling plate respectively, while δ_g and δ_p represent the thickness of the air gap and the cooling plate. h_c is the heat transfer coefficient of condensate, h_{cf} is the heat transfer coefficient of cooling fluid, $T_{c,p}$ is the condensate temperature at the permeate side, $T_{p,p}$ is the cooling plate temperature at the permeate side, $T_{p,cf}$ is the cooling plate temperature at the cooling fluid side, and T_{cf} is the cooling fluid temperature.

The heat transfer resistance in vacuum membrane distillation can be simply neglected due to the presence of a vacuum. A thermal boundary layer resistance at the permeate side is poor due to the vacuum applied.

Thus, the steady-state heat transfer process in membrane distillation is presented in Equation 51. The energy balance can then be used to calculate the membrane surface temperature on both sides which cannot be determined experimentally. The resulting equations to find the membrane surfaces can then be used to calculate using the iterative method.

$$Q_f = Q_m = Q_p \tag{51}$$

Prediction of Heat Transfer Coefficients

The values of heat transfer coefficients, h can be calculated from the corresponding Nusselt number (Nu) using the equation:

$$Nu = \frac{h d_h}{k} \tag{52}$$

where d_h is the hydraulic diameter of the pores, and k is the thermal conductivity.

The Nusselt number can be determined based on some examples of heat transfer correlations in Table 2 that fits the characteristics of the membrane system (see description). The term $\left(\frac{RePrd_h}{L}\right)$ represents the Graetz number (Gz), where Re is the Reynolds number, Pr is the Prandtl number, d_h is the hydraulic diameter, and L is the length of the channel.

The Reynolds number and Prandtl number can be calculated using the formulas as shown in Equation 59 and 60 respectively.

$$Re = \frac{\rho v d_h}{\mu} \tag{59}$$

where ρ is the fluid density, v is the fluid velocity, d_h is the hydraulic diameter, and μ is the fluid viscosity.

Table 2. Heat transfer correlations

Heat Transfer Correlations	Conditions	Equation
Laminar flow $(Re < 2,100)$		
$Nu = 1.86\left(RePrd_h / L\right)^{1/3}$ • Flat-sheet module • Inside/out capillary module • Inside/out hollow fibre module • Inside/out tubular module	Re < 2,100	(53)
$Nu = 3.66 + \frac{0.0668\left(RePrd_h / L\right)}{1 + 0.045\left(RePrd_h / L\right)^{2/3}}$ • Inside/out hollow fibre module	Constant wall temperature	(54)
$Nu = 1.04Re^{0.4}Pr^{0.36}\left(\frac{Pr}{Pr_w}\right)^{0.25}F_c$ • Outside/in hollow fibre module	10 < Re < 500 Pr_w : Prandtl number evaluated at the tube-wall temperature F_c : tube-row correction factor	(55)
Transitional flow $(2,100 < Re < 10,000)$		
$Nu = 0.116\left(Re^{2/3} - 125\right)Pr^{1/3}\left[1 + \left(d_h / L\right)^{2/3}\right]\left(\mu_b / \mu_I\right)^{0.14}$ • Flat-sheet module • Inside/out capillary module	2,100 < Re < 10,000	(56)
Turbulent flow $(Re > 10,000)$		
$Nu = 0.023Re^{0.8}Pr^{n}$ • Flat-sheet module • Inside/out tubular module	$n = 0.4$ (heating); $n = 0.3$ (cooling); Re > 10,000	(57)
$Nu = 0.04Re^{0.75}Pr^{0.33}$ • Flat-sheet module	For a rectangular pipe of height (H) and width (W) with $W \gg H$ and $d_h = 2\text{H}$	(58)

Source: (Chiam & Rosalam, 2014)

$$Pr = \frac{\text{C}_\text{p}\mu}{k} \tag{60}$$

where C_p is the heat capacity, μ is the fluid viscosity, and k is the thermal conductivity.

Figure 15. Mass transfer resistance in membrane distillation based on the Dusty Gas model

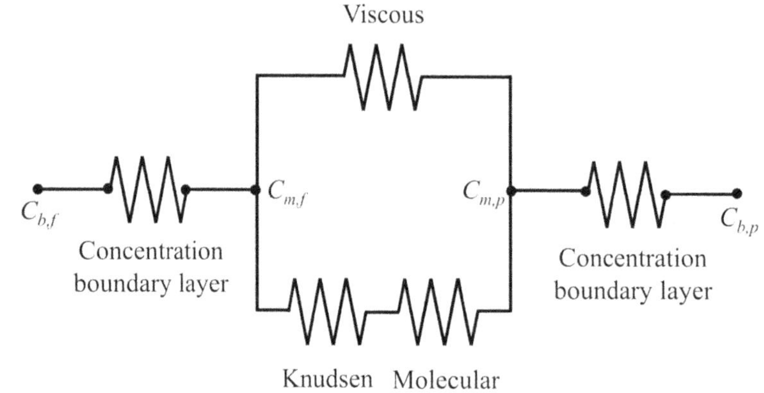

Mass Transfer

Mass Transfer Resistances in Membrane Distillation

There are several established mass transfer models for membrane distillation such as the Dusty Gas model, Fick's las model, Schofield model, Knudsen-molecular diffusion-Poiseuille transition(KMPT) model, and Knudsen flow molecular transition (KMT) model. The former model is the most widely used and studied for membrane distillation and will be topic emphasized in this section. The dusty Gas model uses the assumptions of average pore size to predict the mass transfer of volatile water molecules at an average temperature of the membrane. In this model, the transport mechanisms due to Knudsen diffusion and molecular diffusion are combined as resistance in series which is parallel to the viscous flow. Mass transfer in membrane distillation focuses on the transport of water vapours from the feed side to the permeate side across a porous membrane. Commonly, the mass transfer phenomenon occurs in 3 steps, which are:

Step 1: The water flows from the hot liquid feed bulk to reach the membrane surface to evaporate;
Step 2: The water vaporizes at the vapour-liquid interface and transferred through the membrane pores; and
Step 3: The water vapour diffuses out from the membrane pores into the vacuum compartment at the permeate side and condensates back to the water.

The mass transfer in membrane distillation is controlled by several resistances as depicted in Figure 15.

Mass Transfer on the Feed Side

The mass transfer resistance on the feed side is attributed by the concentration boundary layer due to the concentration polarization that forms near the membrane surface. This occurs as more water vapour escapes, the concentration at the membrane surface becomes higher than the bulk liquid feed. The resistance caused by the concentration boundary layer becomes more significant to the mass transfer with time, causing flux decline. The mass balance on the feed side can be described by the film theory:

$$J = K \ln\left(\frac{C_m - C_p}{C_b - C_p}\right) \tag{61}$$

where J is the flux, K is the mass transfer coefficient, C_m is the concentration at the membrane interface, C_b is the bulk concentration, and C_p is the permeate concentration.

The mass transfer coefficient (K) is calculated from the corresponding Sherwood number (Sh):

$$Sh = \frac{K d_h}{D} \tag{62}$$

where d_h is the hydraulic diameter of the feed flow channel, and D is the diffusion coefficient.

The Sherwood number can be determined based on some examples of mass transfer correlations in Table 3.

The Schmidt number (Sc) is calculated using the formulas as shown in Equation 69.

$$Sc = \frac{\mu}{\rho D} \tag{69}$$

where μ is the fluid viscosity, ρ is the fluid density, and D is the diffusion coefficient in the liquid phase.

Mass Transfer Across the Membrane

In general, there are three types of mass transfer mechanism across the membrane, which are Knudsen diffusion, molecular diffusion, and viscous flow as shown in Figure 16. Furthermore, a combination of these mechanisms is also possible depending on the nature of the membrane system. Also, the membrane distillation configuration has a direct impact on the mechanisms of the mass transfer taking place across the membrane.

By solving the resistance in series-parallel in term of flux from Figure 15, the following general equation for flux is obtained:

$$J = \frac{\left(J_{Kn} + J_{mol}\right) J_{vis}}{J_{Kn} + J_{mol} + J_{vis}} \tag{70}$$

Knudsen diffusion in the membrane distillation is mainly based on the collision between the permeating water vapour with the membrane pores walls. It is usually prevailing in systems operating at high temperature and pressure. The flux equation for this transport mechanism is given by Equation 71.

$$J_{Kn} = K_{Kn}\left(p_{m,f} - p_{m,p}\right) \tag{71}$$

Table 3. Mass transfer correlations

Mass Transfer Correlations	Conditions	Equation
Laminar flow ($Re < 2,100)$		
$Sh = 1.86\left(ReScd_h / L\right)^{1/3}$ • Flat-sheet module • Inside/out tubular module	Re < 2,100	(63)
$Sh = 1.62\left(ReScd_h / L\right)^{1/3}$ • Inside/out hollow fibre module	$d_h^2 v / DL$ > 20	(64)
$Sh = 3.66 + \frac{0.0668\left(ReScd_h / L\right)}{1 + 0.045\left(ReScd_h / L\right)^{2/3}}$ • Inside/out hollow fibre module	Constant wall temperature	(65)
Transitional flow $(2,100 < Re < 10,000)$		
$Sh = 0.116\left(Re^{2/3} - 125\right)Sc^{1/3}\left[1 + \left(d_h / L\right)^{2/3}\right]\left(\mu_b / \mu_I\right)^{0.14}$ • Flat-sheet module	2,100 < Re < 10,000	(66)
Turbulent flow $(Re > 10,000)$		
$Sh = 0.023Re^{0.8}Sc^{n}$ • Flat-sheet module • Inside/out hollow fibre module • Inside/out tubular module	$n = 0.4$ (heating); $n = 0.3$ (cooling); Re > 10,000; L / d_h > 60	(67)
$Sh = 0.04Re^{0.75}Sc^{0.33}$ • Flat-sheet module	For a rectangular pipe of height (H) and width (W) with $W \gg H$ and $d_h = 2\text{H}$	(68)

Source: (Chiam & Sarbatly, 2014)

where J_{Kn} is the flux due to Knudsen diffusion, K_{Kn} is the mass transfer coefficient for Knudsen diffusion, $p_{m,f}$ is the vapour pressure at the membrane surface on the feed side, and $p_{m,p}$ is the vapour pressure at the membrane surface on the permeate side.

For Knudsen diffusion mechanism, the mass transfer coefficient (K_{Kn}) is calculated as:

$$K_{Kn} = \frac{2}{3}\left(\frac{8MW_i}{\pi RT}\right)^{1/2}\frac{\varepsilon r}{\delta \tau} \tag{72}$$

Figure 16. Mass transfer mechanisms across the membrane distillation

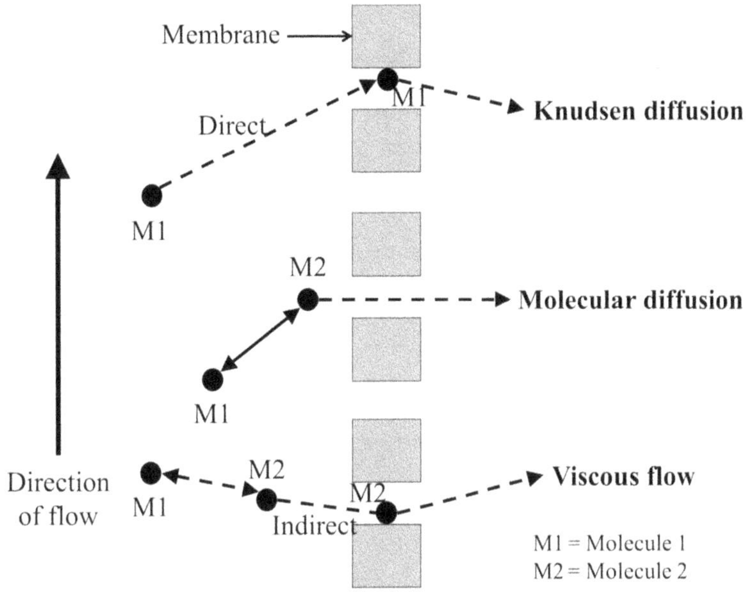

where MW_i is the molecular weight of the transporting component *i*, R is the gas constant, T is the temperature, ε, r, δ, and τ are the membrane porosity, pore radius, thickness, and tortuosity respectively.

The molecular diffusion is attributed by the collision between molecules which are significant in systems having intermediate temperature and pressure. The flux due to molecular diffusion is given by:

$$J_{mol} = K_{mol}\left(p_{m,f} - p_{m,p}\right) \quad (73)$$

where J_{mol} is the flux due to molecular diffusion, and K_{mol} is the mass transfer coefficient for molecular diffusion.

The mass transfer coefficient (K_{mol}) due to molecular diffusion can be calculated as:

$$K_{mol} = \frac{MW_i D_{v/a}}{RT}\frac{P_T}{\bar{P}_a}\frac{\varepsilon}{\delta\tau} \quad (74)$$

where $D_{v/a}$ is the diffusion coefficient of water vapour in the air, P_T is the total pressure inside the pores, $\bar{P}_a$ is the average air pressure within the membrane pores.

For viscous flow, the flux occurs due to the combined collisions of molecules-walls and molecules-molecules. It is important in membrane systems with low temperature and pressure. The flux equation in viscous flow is shown in Equation 75.

$$J_{vis} = K_{vis}\left(p_{m,f} - p_{m,p}\right) \tag{75}$$

where J_{vis} is the flux due to viscous flow, and K_{vis} is the mass transfer coefficient in viscous flow.

For the mass transfer coefficient in viscous flow, Equation 76 gives the formula of calculation.

$$K_{vis} = \frac{1}{8n_v}\frac{\varepsilon r^2}{\delta\tau}\frac{MW_i P_m}{RT} \tag{76}$$

where n_v is the vapour viscosity of the permeation component, and P_m is the average pressure within the membrane pores.

In the transition regions or the combination of two or all three mass transfer mechanisms, the mass transfer coefficient can be calculated using the resistance model, as shown in Equation 77, 78, and 79.

$$K_{Kn-mol} = \left(\frac{1}{K_{Kn}} + \frac{1}{K_{mol}}\right)^{-1} = \left[\frac{3\delta\tau}{2\varepsilon r}\left(\frac{\pi RT}{8MW_i}\right)^{1/2} + \frac{\delta\tau\bar{P}_a RT}{\varepsilon P_T MW_i D_{v/a}}\right]^{-1} \tag{77}$$

$$K_{Kn-vis} = K_{Kn} + K_{vis} = \frac{2}{3}\left(\frac{8MW_i}{\pi RT}\right)^{1/2}\frac{\varepsilon r}{\delta\tau} + \frac{1}{8n_v}\frac{\varepsilon r^2}{\delta\tau}\frac{MW_i P_m}{RT} \tag{78}$$

$$K_{Kn-mol-vis} = \left(\frac{1}{K_{Kn}} + \frac{1}{K_{mol}}\right)^{-1} + K_{vis}$$
$$= \left\{\left[\frac{2}{3}\left(\frac{8MW_i}{\pi RT}\right)^{1/2}\frac{\varepsilon r}{\delta\tau}\right]^{-1} + \left[\frac{D_{v/a}P_T}{\bar{P}_a}\frac{\varepsilon}{\delta\tau}\frac{MW_i}{RT}\right]^{-1}\right\}^{-1} + \frac{1}{8n_v}\frac{\varepsilon r^2}{\delta\tau}\frac{MW_i P_m}{RT} \tag{79}$$

The mass transfer for each of the four membrane distillation configurations is modelled based on the resistance model of Figure 15. Knudsen number analysis is used to determine the mass transfer mechanism(s) that take place in each configuration. Knudsen number is a quantity to provide a pointer to decide the operative mechanism in a given pore diameter under given experimental process conditions. Combinations of two or all of the mechanisms are possible in membrane distillation. For each respective mass transfer mechanism, the mass transfer coefficient can be determined to subsequently calculate the flux. Mechanism of the mass transfer mode can be determined using the Knudsen number (Kn) using the Equation 80.

Table 4. Mass transfer mechanism based on the Knudsen number

Knudsen Number	Mass Transfer Mechanism
$Kn > 1$	Knudsen diffusion mechanism
$Kn < 0.01$	Molecular diffusion mechanism
$0.01 < Kn < 1$	Knudsen-molecular diffusion transition mechanism

$$Kn = \frac{\lambda}{d_p} \tag{80}$$

where λ is the mean free path of water vapour, and d_p is the pore diameter of the membrane.

The mean free path of water vapour can be calculated as follows:

$$\lambda = \frac{k_B T}{\sqrt{2}\pi \bar{P}_m \sigma_i^2} \tag{81}$$

where k_B is the Boltzmann constant, T is the temperature, $\bar{P}_m$ is the mean pressure within the membrane pores, and σ_i is the collision diameter.

Table 4 and Table 5 show the relationship between the calculated Knudsen number with the mass transfer mechanism in general and at low pressure (<400 mbar) operation respectively (Mannella, La Carrubba & Brucato, 2012). When the mean free path for water molecules is greater than the membrane pore diameter ($\lambda > d_p$ or $Kn > 1$) in the absence of air, water molecules will collide with the membrane pores walls. On the other hand, the presence of air inside the membrane pores accompanied by the pore diameter that is much bigger than the mean free path for water molecules ($100\lambda < d_p$ or $Kn < 0.01$) will cause collision between the vapour molecules, thus molecular diffusion becomes prevalent. At the intermediate between these two values ($\lambda < d_p < 100\lambda$ or $0.01 < Kn < 1$), both types of collision occurs, at the Knudsen-molecular diffusion transition region.

At low-pressure operation, the molecular flow is negligible as it is in direct proportionality to pressure. Viscous flow becomes prevalent in low pressure and should be considered alongside the Knudsen diffusion. At this condition, when then Knudsen number is found to exceed 10 (or $\lambda > 10 d_p$), only Knudsen diffusion is the major transport mechanism as fewer molecules are available per unit volume and thus collision with pores walls is more frequent. If $Kn < 0.01$ or $100\lambda < d_p$, only viscous flow is considered as a continuous flow of water vapours driven by the pressure gradient. At the transition region ($0.1\lambda < d_p < 100\lambda$ or $0.01 < Kn < 10$), both molecular-molecular and molecular-wall collisions are considered which is represented by the Knudsen-viscous transition model.

For direct contact membrane distillation and sweeping gas membrane distillation, the mechanism of mass transfer is the same since the fluids from the feed and permeate sides are in direct contact with both sides of the membrane. Thus, Knudsen diffusion, molecular diffusion or combined mechanisms are

Table 5. Mass transfer mechanism based on the Knudsen number at low pressure (<400 mbar)

Knudsen Number	Mass Transfer Mechanism
$Kn > 10$	Knudsen diffusion mechanism
$Kn < 0.01$	Viscous flow mechanism
$0.01 < Kn < 10$ (Vacuum pressure < 15 kPa)	Knudsen-viscous transition mechanism
$0.01 < Kn < 10$ (Vacuum pressure > 40 kPa)	Knudsen-viscous-molecular mechanism

significant in both of these configurations, assuming that the viscous flow is negligible, as expressed in Equations 72, 74, and 77. For air gap membrane distillation and vacuum membrane distillation, that is not the case due to the air gap and vacuum existence at the permeate side of the membrane distillation module. In air gap membrane distillation the viscous flow is also assumed to be negligible, but the air gap thickness is considered into the mass transfer coefficient and is modified for Knudsen diffusion, molecular diffusion and the transition mechanisms as expressed in Equations 82, 83, and 84 respectively. As for vacuum membrane distillation, there are only traces of air present within the membrane pores and molecular diffusion is assumed to be negligible. Thus, the mechanism of mass transfer is either the Knudsen diffusion model, viscous flow model, or both of them together as depicted in Equations 72, 77, and 78. However, vacuum pressure above 40 kPa may allow the molecular diffusion mechanism to occur inside the vacuum membrane distillation, thus the transition region considers all three mass transfer mechanisms as shown in Equation 79.

$$K_{Kn} = \frac{2}{3}\left(\frac{8MW_i}{\pi RT}\right)^{1/2} \frac{\varepsilon r}{\left(\delta\tau + \delta_g\right)} \tag{82}$$

$$K_{mol} = \frac{MW_i D_{v/a}}{RT} \frac{P_T}{\bar{P}_a} \frac{\varepsilon}{\left(\delta\tau + \delta_g\right)} \tag{83}$$

$$K_{Kn-mol} = \left(\frac{1}{K_{Kn}} + \frac{1}{K_{mol}}\right)^{-1} = \left[\frac{3\left(\delta\tau + \delta_g\right)}{2\varepsilon r}\left(\frac{\pi RT}{8MW_i}\right)^{1/2} + \frac{\delta\tau \bar{P}_a RT}{\varepsilon P_T MW_i D_{v/a}}\right]^{-1} \tag{84}$$

REFERENCES

Abu-Zeid, M. A. E., Zhang, Y., Dong, H., Zhang, L., Chen, H. L., & Hou, L. (2015). A comprehensive review of vacuum membrane distillation technique. *Desalination*, *356*, 1–14. doi:10.1016/j.desal.2014.10.033

Bandini, S., Gostoli, C., & Sarti, G. C. (1992). Separation efficiency in vacuum membrane distillation. *Journal of Membrane Science*, *73*(2–3), 217–229. doi:10.1016/0376-7388(92)80131-3

Bazzarelli, F., Giorno, L., & Piacentini, E. (2016). Dense Membranes. In E. Drioli, & L. Giorno (Eds.), *Encyclopedia of Membranes*. Berlin, Germany: Springer. doi:10.1007/978-3-662-44324-8_2222

Bazzi, A., Hammond, S., Redd, K., Sarullo, M., Aflaki, R., & Guendert, D. (2011). Microfiltration and Reverse Osmosis Membrane Replacement Understanding Operating Process Data and Autopsy Data Projection of Useful Remaining Life. In *Proceedings of the Water Environment Federation, WEFTEC 2011: Session 31 through Session 40,* (23), 2097-2119. 10.2175/193864711802713513

Chiam, C. K., & Sarbatly, R. (2014). Vacuum membrane distillation processes for aqueous solution treatment – a review. *Chemical Engineering and Processing: Process Intensification*, *74*, 27–54. doi:10.1016/j.cep.2013.10.002

El-Bourawi, M. S., Ding, Z., Ma, R., & Khayet, M. (2006). A framework for better understanding membrane distillation separation process. *Journal of Membrane Science*, *285*(1–2), 4–29. doi:10.1016/j.memsci.2006.08.002

Grandison, A. S. (1996). *Separation Processes in the Food and Biotechnology Industries*. Boca Raton, FL: CRC Press. doi:10.1533/9781855736573

Lagana, F., Barbieri, G., & Drioli, E. (2000). Direct contact membrane Distillation: Modelling and concentration experiments. *Journal of Membrane Science*, *166*(1), 1–11. doi:10.1016/S0376-7388(99)00234-3

Mannella, G. A., La Carrubba, V., & Brucato, V. (2012). Evaluation of vapour mass transfer in various membrane distillation configurations: An experimental study. *Heat and Mass Transfer*, *48*(6), 945–952. doi:10.100700231-011-0946-x

Olatunji, S. O., & Camacho, L. M. (2018). Heat and mass transport in modelling membrane distillation configurations: A review. *Frontiers in Energy Research*, *6*(130), 1–18.

Singh, R. (2005). *Hybrid membrane systems for water purification: Technology, systems design and operation*. USA: Elsevier.

Srisurachan, S., Jiraratananon, R., & Fane, A. G. (2006). Mass transfer mechanisms and transport resistances in direct contact membrane distillation process. *Journal of Membrane Science*, *277*(1–2), 186–194. doi:10.1016/j.memsci.2005.10.028

Strathmann, H. (2011). Membranes and Membrane Separation Processes, 1. Principles. In Ulmann's Encyclopedia of Industrial Chemistry, 22, 413-456.

Verweij, H. (2012). Inorganic membranes. *Current Opinion in Chemical Engineering*, *1*(2), 156–162. doi:10.1016/j.coche.2012.03.006

Yoon, S. H. (2016). *Membrane Bioreactor Processes: Principles and Applications*. Boca Raton, FL: CRC Press.

Zhu, X., & Bai, R. (2017). Separation of biologically active compounds by membrane operations. *Current Pharmaceutical Design*, *23*(2), 218–230. PMID:27799041

ADDITIONAL READING

Attia, H., Osman, M. S., Johnson, D. J., Wright, C., & Hilal, N. (2017). Modelling of air gap membrane distillation and its application in heavy metal removal. *Desalination*, *424*, 27–36. doi:10.1016/j.desal.2017.09.027

Bi Bella, G., & Di Trapani, D. (2019). A brief review on the resistance-in-series model in membrane bioreactors (MBRs). *Membrane*, *9*(2), 24–52. doi:10.3390/membranes9020024

Nagy, E. (2012). *Basic Equations of the Mass Transport Through a Membrane Layer*. London, UK: Elsevier.

Rochd, S., Zerradi, H., Mizani, S., Dezairi, A., & Ouaskit, S. (2016). Modelisation of membrane distillation: Mass and heat transfer in air gap membrane distillation. *Journal of Membrane Science & Technology*, *6*(2), 1–9.

Strathmann, H. (2011). Membranes and Membrane Separation Processes, 1. Principles. In Ulmann's Encyclopedia of Industrial Chemistry, 22, 413–456.

Theodore, L., & Ricci, F. (2010). *Mass Transfer Operations for the Practicing Engineer*. New Jersey, USA: Wiley. doi:10.1002/9780470602591

Uragami, T. (2017). *Science and Technology of Separation Membranes*. USA: Wiley. doi:10.1002/9781118932551

KEY TERMS AND DEFINITIONS

Concentration Polarization: Phenomenon which occurs naturally due to the build-up of concentration gradient at the membrane-solution interface due to the membrane's permselective property.

Convective Transport: The transport of heat/mass through the porous membrane which is induced by the bulk motion of fluid-driven by the applied driving force.

Diffusive Transport: Transport of fluid down the concentration gradient.

Mass Transfer: In the context of the membrane, mass transfer is the net movement of mass from the feed side to the permeate side.

Membrane Fouling: Accumulation of solutes and/or other materials on the membrane surface and/or inside the membrane pores causing reduced in membrane flux.

Membrane Separation: A process which utilizes a membrane that selectively separates components based on size exclusion or other selectivity mechanisms.

Modelling: In the mathematical context, it is the process of translating a phenomenon from qualitative explanation into the mathematical language which can describe the phenomenon quantitatively.

Transport Resistance: Obstacle to the transport of heat/mass across the membrane.

Chapter 3
Membrane Fabrications

ABSTRACT

The advancement in membrane technologies is driven by the development of membrane fabrication techniques and material studies. Nowadays, there are many fabrication techniques and materials available to produce membranes with characteristics that are ideal for specific applications. In this chapter, some fabrication methods are discussed for the three most widely used membrane types, i.e. the polymeric, ceramic, and metallic membranes. Each type of membrane possesses certain advantages and disadvantages attributed to the membrane structural properties as a result of the fabrication method. The choice of method and material are the two important aspects that determine membrane performance and efficiency. The membrane performance indicator is generally measured from the flux, rejection factor, and ability to withstand extreme chemical and physical conditions. Furthermore, the fabricated membranes are further assembled into a complete unit of membrane system called the membrane modules, namely tubular, plate-and-frame, spiral-wound, capillary, and hollow fibre.

INTRODUCTION

Membrane fabrication is a sophisticated area in the membrane field comprising of various techniques that gives the different characteristics of the membrane. Most synthetic membranes are made from polymers, some incorporated ceramics, metals etc. to enhance the properties of the membrane for a specific application. In the scope of water and wastewater treatments, microfiltration, ultrafiltration, nanofiltration, reverse osmosis, and membrane distillation are the most widely used membrane technologies in these areas. Table 1 shows some examples of the fabrication methods and polymers used for the preparation of polymeric membranes for membrane-based water and wastewater treatment processes.

Typically, membrane fabrication controls the membrane's structural characteristics that will be obtained. Membrane fabrication usually targets certain characteristics of the membrane that is desirable for a specific purpose. Some examples are the membrane pore size, pore size distribution, porosity, and thickness. The characteristics are attributed by the material selection and also the technique of fabrica-

DOI: 10.4018/978-1-7998-2645-3.ch003

Table 1. Examples of fabrication methods and polymers used for the preparation of polymeric membranes widely used in water and wastewater treatments

Membrane Technology	Fabrication Methods	Polymers Used in Fabrication Process	Average Membrane Pore Size
Microfiltration	• Phase inversion • Stretching • Track-etching	• Polyvinylidene fluoride (PVDF) • Poly(tetrafluoroethylene) (PTFE) • Polypropylene (PP) • Polyethylene (PE) • Polyethersulfone (PES) • Polyetheretherketone (PEEK)	0.1 – 10 μm
Ultrafiltration	• Phase inversion • Solution wet-spinning	• Polyacrylonitrile (PAN) • PES • Polysulfone (PS) • Polyethersulfone (PES) • Poly(phthazine ether sulfone ketone) (PPESK) • Poly(vinyl butyral) • PVDF	1 – 10 nm
Nanofiltration	• Interfacial polymerization • Layer-by-layer deposition • Phase inversion	• Polyamides • Polysulfones • Polyols • Polyphenols	1 – 10 nm
Reverse osmosis	• Phase inversion • Solution casting	• Cellulose acetate/triacetate • Aromatic polyamide • Polypiperzine • Polybenziimidazoline	3 – 5 Å
Membrane distillation	• Phase inversion • Stretching • Electrospinning	• PTFE • PVDF	0.1 – 1 μm

Source: (Lalia et. al, 2013)

tion. Furthermore, the method of membrane fabrication has significant effects on its effectiveness and performance. The membrane performance efficiency can be measured in terms of the flux, retention/rejection, chemical and mechanical stability which should be economically feasible. The membrane that possesses high quality can greatly increase the performance of the membrane operation.

Although polymers are the most used material to produce synthetic polymeric membrane, they have some disadvantages that make them unsuitable for some separation processes. For example, the polymeric membrane is prone to be damaged from chemical and physical stresses. To solve this issue, many researchers have succeeded in fabricating membranes made from inorganic materials. Inorganic materials that have been utilized in membrane fabrication are ceramics, metals, and zeolites. Inorganic membrane possesses several attractive advantages over polymeric membranes in terms of robustness, ability to operate at extreme temperature and pressure conditions, as well as ability to withstand some chemical attacks. However, some drawbacks are also associated with this membrane as they are expensive to produce and also complexity during fabrication and operation for it being relatively new to the membrane society.

Membranes used in separation operations are generally further packaged as modules of various choices. A membrane module is a unit of membrane 'cell' which are made from membranes, membrane housing compartment, an inlet(s) and outlet(s). Membranes are usually produced in two forms which are the flat sheet and tubular shapes. Flat sheet membranes are commonly packaged in plate-and-frame

Table 2. General comparison between polymeric, ceramic, and metallic membranes

Aspect	Polymeric Membrane	Ceramic Membrane	Metallic Membrane
Nature	Synthetic	Synthetic	Synthetic
Type of membrane	Organic	Inorganic	Inorganic
Resistance to extreme environments (temperature, pressure, chemical attack)	Low	High	High
Fabrication technology development	Developed	Relatively new	Relatively new
Cost	Low	High	High

or spiral-wound, whereby the tubular membranes are made into hollow-fibre membranes. The selection of the optimum membrane module is influenced by a variety of process conditions.

In this chapter, some fabrication techniques are discussed for three main types of membranes, which are the polymeric, ceramic, and metallic membranes. These three types of membranes are commonly found in many membrane operations. To introduce these membranes, the general comparison between the three types of membranes are discussed in Table 2. Also, the types of membrane modules are discussed which consist of five modules, i.e. tubular, plate-and-frame, spiral-wound, capillary, and hollow fibre.

POLYMERIC MEMBRANE FABRICATION

Membrane utilization is expected to dramatically increase in various application due to the facts that it works without chemical additives, requires relatively low energy usage and easy operation compared to the conventional techniques. Membrane technology in the separation process is becoming more competitive to the conventional methods particularly for the creation of domestic and process water from surface water, groundwater or wastewater. In this section, the most commonly used techniques of polymeric membrane fabrication are discussed which includes phase inversion, interfacial polymerization, stretching, track-etching, electrospinning, and meltblown techniques.

Phase Inversion

Phase inversion is a process of transforming a polymer solution from liquid to solid state in a controlled manner, by wet and/or dry process. The wet process is performed by immersing the polymer solution into a non-solvent coagulation bath, while a dry process involves the polymer solution being exposed to a non-solvent atmosphere. The process of solidification is very often initiated by the transition from one liquid state into two liquids (liquid-liquid demixing). Demixing is a process of separation of components of a mixture. At a certain stage during demixing, one of the liquid phases (the high polymer concentration phase) will solidify so that a solid matrix is formed. By controlling the initial stage of phase transition, the morphology can be controlled, i.e. porous as well as nonporous membranes can be prepared. The transformation of phase inversion membrane can be accomplished by four techniques which are: immersion precipitation, thermally induced phase separation, evaporation-induced phase separation, and vapour induced phase separation.

Figure 1. Schematic drawing depicting the preparation of flat membranes

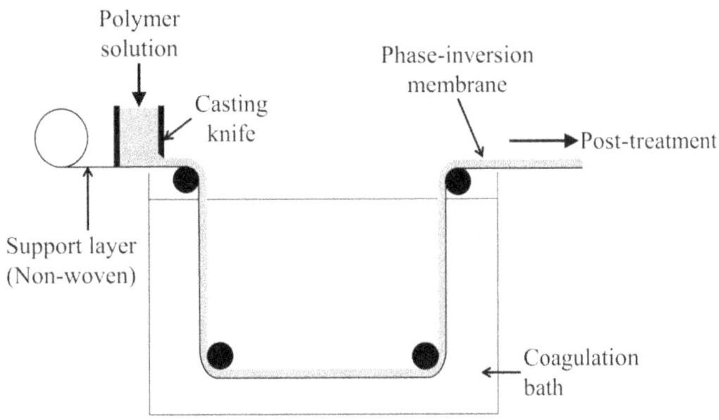

Immersion precipitation is a technique most widely used commercially to prepare polymeric membrane. In this method, a polymer solution (polymer + solvent) is cast on a suitable support and immersed in a coagulation bath containing a non-solvent. Precipitation occurs due to the exchange of solvent and non-solvent. The membrane structure ultimately obtained results from a combination of mass transfer and phase separation. Phase inversion membranes can be prepared from a wide variety of polymers, with the condition that the polymer must be soluble in a solvent or solvent mixture. Membranes formed using this technique can be prepared in two configurations: flat and tubular. Schematic presentation of the process to prepare flat membranes is shown in Figure 1.

In this process, the polymer solution (often referred to as a casting solution or dope solution) is cast directly upon a supporting layer (a non-woven polyester) by a casting knife. The casting thickness can vary from 50 to 500 µm. The cast film is then immersed in a non-solvent bath where exchange occurs between the solvent and non-solvent and eventually the polymer precipitates. Water is commonly used as a non-solvent, however, other kinds of non-solvents can also be utilized. The parameters of the pairing of solvent/non-solvent, polymer concentration, evaporation time, humidity, temperature, and composition of the casting solution (e.g. additives) are very important to the membrane performance obtained (flux and selectivity) to suits the intended application. The membranes obtained after precipitation can be used directly or undergo post-treatment (e.g. heat treatment) before being used. Since flat membranes are relatively simple to prepare, they are very useful for testing on a laboratory scale. For very small membrane surface areas (less than 1000 cm^2), the membranes are cast mostly by hand or semi-automatically, not on a non-woven but often on a glass plate.

Figure 2 illustrates the process after the polymer solution is immersed in a non-solvent. The non-solvent diffuses into the cast film at a flux of J_1, while the solvent diffuses into the coagulation bath at a flux of J_2. After a certain time of solvent/non-solvent exchange takes place, the solution comes to be thermodynamically unstable and demixing occurs which subsequently form a solid polymeric film with an asymmetric structure. Generally, at $J_2 \gg J_1$, ultrafiltration membrane is fabricated with a pore size of 10 – 300 Å, whereas at $J_2 = J_1$, a microfiltration membrane is formed with a pore size of 0.2 – 0.5 µm (Lalia et. al, 2013).

* J_1 is the non-solvent flux, J_2 the solvent flux, X is the position of the interface between the film and the coagulation bath, x is the spatial position coordinate normal to the membrane surface, y = −x

Figure 2. Schematic diagram of film/bath interface
Source: Loddo et al., 2018

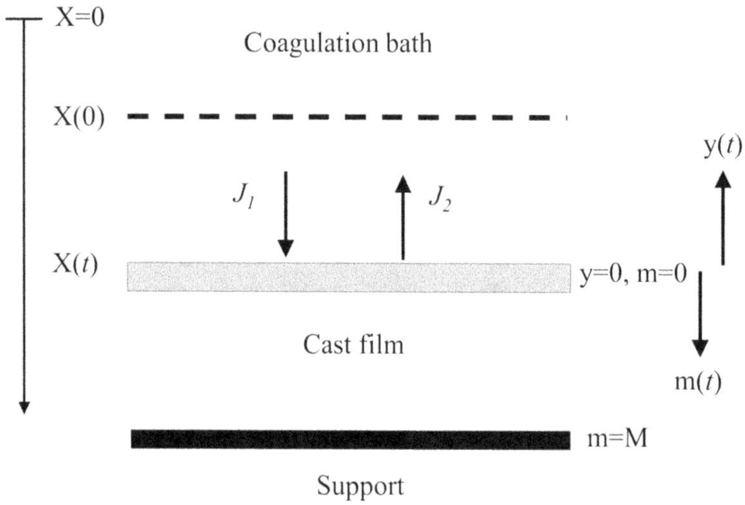

– X(*t*) is the position coordinate that moves with the interface, m is the position coordinate in the polymer-fixed frame of reference, and M is a support (Lalia et. al, 2013).

Unlike flat sheet membranes, tubular, hollow fibre, and capillary membranes are fabricated using a different technique. The classification of these three types of membrane modules is based upon their diameter which is: hollow fibre membrane (diameter: <0.5 mm), capillary membrane (diameter: 0.5 – 5.0 mm), and tubular membrane (diameter: > 5mm). Tubular membranes are so large that they have to be supported, whereas hollow fibres and capillaries are self-supporting. Hollow fibres and capillaries can be prepared via three different methods, namely wet-spinning (or dry-wet spinning), melt-spinning, and dry-spinning.

In the fabrication of hollow fibres and capillaries membranes, demixing takes place from the boring side (lumen) and from the shell (outside) which differs from flat membrane which undergoes demixing on only one side. Figure 3 shows the schematic drawing of a dry-wet spinning process. Initially, the polymer solution which contains a polymer, solvent, and sometimes additives, is filtered prior to being pumped through a spinneret. The viscosity of the polymer solution is desired to be high (in general above 100 Poise) to produce a membrane with small pore size. The bore injection fluid is pumped through the inner tube of the spinneret. After a short residence time in the air or a controlled atmosphere (the term dry originates from this step), the fibre is immersed in a non-solvent bath where coagulation occurs. These lead to the phase inversion and allow the fibre to solidify which is then collected upon a godet. The parameters affecting the fibre prepared from the spinning process are the extrusion rate of the polymer solution, the bore fluid rate, the 'tearing-rate', the residence time in the air-gap, and dimensions of the spinneret. These spinning parameters interfere with the membrane-forming parameters which are the composition of the polymer solution, composition of the coagulation bath, and the coagulation temperature.

The spinneret is a moulding devise used to extrude polymer solution/melt to form fibre membranes. There are two main types of spinnerets depending on the processes which are either to form fibres from solution or from the melt Table 3 lists the distinct differences between the two types of spinnerets: wet spinning & dry-wet spinning spinneret (Figure 4), and melt spinning & dry spinning spinneret (Figure 5).

Figure 3. Schematic diagram of a dry-wet spinning process

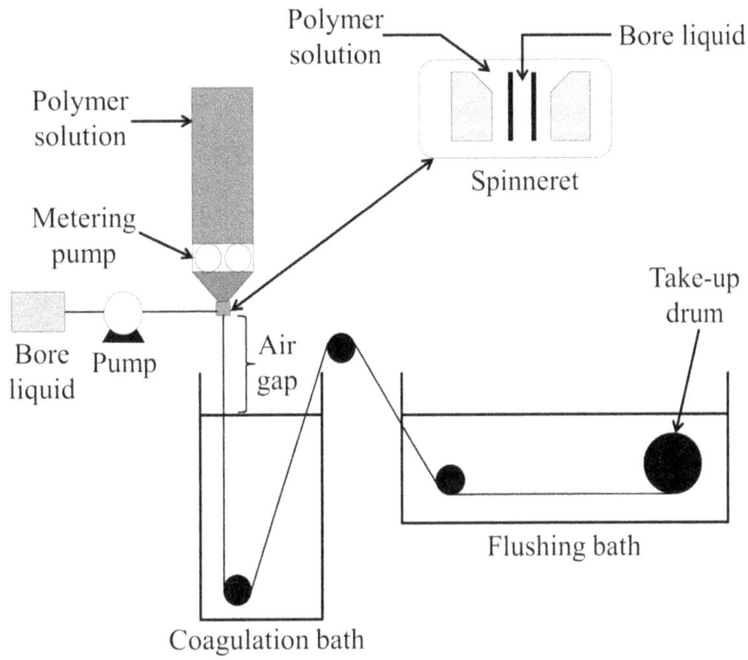

Table 3. Types of spinnerets

Wet Spinning & Dry-wet Spinning	Melt Spinning & Dry Spinning
Spinneret dimensions are very important because the fibre dimensions are mainly determined by the membrane-forming parameters.	Spinneret dimensions are not so crucial because the fibre dimensions are mainly determined by the extrusion rate and tearing rate.
Spinning rate: meters per minute.	Spinning rate: thousands of meters per minute.

Figure 4. Cross-section view of wet spinning & dry-wet spinning spinneret

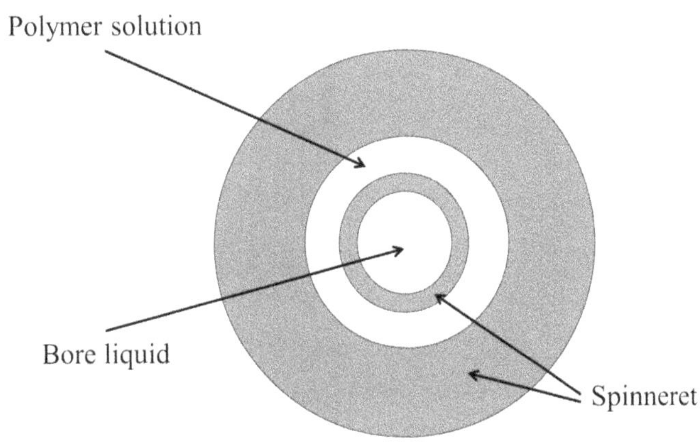

Figure 5. Cross-section view of melt spinning & dry spinning spinneret

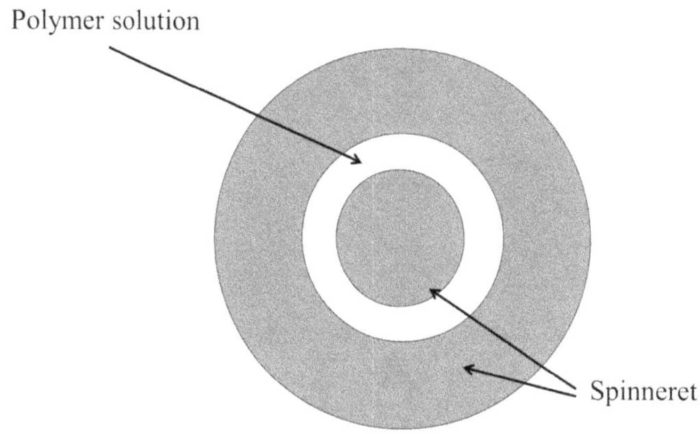

For tubular membranes, the fabrication is slightly different than that of hollow fibres and capillaries membranes since it is not self-supporting. Thus, the casting of the polymer solution has to be carried out on a supporting tubular material (for example a non-woven polyester or porous carbon tube). The laboratory set-up for the preparation of tubular membrane is depicted in Figure 6. In this method, the pressure is applied to a reservoir filled with a polymer solution so that the solution is transported through a hollow pipe. At the end of the pipe is 'casting bob' with small holes through which the polymer solution is forced (Figure 6(a)). If the porous tube is moving vertically, either mechanically or by gravity, a film is cast upon its inner wall (Figure 6(b)). The pipe is then immersed in a coagulation bath where precipitation of the cast polymer solution leads to the formation of a tubular membrane (Figure 6(c)).

Thermally-induced phase separation is a method using the principle of temperature; as temperature decrease, the solvent quality is usually decreased. After demixing is induced, the solvent is eliminated by the methods of evaporation, extraction, or freeze-drying.

Figure 6. Laboratory set-up for tubular membrane preparation

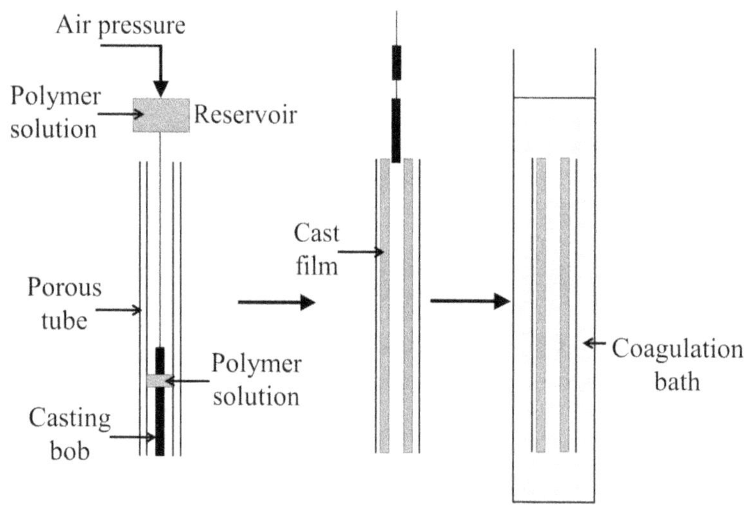

Figure 7. Fabrication of polyamide thin film membrane using the interfacial polymerization method

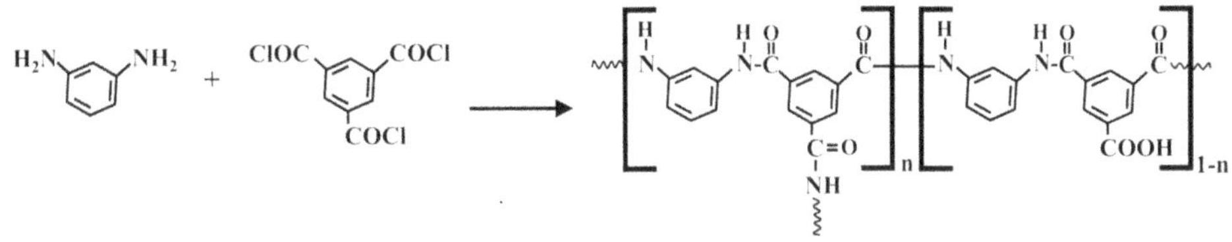

Evaporation-induced phase separation (or solution casting) is a technique in which polymer solution is formulated in a solvent or in a volatile non-solvent mixture. The solvent is then allowed to evaporate which leads to the precipitation or demixing-precipitation of the polymer solution.

Vapour-induced phase separation allows the polymer solution to be exposed to an atmosphere comprising of a non-solvent which is typically water, in which the non-solvent absorption leads to the demixing-precipitation reaction.

Interfacial Polymerization

Interfacial polymerization method is generally used to fabricate commercial reverse osmosis and nanofiltration membranes. The original fabrication method involves soaking microporous polysulfone support in a solution of aqueous polymeric amine. Then, the membrane saturated with amine is immersed in a di-isocyanate in hexane solution. Next, the membrane undergoes heat treatment at 110 °C to allow the monomers to cross-link. The resultant membrane obtained typically has good salt rejection and high water flux.

In terms of its morphological structure and membrane layer composition, various factors contribute to the results, such as the monomers concentration, reaction duration, type of solvent used, and post-treatment operation. With the significant benefits of this technique, particularly in the optimization of the membrane skin layer and microporous substrate properties, various membranes have been successfully fabricated. For example, Figure 7 shows the polymeric linkage reaction of m-phenylenediamine in water with trimesoyl chloride in hydrocarbon to fabricate the polyamide thin film membrane. In recent time, novel monomers have been proposed for membrane fabrication using this method. The novel monomers have a more functional or polar group, thus the resultant membrane displays smoother membrane surface or good hydrophilicity. These characters are important to the improvement of antifouling property for the membrane.

Stretching

The fabrication of polymeric membrane using the stretching technique was first developed in the 1970s. In this method, the polymer is melted by heating and then extruded into a thin sheet layer. The thin sheet is then stretched to create the membrane pores. Unlike the methods discussed earlier, the stretching technique does not use any solvent to create the pores, but instead, it is achieved through a mechanical means.

This method is especially suitable for membrane having high crystallinity, as the crystalline area offers strength and support, whereas the amorphous area creates the membrane pores. Stretching is typically performed in two steps: cold stretching accompanied by hot stretching. Cold stretching is conducted to

Figure 8. Track-etching membrane fabrication process

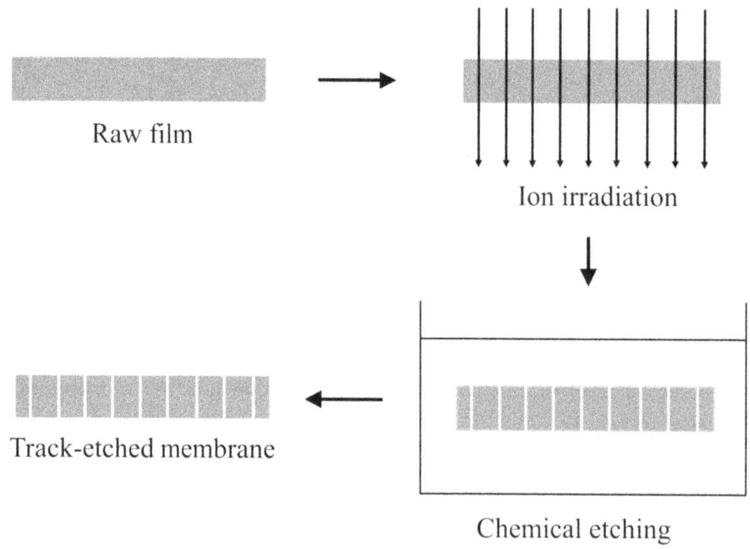

initiate the micro-pores formation on the membrane film, while the subsequent hot stretching is carried out to increase or control the resulting pore structure of the membrane. Eventually, the final pores formed are affected by the physical properties of the material used, such as the melting point, crystallinity, and tensile strength, as well as the processing parameters being employed.

Track-etching

Track-etching method uses energetic ion irradiation on a nonporous polymeric film which causes the formation of linear damaged (etched) track through the polymeric film (i.e. the membrane pores). The main advantage of this technique is its precise determined control on the membrane pore size distribution. Both the pore size and pore density are independent parameters that can be varied in a controlled manner, typically in the range of 10^{-9} – 10^{-5} m and 1 – 10^{10} cm^{-2} respectively. These properties make it easy to achieve a good relationship connecting the required water transport characteristics with the membrane pores size/shape. Besides, the membrane pores formation is essentially controlled by the etching duration and operated temperature.

The next stage after the polymeric film undergoes ion irradiation is the chemical etching process. Chemical etching is a process of soaking the polymer films in a temperature-regulated etching chemical to formulate the porous structure. This process is fundamental for the pore-size and pore-shape determination, whereby the damaged linear tracks are removed leaving hollow channels. The simplified process flow for the making of the track-etched membrane is shown in Figure 8.

Electrospinning

Electrospinning is a relatively new method to fabricate porous membrane. This technique uses the application of high-intensity electric field on polymer solutions or melts to generate micro- to nano-scale fibres which then deposited on a grounded collector. A typical electrospinning set-up is illustrated in

Figure 9. Schematic of the electrospinning process

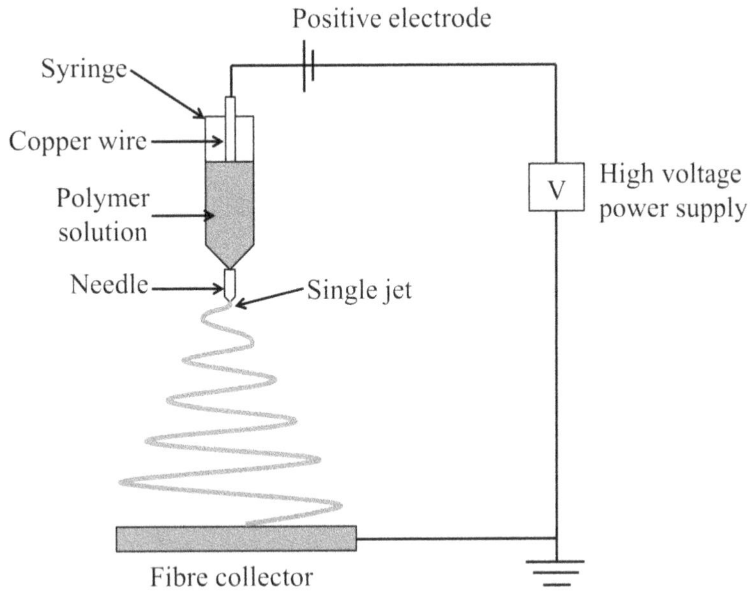

Figure 9. Electrospinning process set-up consists of three main elements, which are (1) high-voltage power supply ranging from 1 – 50 kV, (2) container with a metallic needle containing the polymer solution or melts, and (3) grounded fibre collector (typically a flat plate or collection drum).

Succinctly, the syringe container is filled with a polymer solution which is then fed at a precise rate to keep a droplet at the end of a conductive capillary (i.e. the metallic needle). The needle directly connects to the high voltage power supply and functions as an electrode to charge the polymer solution. A jet of the solution is then formed which moves across the air gap between the needle and collector (typically spaced in the range of 1 – 30 cm away). This allows the jet fibres to dry and deposited on the grounded collector in a random positioned manner.

Meltblown Technique

Meltblowing is a distinctive one-step process to produce self-bonding fibrous nonwoven membranes directly from a polymer resin, with fibre diameter ranging from micrometre to nanometer in size. It is a common technique used to produce nanofibre membrane which has been widely applied in filtration, medical fabrics, sanitary products, oil absorbents and electronic specialities. The main advantage of membrane fabricated from the melt-blown technique is that a relatively high production rate can be achieved. Unlike the electrospinning technique, the melt-blown technique does not require the massive scale solvent recovery from the dilute air stream. This process involves: (1) heating the polymer pellets at a temperature exceeding the melting temperature of the polymer, (2) the molten polymer and compressed hot air are sprayed through the nozzle, (3) nanofibres formed instantaneously in the air, whereby a rotating cylinder is designed to collect the nanofibres. This process can be illustrated as shown in Figure 10.

Figure 10. Schematic of a meltblown process

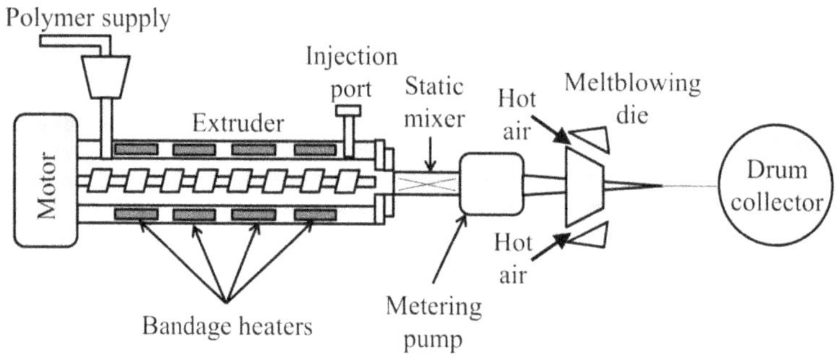

CERAMIC MEMBRANE FABRICATION

The ceramic membrane was first invented in the 1960s for the application of beer filtration and extract recovery and gas separation. It is prepared by the combination of metal (aluminium, silicium, iron, zinc, zirconium) with non-metal in oxide, nitride, or carbide forms, for examples aluminium oxide (Al_2O_3), zinc oxide (ZnO) and zirconium dioxide (ZrO_2). Nowadays, the ceramic membrane has gain popularity due to its ability to withstand severe operation for instances in a high-temperature environment and harsh chemical condition. There is a various method of ceramic membrane fabrication, such as slip casting, pressing, tape casting, phase inversion, sol-gel method, chemical vapour deposition, and extrusion. In this section, several production processes of the ceramic membrane are discussed which includes the phase inversion/sintering method, tape casting, and slip casting.

Phase Inversion/Sintering Method

The phase inversion method was first invented for the application of polymeric membrane fabrication in the early 1960s by Loeb and Sourirajan, which then practised for other material like ceramic. The fundamental principle of the phase inversion technique has been discussed in Section 3-2.1. The early ceramic membrane developed through phase inversion process was conducted by Luyten et al. (2000) for the formulation of LaSr-$CoFeO_{3-x}$ membrane. The ceramic membrane was prepared by using a mixture of ceramic powders as the main component with a polymer which acts as a binder. The mixture is typically referred to as ceramic suspension, dope suspension, or powder suspension.

For the typical preparation procedure, initially, the ceramic powder with the solvent is milled for over 1 day. Then, the mixture is added with polymer and further mixed for more than 1 day. After that, the ceramic suspension is extruded through a spinneret nozzle (for hollow fibre configuration) that is positioned near to the non-solvent bath. The ceramic hollow fibre is then cut into the preferred length and dried at room temperature. Next, it undergoes sintering process at a temperature of 500 – 600 °C to burn off the polymer, and further sintered at a higher temperature of 1,000 – 1,500 °C to allow the growth between ceramic grains. Similarly, the flat sheet ceramic membrane can be prepared by using a glass plate and cast using a casting knife and prepared under a similar procedure. The membrane pore size formation is generally affected by the sintering temperature, at which increasing the sintering temperature causes a decrease in the membrane pore size.

Figure 11. Tape casting process

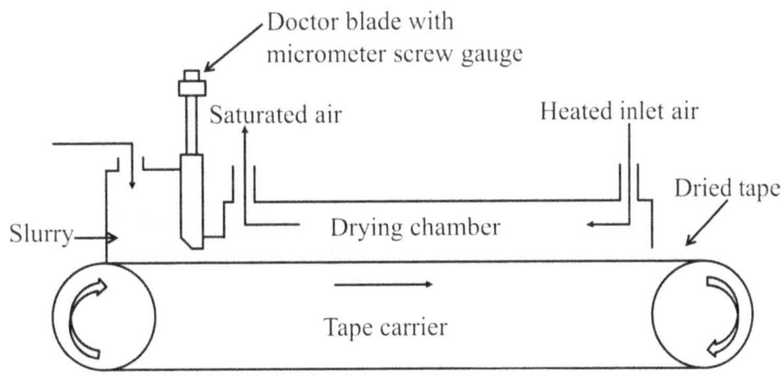

Tape Casting

Tape casting technique as shown in Figure 11 was first introduced in the mid-1940s by G. N. Howatt to make thin piezoelectric material. This method is widely known for ceramic membrane production. Tape casting is a membrane fabrication method to produce thin sheet ceramic membrane from the ceramic slurry. The fabrication process involves three steps: (1) preparation of ceramic slurry of the right viscosity (typically within 100–130 Poise (10^{-1} Ns/m^2)), (2) tape casting process to produce thin-film ceramic membrane, and (3) sintering at high temperature.

In this process, ceramic slurry in aqueous or solvent is continually transferred to a slurry chamber and dispensed under the doctor blade to an adjustable desired thickness. The ceramic cast film is brought along a carrier tape to a drying chamber/oven to be dried. In general, the cast film undergoes two heating sequence: (1) dried at room temperature followed by drying in a hot air oven at low heating (100 – 250 °C), and (2) sintering at a temperature of around 1,000 °C. The resulting ceramic membrane can then be cut into different shapes such as rectangular, circular, or square. The downside of tape casting method is, however, the indistinctness in shape due to corrosion of the plaster mould. Besides, tape casting requires a long duration when ceramic slurry with fine powders is used. The problem was tried to be fixed by pressing, yet this only adds to the processing cost. In another attempt, the solution is to change the ratios of the slurry mixture, or adding pore-forming agents and activated carbon to enhance porosity during the sintering process.

Slip Casting

The slip casting method is a well-known method to produce pottery and ceramic products, especially when it involves a complex shape that is irregular and hollow. The mixture/solution used is typically termed as slurry or slip, which is normally made of kaolin and water. This method was first introduced by Monsieur Tendelle in 1790 for the manufacturing of porcelain.

The principle technique of slip casting is illustrated in Figure 12. In this method, the slurry is prepared which often consists of kaolin, water, and viscosity reducing additive such as tripolyphosphate. It is important to ensure that the slurry viscosity is in the right consistency to prevent it from getting stuck to the mould. Usually, the viscosity density fall around 1.5 g/cm^3. The slurry is then poured into the casting mould and let to set. After casting for a period of time, the excess slurry is poured out. The cast

Figure 12. The principle of slip casting: (1) slurry is filled into the mould, (2) casting, (3) remove the excess slurry after a period of time, and (4) resulting cast is separated from the mould

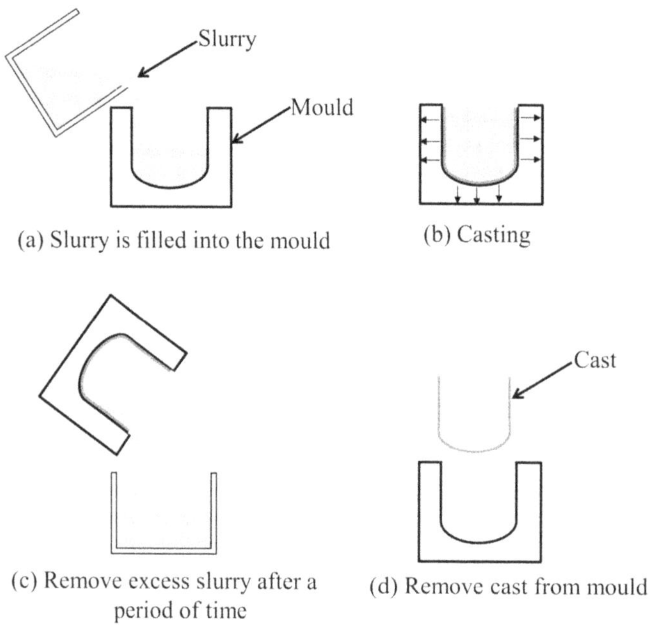

obtained is then dried at room temperature followed by oven heating at a temperature around 100 °C, and finally by sintering up to temperatures of 800 – 1,200 °C. The sintering process affects the pore size formation in a way that increased temperature results in increased membrane porosity, which in turns increase the membrane flux. Nonetheless, this causes low performance in terms of separation factor due to the large pore size of the membrane.

METALLIC MEMBRANE FABRICATION

The fabrication of porous metallic membrane is mainly prepared from metal-base powder, and sometimes with the addition of a solvent or polymeric binder. The most widely used metallic membrane is a membrane of palladium (Pd), which is principally used for hydrogen (H_2) recovery from gas mixtures. Pd is characterized by high permeability, solubility, and selectivity towards H_2 gas, thus widely used to obtain highly purified of H_2. This membrane has both advantages and disadvantages. The advantages are good chemical and thermal stability and better antifouling property than the polymeric membrane. The limitations include high cost due to low elementary Pd availability in nature and surface poisoning. Surface poisoning occurs when sulfur bonds to Pd, blocking the sites for H_2 adsorption and ultimately restructures the metal surface. There are plenty methods that can be utilized to prepare metallic membranes, such as tape casting, phase inversion, electroplating, chemical powder deposition, thermal sintering, electrospinning, de-alloying, electrical sintering, template-directed synthesis, sol-gel method, and block copolymer method. In this section, several methods of metallic membrane fabrication are discussed, which includes phase inversion, particle sintering, and the sol-gel method.

Phase Inversion

Like polymeric and ceramic membrane fabrication, phase inversion can also be utilized to produce a metallic membrane in a similar fashion. This method is mostly used to make hollow fibre membrane by extruding a suspension of metal powders (e.g. stainless steel, nickel) in a polymer solution through a spinneret into a coagulation bath. After the phase inversion takes place, the hollow fibre precursors are sintered at high temperature in the range of 1,100 – 1,500 °C. Combining both phase inversion and sintering into the fabrication of metallic membrane allows a large amount of hollow fibre to be produced in one step. The microstructure, as well as macrostructure of the membrane, can then be adjusted for the intended applications.

Particle Sintering

Particle sintering is a method which involves high-temperature compression of finely-sized metal powders or fibres at its softening temperature to produce porous metal membrane. This method consists of two main steps, which are (1) dispersing of inorganic metal powder into an organic solvent (also contains a binding agent, additive, and dispersant), and (2) sintering. Upon mixing of the ingredients, the slurry is allowed to stabilize and subsequently shaped via special moulding. The resulting membrane film is then heated at a temperature of about 100 ° to burn off the solvent, to 400 – 500 °C to burn off the polymer, and finally sintered at a high temperature of around 1,000 – 1,500 °C. The principle of sintering is to join together the metal particles, forming links of the soften metals at high temperature. The microstructure of the membrane formed in terms of porosity and pore size is dependent on the size, shape, distribution of particles, ratio of ingredients used, and the sintering technology parameters. Some elementary metals that can be used to fabricate porous metallic membrane through this method are aluminium (Al), titanium (Ti), stainless steel (*SS*), gold (Au) and copper (Cu)).

Sol-gel Method

The sol-gel method is one of the conventional methods used to coat a thin layer of ultrafine metal oxides onto a porous metal substrate. Usually, the metal oxides used are TiO_2, SiO_2, ZrO_2, and Al_2O_3, while the porous metal substrate is typically from stainless steel. The morphology and layer obtained are affected by other parameters such as the binder used, the concentration of casting solution, and sintering condition. This process has two routes, which are colloidal gel route and polymer gel route as shown in Figure 13. Principally, the sol-gel method comprises of three steps: (1) hydrolysis of precursor, (2) condensation of precursor, and (3) drying and sintering.

MEMBRANE MODULE

Membrane module (cell) is a manifold assembly containing a membrane or membranes to separate the streams of feed, permeate, and the retentate. Flat and tubular are the two general configurations of the membrane, and fabrication of membrane usually takes place in these shapes before arranged into different types of the module to suit the application required. These configuration and module are illustrated in Figure 14. The selection of suitable module configuration is usually based upon the type of separation

Figure 13. Sol-gel method routes

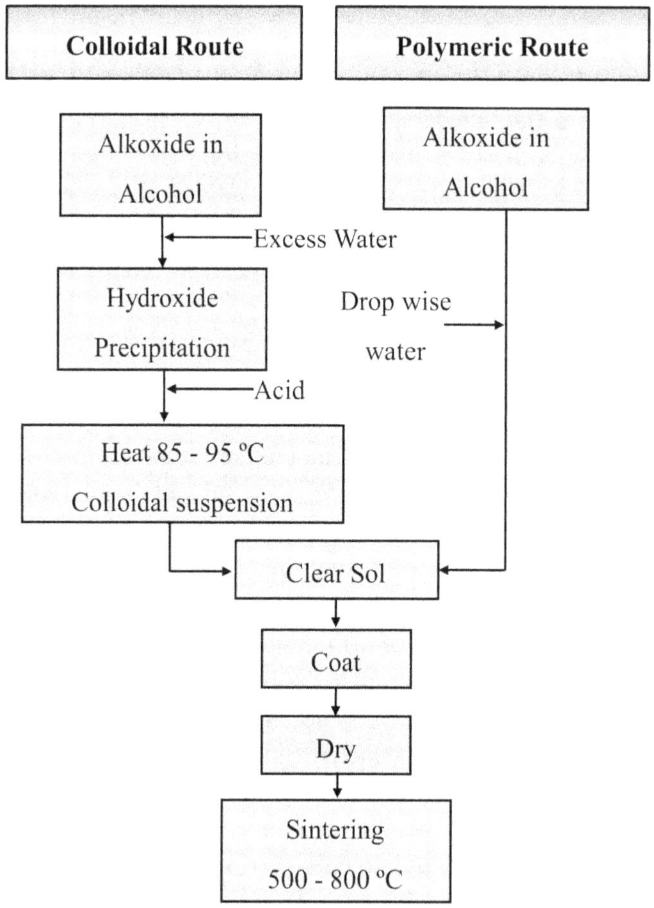

problem, cleaning, maintenance and operation costs, compactness of the system, scale, and possibility of membrane replacement. Table 4 shows the qualitative comparison of the flat and tubular membrane modules.

Flat Membrane Modules

Plate-and-frame Module

The plate-and-frame module is arranged in a way that sets of two membranes are placed in a sandwich-like fashion with their feed sides facing each other. A suitable spacer is placed between each set to provide a compartment for the feed flow. The number of membrane sets required for a certain operation is determined by the required membrane area. An example is shown below.

EXAMPLE 1: Number of Set for Plate-and Frame Module

Figure 14. Two types of membrane configurations

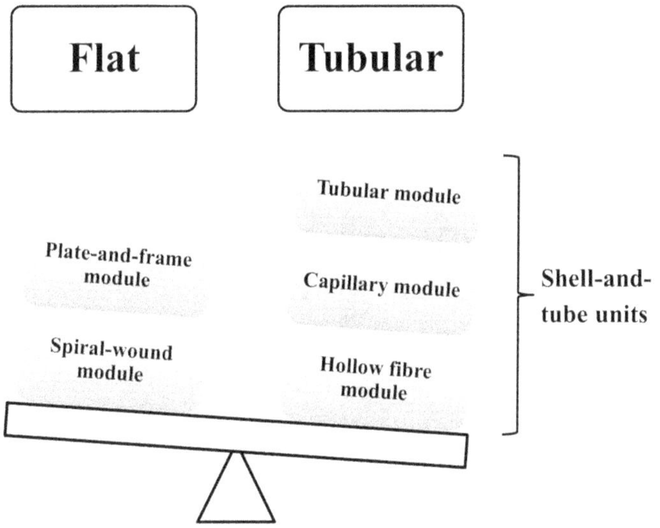

Problem. Calculate the number of membrane set required to fabricate a plate and frame module which requires 100 m^2 of membrane area. One set of membrane uses 2 m^2 of the membrane.

Solution. The number of sets required can be calculated directly by dividing the total membrane are required with the membrane area per one set of the membrane, given by:

Total membrane area required = 100 m^2

Area of one set of membrane = 2 m^2 (value usually provided by suppliers)

Then, we need 50 sets of membranes.

The membrane sets are furnished with sealing rings and two end plates and builds up to a plate-and-frame stack. The packing density (membrane surface are per module volume) is typically in the range of 100 – 400 m^2/m^3. Figure 15 depicts the schematic diagram of a plate-and-frame module.

Table 4. Qualitative comparison of the various membrane module

Membrane Module	Tubular	Plate-and-frame	Spiral-wound	Capillary	Hollow Fibre
Packing density	Low---> Very high				
Investment	High---> Low				
Fouling tendency	Low---> Very high				
Cleaning	Good---> Poor				
Operating cost	High---> Low				
Membrane displacement	Yes/no	Yes	No	No	No

Source: (Abbasi-Garravand, 2012)

Figure 15. Plate-and-frame module

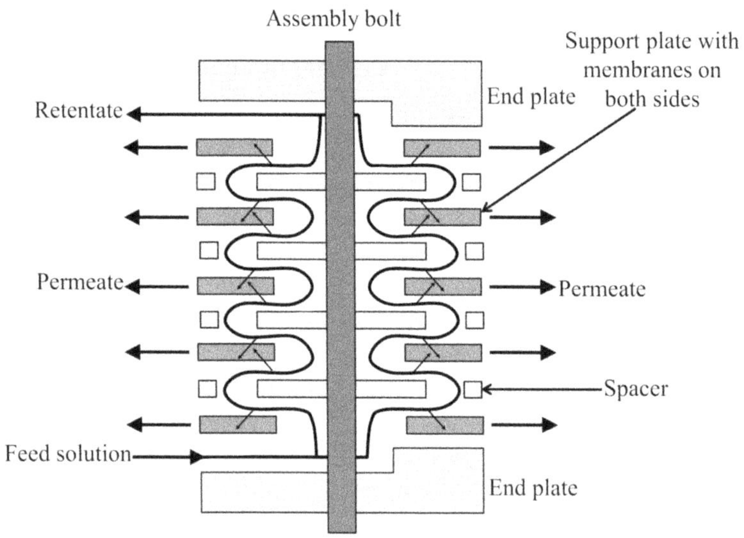

Spiral-wound Module

The spiral-wound module is essentially a plate-and-frame system wrapped around a central collection pipe. Membranes and permeate-side spacer are then glued along three edges to build a membrane envelope. The feed-side spacer separating the top layer of the two flat membranes acts as a turbulence promoter. A schematic diagram of a spiral-wound module is shown in Figure 16.

The feed flows axially through the cylindrical module parallel along the central pipe, whereas the permeate flows radially toward the central pipe. The packing density of the spiral-wound module is between 300 to 1,000 m^2/m^3, which is higher than the plate-and-frame module. The packing density

Figure 16. Spiral-wound module

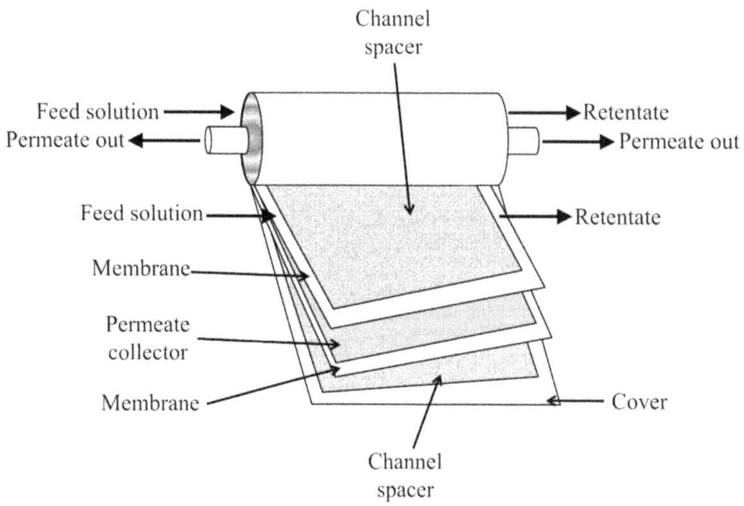

Figure 17. A pressure vessel containing three spiral-wound modules arranged in series

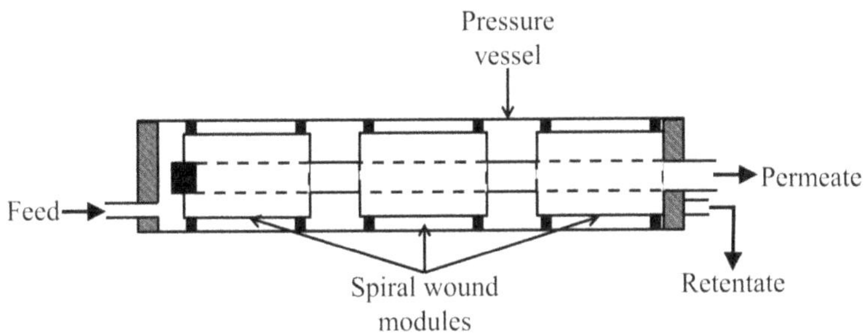

depends heavily on the channel height and is determined by the permeate and feed-side spacer material. In application, a number of spiral-wound modules are assembled in one pressure vessel, connected in series via the central permeate tubes as shown in Figure 17.

Tubular Membrane Modules

Tubular Module

The tubular module as shown in Figure 18 comprised of non-self-supporting tubular membrane placed inside porous stainless steel, ceramic or plastic tube with a diameter more than 10 mm. the number of tubes in the module is normally in the range of 4 – 19 tubes, though not limited by this number.in this module, the feed solution always flows through the centre of the tubes, the permeate flows across the porous supporting tube into module housing. Ceramic membranes are mostly assembled in such tubular module configuration. The packing density is usually less than 300 m^2/m^3.

Capillary Module

The capillary module consists of a large number of self-supporting capillary membranes assembled together in a module. The free ends of the fibres are potted with agents such as epoxy resins, polyure-

Figure 18. Tubular module

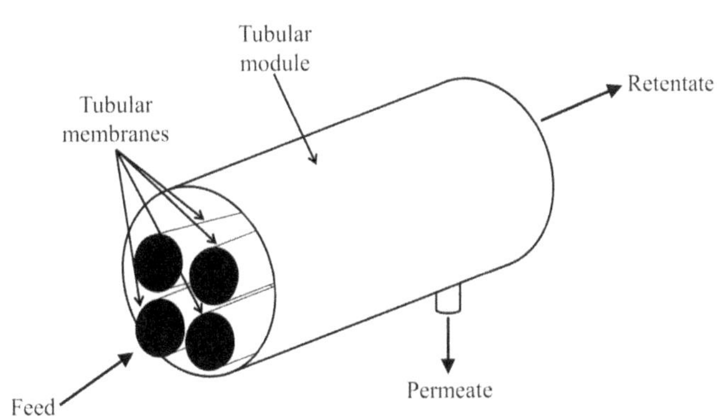

Figure 19. Capillary module

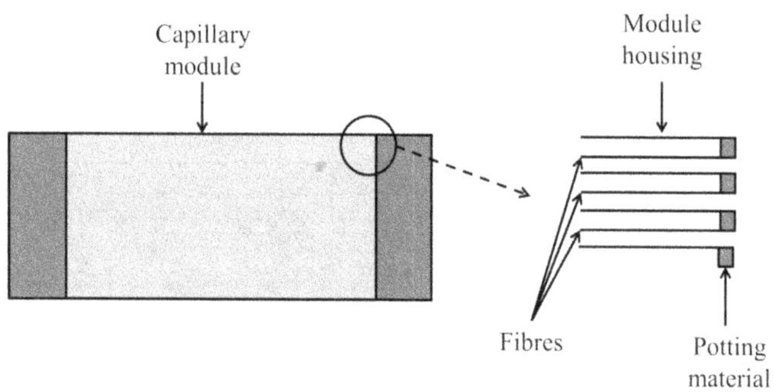

Figure 20. Capillary module modes, (a) inside-out, and (b) outside-in

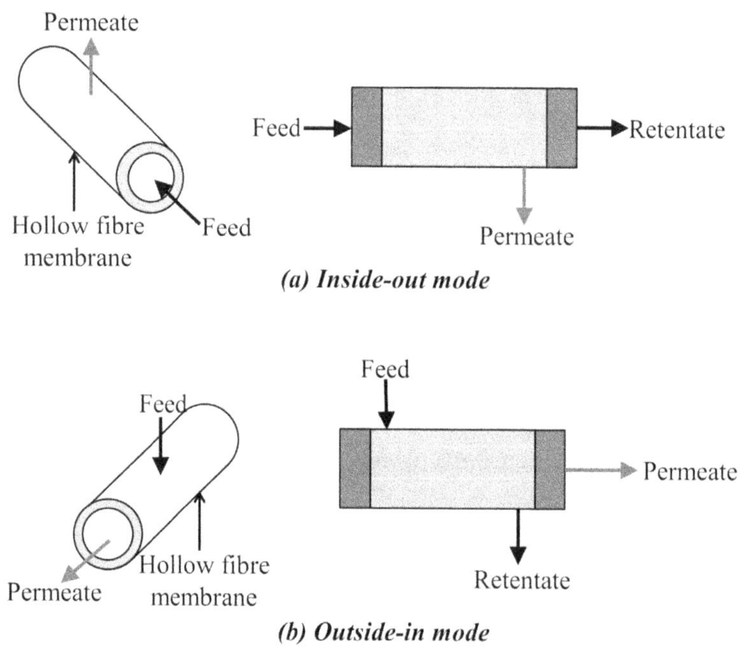

thanes, or silicone rubber as shown in Figure 19. This module has a packing density in the range of 600 – 1,200 m^2/m^3.

There are two modes of the capillary membrane: inside-out and outside-in. In inside-out, the feed solution passes through the bore of the capillary (lumen) whereas the permeate is collected on the outside of the capillaries. On the other hand, the feed solution enters the module on the shell side of the capillaries (external) and the permeate passes into the fibre bore in the outside-in mode. Figure 20 shows the schematic representation of the modes of the capillary module.

Hollow Fibre Module

The hollow fibre module has a similar concept with the tubular and capillary module. The hollow fibre membrane bundle are packed together in a housing chamber, which is commonly arranged in parallel,

Figure 21. Schematic diagram of a typical hollow fibre module

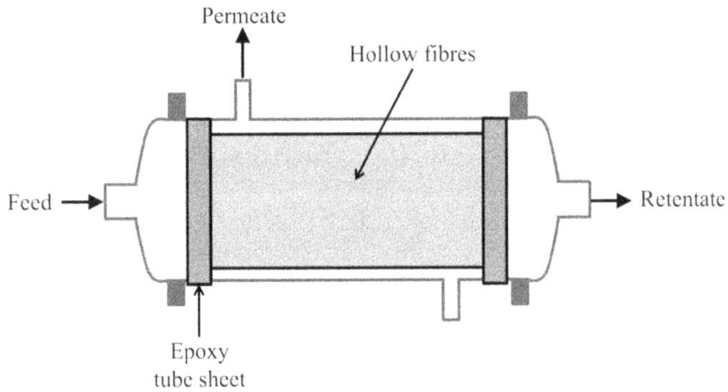

crisscross or other controlled geometric. Additionally, this module can be arranged in inside-out or outside-in modes. The module has a high packing density of around 30,000 m^2/m^3, meaning that it has high resistance to flow which could lead to pressure drops in the hollow fibres flow externally. This makes a hollow fibre module suitable for gas separation application. Figure 21 depicts the schematic diagram of a typical hollow fibre module.

REFERENCES

Abbasi-Garravand, E. (2012). *Removal of Cr (VI) and Cr (III) From Water by Reduction and Micellar Enhanced Ultrafiltration Techniques* (Master's dissertation). Concordia University, Canada.

Lalia, B. S., Kochkodan, V., Hashaikeh, R., & Hilal, N. (2013). A review on membrane fabrication: Structure, properties and performance relationship. *Desalination*, *326*, 77–95. doi:10.1016/j.desal.2013.06.016

Loddo, V., Bellardita, M., Camera-Roda, G., Parrino, F., & Palmisano, L. (2018). Heterogeneous Photocatalysis: A Promising Advanced Oxidation Process. In A. Basile, S. Mozia, & R. Molinari (Eds.), *Current Trends and Future Developments on (Bio-) Membranes: Photocatalytic Membranes and Photocatalytic Membrane Reactors* (pp. 1–43). New York: Elsevier. doi:10.1016/B978-0-12-813549-5.00001-3

Luyten, J., Buekenhoudt, A., Adriansens, W., Cooymans, J., Weyten, H., Servaes, F., & Leysen, R. (2000). Preparation of $LaSrCoFeO_{3-x}$ membranes. *Solid State Ionics*, *135*(1-4), 637–642. doi:10.1016/S0167-2738(00)00425-2

ADDITIONAL READING

Abd. El-Ghaffar, M. A., & Tieama, H. A. (2017). A review of membranes classifications, configurations, surface modifications, characteristics and its applications in water purification. *Chemical and Biomolecular Engineering*, *2*(2), 57–82.

Abitha, V. K., & Thomas, S. (2018). Fabrication Methods: Polymer Membranes for Liquid Transport. In S. Thomas, W. Runcy, S. Anil Kumar, & S. George (Eds.), *Transport Properties of Polymeric Membranes* (pp. 35–50). United States: Elsevier. doi:10.1016/B978-0-12-809884-4.00003-3

Belfort, G. (1988). Membrane modules: Comparison of different configurations using fluid mechanics. *Journal of Membrane Science*, *35*(3), 245–270. doi:10.1016/S0376-7388(00)80299-9

Hilal, N., Ismail, A. F., & Wright, C. J. (Eds.). (2015). *Membrane Fabrication*. Florida, USA: CRC Press. doi:10.1201/b18149

Johnson, D. J., & Hilal, N. (2018). State of the art review on membrane surface characterization: Visualization, verification and quantification of membrane properties. *Desalination*, *434*, 12–36. doi:10.1016/j.desal.2017.03.023

KEY TERMS AND DEFINITIONS

Ceramic: Solid material made from clay which is used to produce a thin sheet of the membrane by sintering at high temperature.

Configurations: Particular arrangement of membrane's geometric module.

Fabrication Technique: Methods used to produce membrane which comprised of several steps which can be achieved through mechanical, chemical, or combination of both processes.

Membrane: A thin sheet of permselective barrier that separates two phases.

Membrane Material: Material or 'ingredient' used to produce membrane, which is primarily from the polymer (organic), ceramic (inorganic), metal (inorganic), or combinations of materials in the fabrication of composite membrane.

Metal: Solid material typically having the characteristics of hard, malleable, shiny, ductile, and good electrical and thermal conductivity, such as palladium, silver, gold etc.

Module: A unit of membrane 'cell' which are made from membranes, membrane housing compartment, an inlet(s) and outlet(s).

Polymer: Chemical compound of large molecules formed by polymerization of repeating structural units known as the monomers.

Chapter 4
Membrane Characterizations

ABSTRACT

Membrane characterization is a critical necessity throughout the membrane's lifecycle. It mainly provides the connection between the fabrication and performance of the membrane. At the fabrication stage, membrane characterization allows us to study the membrane's characteristics in terms of its structure, morphology, chemical and physical properties, transport properties, etc. Membrane operation allows the determination of whether membrane cleaning is required or replacement is necessary. Finally, characterization at the end stage of membrane usability provides the causes of membrane failure which are significant data for future references. In that essence, this chapter discusses several methods that are used in membrane characterization processes, which are mainly categorized into the transport/flow, electron microscopy, scattering, and spectroscopy characterization methods.

INTRODUCTION

Membrane characterization provides the connection between its fabrication and performance with relation to the membrane structure, morphology, chemical and physical properties, transport/flow properties etc. The membrane field is continuously developing owing to its branches of the study of membrane fabrication and characterization. As more and more researches are conducted on membrane production methods, an improvement on the membrane can be achieved when better understanding on how to create the best membrane is developed. Various membrane characterization techniques have been established and practised to obtain the parameters that make out the membrane characteristics. Membrane characterization can be carried out throughout the lifecycle of membrane particularly after fabrication, during the operation, and lastly at the end of membrane lifecycle. They are performed at these different periods for different reasons such as obtaining the membrane characteristics, determining whether cleaning or membrane replacement is needed, and gathering information on the factors causing membrane failure respectively.

DOI: 10.4018/978-1-7998-2645-3.ch004

Table 1. Common methods of membrane characterization

Transport/Flow Methods	Equilibrium Methods	Advanced Methods
• Bubble gas transport • Gas permeability • Solute rejection/ molecular weight cut-off (MWCO) • Liquid-liquid displacement • Mercury intrusion porosimetry	• Gas adsorption/desorption • Thermometry • Permporometry • Bubble point	• Spectroscopy methods (e.g. RS, FTIR, PALS, RBS) • Scattering methods (e.g. XRD, SAXS, and SANS) • Electron microscopy methods (e.g. TEM, SEM, and ESEM)

Source: (Bernstein et al., 2013)

The performance of a membrane is influenced by its characteristics attributed to the chosen fabrication technique. This mainly includes membrane pore structure: porosity, pore size distribution and pore tortuosity, crystallinity, and surface property: hydrophilic/hydrophobic property, surface charge and surface roughness. These membrane properties can be measured by methods of characterization that are shown in Table 1, which are categorized into transport/flow techniques, equilibrium techniques, and advanced techniques. In this section, several methods of membrane characterization are discussed which includes transport/flow, electron microscopy, scattering, and spectroscopy characterization methods. These methods can be used individually to characterize a certain membrane property, or collectively to obtain whole characteristics of the membrane in terms of its chemical, physical, and mechanical properties. Consequently, the correlation obtained between the fabrication method and the membrane characteristics would assist researchers in the further development of membrane with better properties.

MEMBRANE PORE SIZE AND HYDROPHILICITY/ HYDROPHOBICITY CHARACTERIZATION METHODS

Bubble Point Method

The bubble point method is a technique which utilized the displacement of air with an original liquid inside the membrane pores to determine the pore size and distribution. This technique is based upon the capillary flow principle, whereby the air will pass through the membrane pores if the applied pressure surpasses the capillary attraction of liquid inside the largest pore. Figure 1 shows the principle of the bubble point method.

This method is carried out by first completely wetting the membrane with liquid which will be retained inside the pores by capillary force. Next, a gas is introduced at one side of the membrane, which subsequently forms a gas-liquid interface with surface tension γ. When the gas pressure is increased, the liquid inside the pores is forced out and displaced by the gas. Cantor equation can be used to measure the pressure difference, ΔP needed to force out the liquid from the membrane pore with radius, r as shown in Equation 1. This equation shows that the mechanism of gas bubble formation and release is governed by the gas pressure applied, the surface tension of the liquid, and the membrane pore size.

$$\Delta P = \frac{2\gamma}{r} \quad (1)$$

Figure 1. Principle of bubble point method, (a) mechanism of gas bubble formation and release, (b) theoretical model of a meniscus in a cylindrical pore

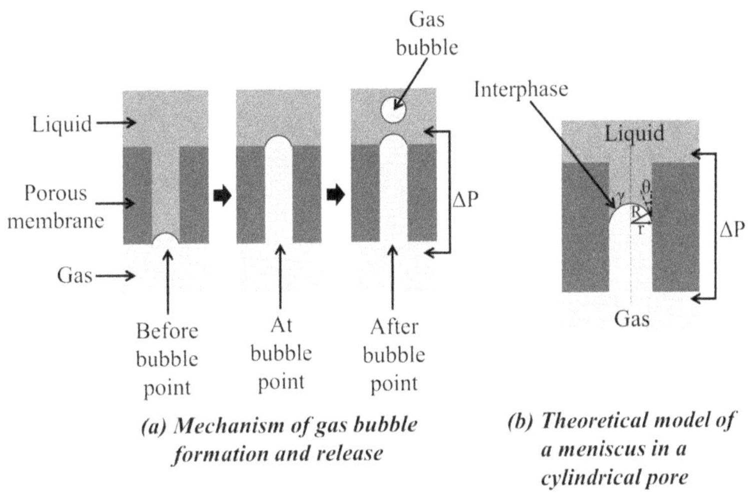

where ΔP = applied transmembrane pressure, γ = liquid surface tension, r = radius of pore.

Solute Rejection/Molecular Weight Cut-off (MWCO) Method

Molecular weight cut-off (MWCO) is a method which studies the characteristics of membrane separation of various solutes as it measures the solute rejection using non-ionic solutes with identified molecular weight. A membrane's MWCO is determined from a filtration experiment which identifies the molecular weight value of the solute molecule that provides 90% rejection (Figure 2) when various different molecular weight solutes are filtered in the target solvent. The MWCO of a membrane characterizes its selectivity for solute of different molecular weight. For that purpose, non-ionic polymeric solutes are used such as polyethene glycol (PEG), alkanes, dextrans, and oligostyrenes. The most widely used solute for the determination of MWCO is PEGs due to their colloidal properties which cause low fouling of the membrane pores. The PEG detection is usually carried out using sophisticated equipment like high-performance liquid chromatography (HPLC) and low-end colourimetric method.

The MWCO of a membrane is dependent on the pore size of the membrane. Table 2 shows the typical MWCO of some membrane processes which includes microfiltration, ultrafiltration, nanofiltration, and reverse osmosis.

Hydrophilicity/Hydrophobicity Analysis

Hydrophilicity and hydrophobicity are used to explain the relative affinity of water molecules dispersion on any surface material. The affinity of water molecules with the surface material is caused by the uneven secondary force interaction between the water droplet and material. They are usually defined by the geometry of water of the surface material, specifically on the contact angle (θ), i.e. the angle between water droplet's edge and the surface underneath it. The magnitude of the contact angle is connected to the material's surface energy.

Figure 2. Typical graph of rejection versus the molar mass of solutes to determine the 90% MWCO of a membrane

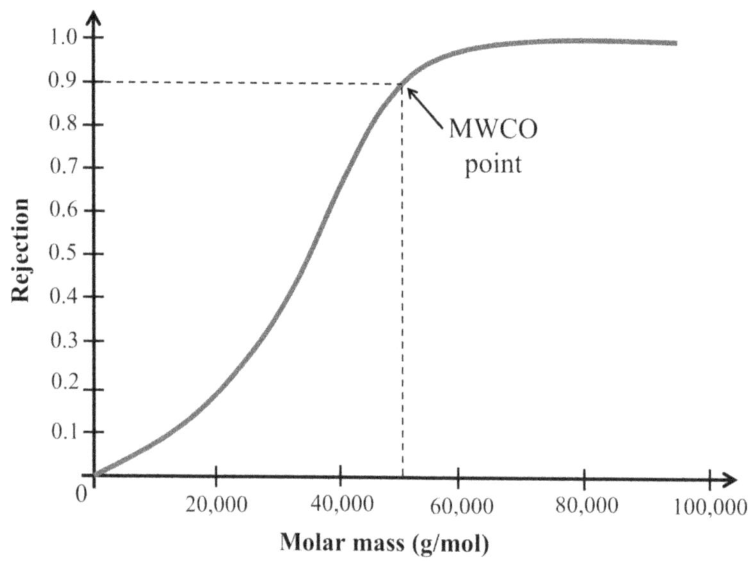

Generally, the magnitude of θ below 90° signifies high affinity of surface material towards the water molecules and categorized as hydrophilic (water-loving). Hydrophilic can be defined as the tendency of surface material to wet or form a thin water layer over the surface. On the other hand, θ higher than 90° characterize a non-wetting property of the surface, meaning that it has a low affinity towards water molecule. This type of surface material is called hydrophobic (water rejecting), whereby water molecules tend to form water beads on the surface due to low affinity. The effect of contact angle on the hydrophilicity and hydrophobicity of membrane are illustrated in Figure 3.

ELECTRON MICROSCOPY METHODS

Electron microscopy method is one of the widely used methods for membrane characterization that allows the imaging of the surface of the membrane to show its physical characteristics and morphology. This method applies the interaction of electron and membrane specimen and generates topological images of the specimen surface. Compared to the light microscope, electron microscopy has higher magnification which allows the observation of matter at refined detail beyond the naked human eye.

Table 2. Typical MWCO for various membrane processes

Membrane Process	Pore Size Range	MWCO (Daltons)
Microfiltration	0.1 – 10 μm	-
Ultrafiltration	1 – 100 nm	>1,000 – 10,000
Nanofiltration	1 – 10 nm	>200 – 400
Reverse osmosis	1 – 10 Å	<100s

Figure 3. Factor of contact angle on the hydrophilicity and hydrophobicity of membrane

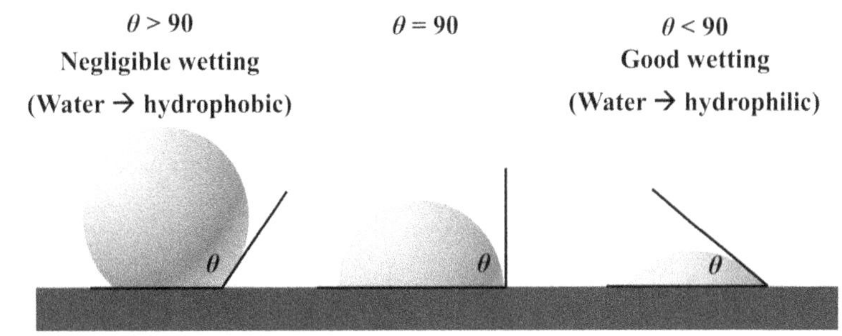

This method utilizes electrons scattered from a specimen hit with a finely focused electron beam. The faster the electrons travel (higher accelerating voltage), the shorter the wavelength which in turn form image that has higher resolution. This instrument usually uses a high vacuum, thus the specimen usually needs to be totally dry or completely frozen (for cryo-electron microscopy application). The electron beam causes heating and charging of the specimen. Nonetheless, these can be reduced using a conductive coating of a few nanometer thicknesses, cryo-technique, or lower beam energy. However, these lead to lower resolution imaging obtained. There are many types of electron microscopy, including transmission electron microscopy (TEM), scanning electron microscopy (SEM), environmental scanning electron microscopy (ESEM), reflection electron microscopy (REM), and scanning transmission electron microscopy (STEM). The first two types aforementioned will be discussed in this section.

Transmission Electron Microscope (TEM)

Transmission electron microscopy (TEM) is the original form of electron microscopy. TEM utilizes an electron beam passing through a specimen and scattered to provide a microscopic image of the specimen which can be observed through its objective lens. The electron beam is produced from an electron gun fitted on the transmission electron microscope as shown in Figure 4. A tungsten filament cathode is usually used as the source of the electron beam, while an anode is utilized to accelerate the electrons. In addition, electrostatic and electromagnetic lenses are used to focus the electron beam.

The limitation of this technique is that a very thin specimen is required, typically below 100 nm. Thus, only nano-objects such as nanoparticles or nanotubes can be directly viewed using TEM. Because of that, most specimens need to be chemically/physically altered and dehydrated before being embedded in a polymer resin prior examination under TEM. For bulk materials embedded in polymeric resin, it needs to be cut into thin cross-sections by using ultra-microtome before it can be viewed. Besides, the limitation with the examination of organic materials is that they usually have very small differences in electron density at the target area. Thus, staining is typically conducted on organic specimen into the selected region. Despite its several limitations, TEM has an extensively high resolution which could reach the subatomic level for inorganic materials and several nanometers for polymers. An example of an image obtained from TEM is shown in Figure 5.

Figure 4. Schematic diagram of a TEM
Source: Wikimedia Commons, 2011 (Licensed under CC BY-SA 3.0, 2.5, 2.0, & 1.0)

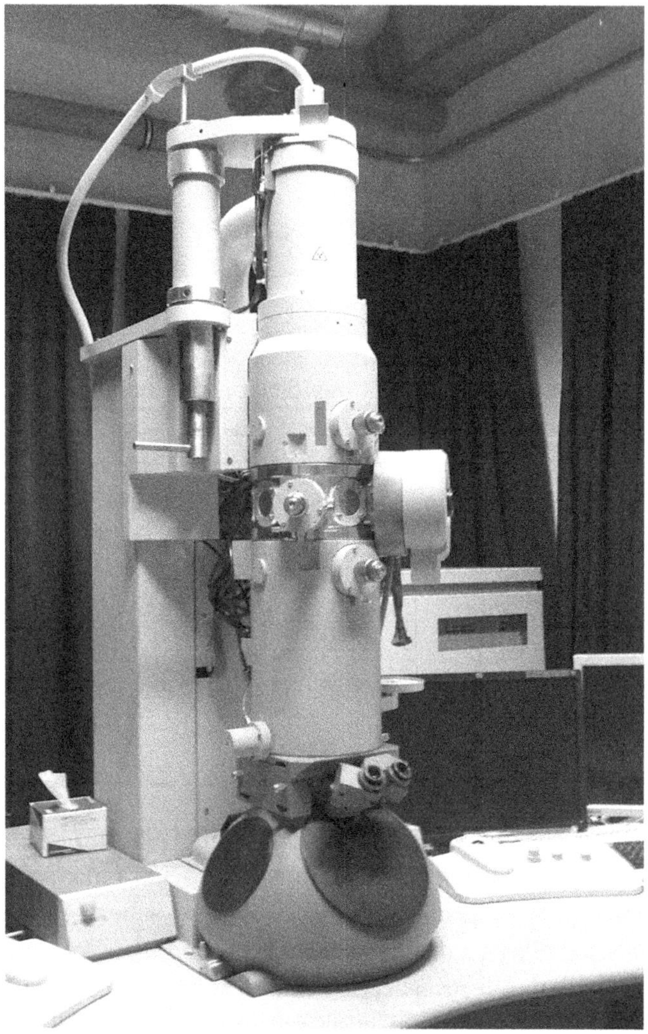

Figure 5. TEM images of MCM-41 and ionic liquid-modified MCM-41: (a) MCM-41, (b) MCM-41-IL1 and (c) MCM-41-IL2
Source: Li, Yan, & Wu, 2019 (Licensed under CC BY 4.0)

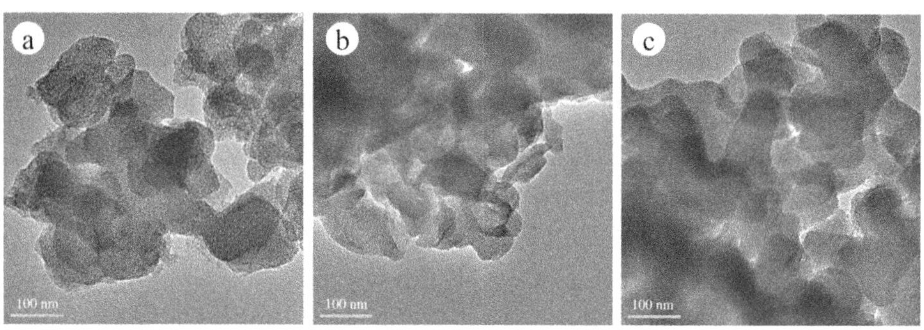

Figure 6. SEM images of MCM-41 and ionic liquid-modified MCM-41: (a) MCM-41, (b) MCM-41-IL1, (c) MCM-41-IL2
Source: Li, Yan, & Wu, 2019 (Licensed under CC BY 4.0)

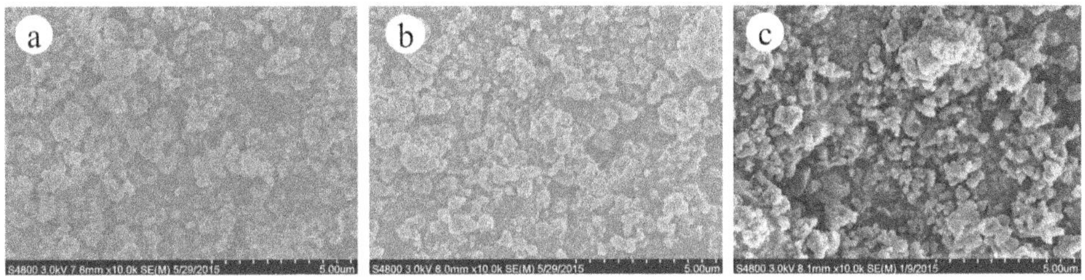

Scanning Electron Microscope (SEM)

Scanning electron microscopy (SEM) technique uses a focused electron beam that scans across the specimen surface to produce a microscopic image of the sample. When the electron beam comes into contact with the specimen, the interaction produces signals of secondary electrons, backscattered electrons, and x-ray. The signals are detected by detectors which are then displayed on the monitor.

The SEM images are usually produced from two types of interactions: elastically backscattered electrons and inelastically scattered secondary electrons. Inelastic interaction, lower energy secondary electrons are emitted due to deflection of primary electrons when it interacts with sample atomic nuclei or electrons of equivalent energy. While in inelastic interaction, lower energy secondary electrons are produced from the sample due to energy exchange between the primary electron with the atom in the sample.

SEM can function at a magnification of up to 100,000. Unlike TEM, SEM can be used for imaging of wide regions, large, and irregular shape sample. Generally, SEM can reach a resolution limit of around 0.1 μm for typical SEM. On the other hand, advanced SEM using a field emission gun (i.e. field emission scanning electron microscopy (FESEM)) can reach a resolution limit of up to 2.5 nm. Figure 6 shows examples of SEM images.

SCATTERING METHODS

X-ray Diffraction (XRD)

X-ray diffraction (XRD) is a technique which utilizes x-rays to examine and measure the crystallinity of a sample by quantifying the x-rays diffraction from the planes of atoms within the sample. XRD can be used to measure the samples' crystalline content; detect the crystalline phases etc. due to its sensitivity to the type of and relative position of atoms in a sample. The membrane sample for analysis can vary from various ranges of types, comprising of polymeric, metal, ceramic, composite, organic, and inorganic materials. Typically, the sample is prepared as finely divided powder, but it can also be in the form of a flat surface, on condition that the surface has relatively low roughness. The typical instrumentation of an XRD method is illustrated in Figure 7, which is known as the x-ray diffractometer. Generally, there are

Figure 7. Instrumentation of XRD
Source: Wikimedia Commons, 2016 (Licensed under CC BY-SA 4.0)

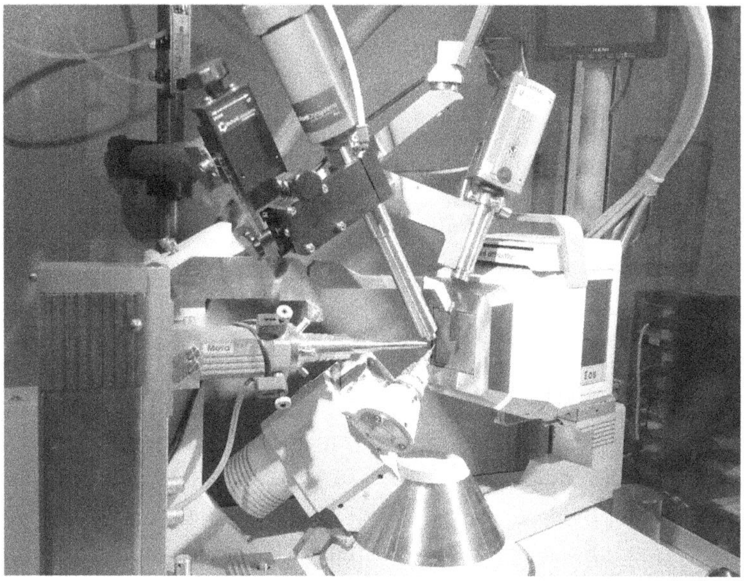

four types of XRD methods namely Laue's photographic method, Bragg's x-ray spectrometer method, and rotating crystal method, and powder method.

In principle, when x-ray interacts with matter, scattering phenomenon will occur. By looking at x-ray behaviour as a wave, we obtain coherent scattering (as oppose to particles behaviour where we deal with incoherent scattering). Incoherent scattering, photons scatter with no energy loss and produce scattered radiation of the same wavelength. When the waves hit electrons, the electrons start to vibrate and become the source of a wave, whose phase is established by the incident wave. Subsequently, all the electrons in the material that the wave encounters form a group of coherent sources whose radiation can interfere constructively or destructively. The data of scattering angles together with the x-ray pattern can then be used to characterize the crystalline structure or polymer characterization of membrane sample.

Figure 8 illustrates an XRD pattern for a polypropylene/titanium dioxide (PP/TiO_2) composite membrane with different TiO_2 composition which can show the crystalline characteristic of the membrane. Apparent peaks centred at angles 14.0°, 16.9 °, 18.5°, and 25.4° of the 2θ are associated with α-form crystallography planes (110), (040), (130), and (060) respectively. The membrane is characterized as to have a very oriented lamellar structure in terms of its crystalline morphology from the observation of weak peaks of the (110/041) planes at angles 21.3°, 21.8° (Wang et al., 2017).

Small-Angle Scattering (SAS)

Small-angle scattering (SAS) is a technique used to probe the structure of a membrane sample up to the length scale in the range of 1 – 100 nm using either neutron or x-ray at small-angle scattering. There are generally two types of SAS technique, which are small-angle neutron scattering (SANS) and small-angle x-ray scattering (SAXS). The schematic representation of SANS/SAXS is shown in Figure 9, while the general comparison between SANS and SAXS are tabulated as in Table 3.

Figure 8. Example of XRD patterns image showing polypropylene (PP) and PP/TiO_2 composite membrane with different TiO_2 content
Source: Wang et al., 2017 (Licensed under CC BY 4.0)

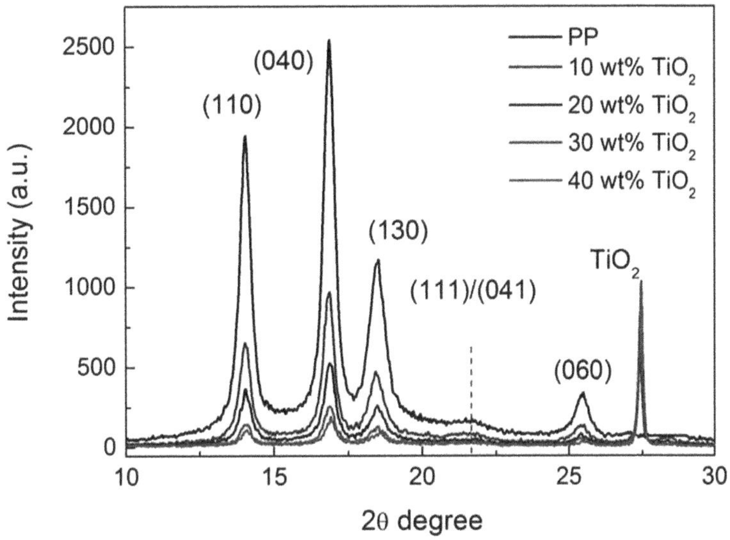

Figure 9. Schematic presentation of SAS technique

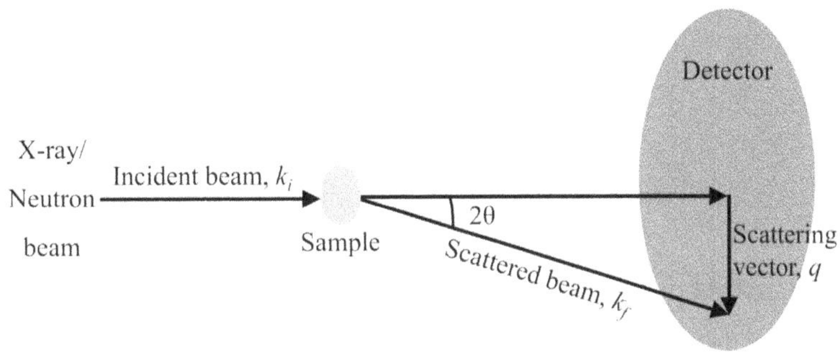

Table 3. Types of SAS technique

Types of SAS Technique	SANS	SAXS
Source	Neutron, few and weak	X-ray, many and strong
Length scale probed (nm)	1 – 100	1 – 100
Scattering angle	Small	Big
q range (Å^{-1})	0.001 – 1.0	0.8 – 1.0
Sample volume (cm^3)	0.05 – 3.5	0.0001 – 0.5
Radiation damage to the sample	Very unlikely	Very likely

From Figure 9, X-rays or neutrons is scattered upon contact with the sample, whereby the scattering profile can be illustrated mainly with the parameters of the scattering intensity, $I(q)$. The scattering vector, q is calculated from the difference between the scattered beam vector, k_f and incident beam vector, k_i as shown in Equation 2. The magnitude of k_f and k_i are principally from the elastic scattering since inelastic scattering is negligible for small-angle scattering.

$$q = k_f - k_i \tag{2}$$

The magnitude of q can be defined as,

$$q = \left|q\right| = \frac{4\pi \sin\left(\theta\right)}{\lambda} \tag{3}$$

where q = scattering vector, $\left|q\right|$ = magnitude of scattering vector, 2θ = scattering angle, λ = wavelength of the incident beam.

The magnitude of scattering intensity, $I\left(q\right)$ is defined as,

$$I\left(q\right) = \varnothing.V.\Delta p^2.P\left(q\right).S\left(q\right) \tag{4}$$

where $I\left(q\right)$ = scattering intensity, $\varnothing$ = density of particles in volume, V = volume, Δp^2 = contrast factor, $P\left(q\right)$ = form factor, $S\left(q\right)$ = structure factor.

SPECTROSCOPY METHODS

This section discusses some vibrational spectroscopy methods which utilize vibrational molecules to study the chemical characteristics of the membrane. This includes the membrane's chemical structure and composition. The vibrational data obtained will result in absorption spectra or specific band which can be translated into information such as chemical bonds, chemical substructures, or chemical groups within a complex structure. The spectroscopy methods which are commonly used and which will be discussed in this section are infrared spectroscopy and Raman spectroscopy.

Fourier Transform Infrared (FTIR) Spectrophotometry

Fourier transform infrared (FTIR) is a characterization method mainly used to determine the functional group in a sample as well as the possible molecular bonds between chemical compounds present in the sample. FTIR spectrophotometer is the instrument that determines the infrared absorption spectrum of a sample. Figure 10 illustrates the basic components of a typical FTIR spectrophotometry. The instrument works in a way that infrared beam is emitted from a source and passes through into interferometer which is responsible for the spectral encoding. In here, the beams with different path lengths recombine

Figure 10. Elements of FTIR spectrophotometry

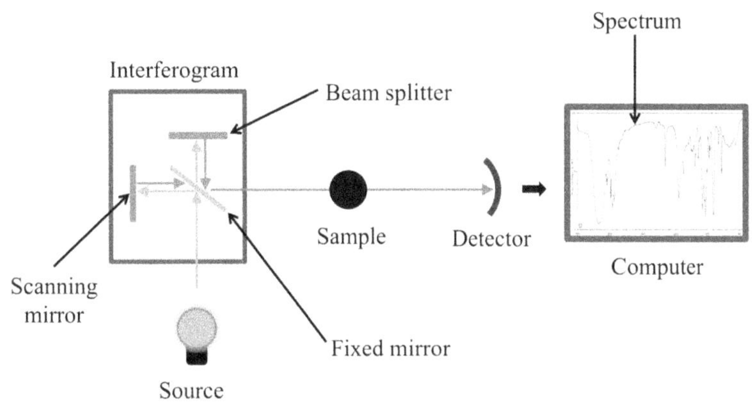

and generate constructive and destructive interference known as an interferogram. The beam then passes through the sample compartment, where the sample absorbs energy with specific frequencies, which corresponds to the sample characteristics from the interferogram. A detector subsequently measures the signal of interferogram in energy versus time for all frequency signals before the infrared spectrum is finally attained.

The FTIR spectrophotometer usually produces an infrared spectrum in the range of 4,000 – 666 cm^{-1}. Typically, many functional groups can be detected in the mid-infrared region of 4,000 – 400 cm^{-1} making FTIR a versatile technique to locate the functional groups which exist within the molecule. Basically, the spectrum can be categorized into four regions of types of bonds. Single bonds, for example, O–H, C–H, and N–H are noticeable in the higher wavenumber in the range of 2,500 – 4,000 cm^{-1}. On the other hand, double bonds such as C=O, C=N, C=C, and N=O; and triple bonds like C≡C, and C≡N are detectable in the range of 1,500 – 2,000 cm^{-1} and 2,000 – 2,500 cm^{-1} respectively. For the lower wavenumber region (fingerprint region) between 1,500 – 600 cm^{-1}, complex pattern are produced that are characterized by the bonds C–C, C–N, C-O, and C–Cl. The wavenumber values of FTIR absorption band which corresponds to the types of chemical bonds are tabulated in Table 4.

Raman Spectroscopy

In membrane application, Raman spectroscopy is used to observe the crystalline structure of macromolecules and also changes in the membrane structure. Principally, Raman spectroscopy examines the resonant vibrational energies within molecules typically using monochromatic laser light. Interaction between the light and vibrational molecules or lattice (phonon) will result in inelastic scattering of photons.

Figure 11 shows the schematic diagram of a typical Raman spectrophotometer operation. Raman scattering is the inelastic scattering of a photon when the laser light interacted with the sample with energy transfer between the excitation beam and sample. Raman scattering is produced when the molecular motion causes change in the molecule polarizability. The scattering light has frequency different from the monochromatic light, which corresponds to the molecular vibrational frequency. Raman spectra are then produces which are often representative of certain bonds or chemical structure. Because of that, this technique is very useful to determine the functional group and chemical substructure of membrane

Table 4. Typical types of bond and wavenumber value of FTIR absorption band

Type of Bonds	Wavenumber Value (cm^{-1})	Intensity*
C–H	3,000 – 2,850 (alkanes stretch) 1,450 & 1,375 (alkanes: $-CH_3$ bend) 1,465 (alkanes $-CH_2-$ bend) 3,100 – 3,000 (alkenes stretch) 1,000 – 650 (alkenes out-of-plane bend) 3,150 – 3,050 (aromatic stretch) 900 – 690 (aromatic out-of-plane bend) 3,300 (alkyne stretch) 2,900 – 2,800 (aldehyde)	s m m m s s s w W
C=C	1,680 – 1,600 (alkene) 1,600 – 1,475 (aromatic)	m–w m–w
C≡C	2,250 – 2,100 (alkyne)	m–w
C=O	1,740 – 1,720 (aldehyde) 1,725 – 1,705 (ketone) 1,725 – 1,700 (carboxylic acid) 1,750 – 1,730 (ester) 1,680 – 1,630 (amide) 1,810 – 1,760 (anhydride) 1,800 (acid chloride)	s s s s s s s
C–O	1,300 – 1,000 (alcohol, ethers, esters, carboxylic acids, anhydrides)	s
O–H	3,560 – 3,600 (free alcohols, phenols) 3,400 – 3,200 (H-bonded alcohols, phenols) 3,400 – 2,400 (carboxylic acid)	m m m
N–H	3,500 – 3,100 (primary, secondary amines and amides: stretch) 1,640 – 1,550 (primary, secondary amines and amides: bend)	m m–s
C–N	1,350 – 1,000 (amines)	m–s
C=N	1,690 – 1,640 (imines and oximes)	w–s
C≡N	2,260 – 2,240 (nitriles)	m
X=C=Y	2,270 – 1,940 (allenes, ketenes, isocyanates, isothiocyanates)	m–s
N=O	1,550 & 1,350 (nitro: $R-NO_2$)	s
S–H	2,550 (mercaptans)	w
S=O	1,050 (sulfoxides) 1,375 – 1,300, 1,350 – 1,140 (sulfones, sulfonyl chlorides, sulfates, sulfoamides)	s s
C–X	1,400 – 1,000 (fluoride) 785 – 540 (chloride) <667 (bromide, iodide)	s s s

*s: small, m: medium, w: wide
Source: (Hilal et al., 2017)

sample; macromolecule chains positioning, and to observe the changes in membrane structural properties after undergoing some chemical or mechanical stresses.

Raman spectroscopy is usually associated as a complementary technique to infrared spectroscopy. In some cases, come vibrational modes only shown in the Raman spectrum and not visible in the infrared spectrum. For example, some symmetric molecules become visible in Raman spectra but inactive in infrared, whereas certain asymmetric molecules are active in infrared spectra but not in Raman. For

Figure 11. Schematic of the Raman spectrophotometer

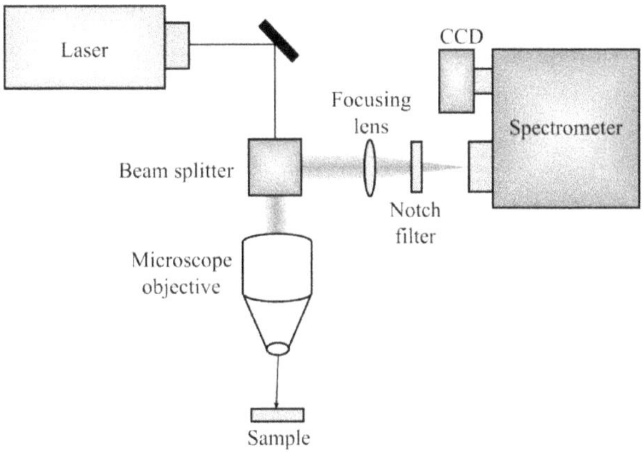

Figure 12. Example of Raman spectra image of synthesized nanopowder from sulfated-titanium dioxide (S-TiO_2) filler and nanocomposite Nafion membranes which illustrates the before (grey) and after (black) storage in high humidity environment
Source: Allodi et al., 2016 (Licensed under CC BY 4.0)

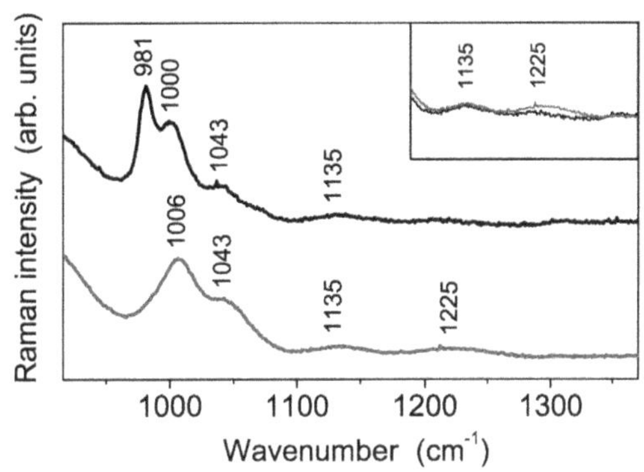

Figure 13. Example of Raman spectra image carried out in high wavenumber region of (a) pure Nafion membrane, and composite membranes of different TiO_2 nanopowders composition of (b) 2%, (c) 5%, and (d) 7%
Source: Allodi et al., 2016 (Licensed under CC BY 4.0)

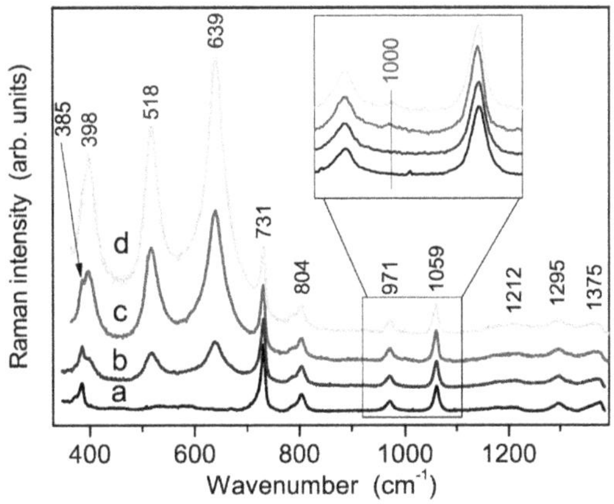

polymeric molecule, the carbon chain vibration can be easily examined using Raman spectroscopy, while the branches vibration is easily studied using infrared spectroscopy.

Figure 12 and Figure 13 illustrate the application of Raman spectroscopy to characterize the membrane in terms of changes in membrane structural properties after being introduced in an environment, and vibrational characteristics of the composite membranes respectively. Figure 12 depicts an example of Raman spectra of a synthesized nanopowder from sulfated-titanium dioxide (S-TiO_2) filler and nanocomposite Nafion membranes which illustrates the changes before and after the sample is stored in high humidity environment. On the other hand, Figure 13 shows an example of Raman spectra of pure Nafion membrane and composite membrane with different TiO_2 nanopowders composition. Higher TiO_2 nanopowders composition corresponds to higher Raman intensity in the spectrum.

REFERENCES

Allodi, V., Brutti, S., Giarola, M., Sgambetterra, M., Navarra, M. A., Panero, S., & Mariotto, G. (2016). Structural and spectroscopic characterization of a nanosized sulfated TiO_2 filler and of nanocomposite Nafion membranes. *Polymers*, *8*(3), 68–80. doi:10.3390/polym8030068 PMID:30979163

Bernstein, R., Kaufman, Y., & Freger, V. (2013). Membrane characterization. In Encyclopedia of Membrane Science and Technology. doi:10.1002/9781118522318.emst063

Hilal, N., Ismail, A. F., Matsuura, T., & Oatley-Radcliffe, D. (2017). *Membrane Characterization*. United States: Elsevier.

Li, Y., Yan, D., & Wu, Y. (2019). Ionic liquid-modified MCM-41-polymer mixed matrix membrane for butanol pervaporation. *Royal Society Open Science*, *6*(7). doi:10.1098/rsos.190291 PMID:31417733

Patterson, D. A., Havill, A., Costello, S., See-Toh, Y. H., Livingston, A. G., & Turner, A. (2009). Membrane characterization by SEM, TEM and ESEM: The implications of dry and wetted microstructure on mass transfer through integrally skinned polyamide nanofiltration membranes. *Separation and Purification Technology*, *66*(1), 90–97. doi:10.1016/j.seppur.2008.11.022

Wang, S., Ajji, A., Guo, S., & Xiong, C. (2017). Preparation of microporous polypropylene/titanium dioxide composite membranes with enhanced electrolyte uptake capability via melt extruding and stretching. *Polymers*, *9*(3), 110–121. doi:10.3390/polym9030110 PMID:30970788

Wikimedia Commons. (2011, January 22). *File: Transmission electron microscope (Philips TEM) Mega View II pl.jpg* [digital image]. Retrieved August 15, 2019 from https://commons.wikimedia.org/wiki/File:Transmission_electron_microscope_(Philips_TEM)_Mega_View_II_pl.jpg

Wikimedia Commons. (2016, July 23). *File:Scxrd.jpg* [digital image]. Retrieved August 15, 2019 from https://commons.wikimedia.org/wiki/File:Scxrd.jpg

ADDITIONAL READING

Ahmad, N. A., Leo, C. P., Ahmad, A. L., & Ramli, W. K. W. (2014). Membranes with great hydrophobicity: A review on preparation and characterization. *Separation and Purification Reviews*, *44*(2), 109–134. doi:10.1080/15422119.2013.848816

Capannelli, G., Vigo, F., & Munari, S. (1983). Ultrafiltration membranes – characterization methods. *Journal of Membrane Science*, *15*(3), 289–313. doi:10.1016/S0376-7388(00)82305-4

Childress, A. E., Brant, J. A., Rempala, P., Phipps, D. W., & Kwan, P. (2012). USA: Water Research Foundation.

Khulbe, K. C. Feng, Evaluation of Membrane Characterization Methods C. Y., & Matsuura, T. (2002). Membrane Characterization. In Water and Wastewater Treatment Technologies: Encyclopedia of Life Support System. Oxford, UK: EOLSS Publishers.

Khulbe, K. C., Feng, C. Y., & Matsuura, T. (2008). *Synthetic Polymeric Membranes: Characterization by Atomic Force Microscopy*. Berlin, Heidelberg: Springer.

Tamime, R., Wyart, Y., Siozade, L., Baudin, I., Deumie, C., Glucina, K., & Moulin, P. (2011). Membrane characterization by microscopic and scattering methods: Multiscale structure. *Membranes*, *1*(2), 91–97. doi:10.3390/membranes1020091 PMID:24957612

Tylkowski, B., & Tsibranska, I. (2015). Overview of main techniques used for membrane characterization. *Journal of Chemical Technology Metallurgy*, *50*, 3–12.

Xu, Z. K., Huang, X. J., & Wan, L. S. (2009). Techniques for Membrane Surface Characterization. In *Surface Engineering of Polymer Membranes* (pp. 5–63). Berlin, Heidelberg: Springer. doi:10.1007/978-3-540-88413-2_2

KEY TERMS AND DEFINITIONS

Characterization: Critical process in membrane research and development to characterize the membrane characteristics/properties in terms of its structure, morphology, chemical and mechanical properties, transport/flow properties, etc.

Electron Microscopy: Technique used in membrane characterization that allows the imaging of the surface of the membrane to show its physical characteristics and morphology.

Membrane Characteristics: Properties that wholly define the membrane which is generally classified into structural, morphological, transport, flow, chemical, physical, and mechanical characteristics.

Membrane Structure: The overall physical construction of the membrane.

Morphology: Covers the form, shape, size, structure, and appearance of the membrane.

Scattering Methods: Techniques used in membrane characterization which typically use light, moving particles, x-ray etc. which are scattered and translated to provide information about the membrane characteristics.

Spectroscopy: Techniques used in membrane characterization which involves the splitting of light or electromagnetic radiation, which are then measured into the constituent wavelength or spectrum. The spectrum data obtained can be translated into information regarding the membrane characteristics.

Transport Property: Parameters that define the rate of transport (of mass, heat, etc.) such as the viscosity, diffusivity, and thermal conductivity.

Chapter 5 Membrane Processes in Water and Wastewater Treatment

ABSTRACT

Membrane technologies play a very important role in water and wastewater treatments. These membrane processes provide key advantages over the conventional processes, such as lower energy requirement, lower footprint, easier to operate, and more effective contaminants removal. This chapter introduces different membrane processes: (1) pressure-driven membrane processes which are the most widely used in water and wastewater treatments, and (2) several advanced membrane processes. These processes perform physical or physicochemical separations. Most of the separations occur between liquid-liquid phases, but liquid-gas and gas-gas separation phases are also performed in the latest membrane development. The contemporary membrane bioreactor is the heart of membrane technologies that are used in various applications. However, fouling is a common phenomenon that reduces the efficiency of the membrane operation. Thus, the concept of critical flux and introduction of some control and preventive mechanism could prevent or reduce the fouling in membrane bioreactors.

INTRODUCTION

A Brief History Development of Membrane Technology

The early study of membrane field can be trailed back to the eighteenth century, when the term "osmosis" was first introduced by Jean-Antoine (Abbé) Nollet in 1748 which signifies the water permeation across a diaphragm. In his study, Nollet experimented using pig bladder as the membrane material. Yet through the 1800s until early 1900s sees no commercial application for membrane, except as laboratory experimentation tool to acquire breakthrough in physical and/or chemical theories in membrane field. These were the golden period to study this relatively new field which has brought to the discovery of van's Hoff equation and the development of the kinetic theory of gases using the model of the selective

DOI: 10.4018/978-1-7998-2645-3.ch005

semipermeable membrane by Maxwell. It wasn't until two centuries later before membrane technologies have seen tremendous development and evolution in terms of their use of materials, designs, operations, as well as their applications.

In terms of membrane material, early membrane researches experimented with available diaphragms such as pig, cow, and fish bladder, and animal gut. Other materials were subsequently used to make membrane such as collodion (nitrocellulose), and other polymers like cellulose acetate which are still extensively used today. However, the membrane has yet been widely used for separation application let alone commercialized due to significant problems aroused with the available membranes. These drawbacks included membrane being very undependable, low flux, unselective and expensive. Industrial application of membrane was made possible at the beginning of 1963 through the Loeb-Sourirajan process. Loeb and Sourirajan's work managed to produce reverse osmosis membrane characterized by fault-free, high flux, asymmetric membrane. The membrane developed was a practical method to desalt water. This innovative commercialization gave birth to the progression of other membrane types such as microfiltration and ultrafiltration.

The peak development of these membranes was in the mid-1990s to treat municipal water and integrated with the membrane bioreactor systems to treat sewage. The swift growing of wastewater treatment application using membrane technology was made possible through scaling up and commercialized installation of the system by leading companies like Mitsubishi Rayon, Kubota and Zenon. Since then, membrane technology utilization in water and wastewater application has been widespread in various countries around the world.

In Singapore for example, membrane technology has gained significant interest in wastewater treatment application. Due to land scarcity, Singapore faced a problem of water shortage as a result of the lack of natural water resources. Strategic planning, investment, and management on its water technology helped Singapore to overcome these problems mentioned beforehand. Currently, Singapore has four major water sources termed as the "Four National Taps" which are water obtained from their local collection area, imported from Malaysia, reclaimed water known as the NEWater, and desalinated water. Membrane bioreactor technology has been extensively studied for the application of water reclamation. In 2003, a pilot study on the membrane bioreactor technology was initiated at Bedok Water Reclamation Plant to analyse on the design and operation suitable to the tropical environment of Singapore. It was able to produce a good feed water quality for the production of NEWater through membrane bioreactor/reverse osmosis system. The result was very satisfactory which leads to the opening of a municipal-scale plant with a capacity of 23 million litres per day (MLD) at Ulu Pandan Water Reclamation Plant in 2006. Domestic wastewater is treated to supply clean process feed water to industries which are coined as the Industry Water. The systems are feasible because of the low economy of scale on the membrane systems. Moreover, another membrane bioreactor plant was planned to be commissioned in 2011 at Jurong Water Reclamation with a capacity of 68 MLD.

Membrane technology researches in Australia were initiated in the late 1990s. In 2002, the first full-scale operation of the membrane bioreactor system was erected at Picnic Bay on Magnetic Island, Australia. The shortage in freshwater supply forces for the water recycling initiative, which subsequently becomes the main driving force for the elevating utilization of membrane technology. The water recycling initiative is an initiative to save water resources via water reuse for not-for-drinking application. For example, reusing domestic water from sinks and showers for agriculture irrigation and toilet flushing. However, domestic sewage and industrial wastewater would need further treatment before they can safely be reused for other purposes. One of the largest applications of membrane bioreactor in Australia is the Gippsland

Water Factory in Australia, commissioned in 2010. The treatment facility treated both domestic sewage and pulp and paper mill wastewater with a capacity of 35 MLD. To date, membrane bioreactor has been extensively used here from a small system at household to a large sewage treatment plant.

In Malaysia, membrane bioreactor technology has yet been implemented for treating manufacturing or municipal wastewater. The reasons are due to:

- Lack of awareness into the insights of this promising technology by the government, policymakers, and treatment facility operators; and with the mindset that the technology is expensive and difficult to install and run;
- High capital investment in commencing the sophisticated and advanced membrane bioreactor technology, which could be out of the financial capacity of private sectors;
- The low economy of scale in membrane bioreactor technology compared to conventional technology in many industries, making the technology hard to be rationalized.

Nonetheless, continuous research and analysis to study the feasibility of membrane technology implementation in Malaysia provide promising insights for the future prospect. In University Malaysia Sabah (UMS), membrane technology development is comprehensively ongoing with some new technology being proposed for various applications. The Membrane Research Group in UMS was first established by Professor Ir Dr Rosalam Sarbatly in 2006, located at the School of Engineering and Information Technology (SKTM). The research evolved from membrane production to utilization in different application, particularly in wastewater treatment and seawater desalination. The membrane fabrication comprised of flat sheet membrane and hollow fibre membrane made from polymeric and ceramic based. Another invention by the research group was on the membrane distillation technology essentially for groundwater treatment. This technology is also predominantly used to cater to the needs mainly in the oil and gas industry. Presently, the research group is extensively researching on the membrane application for freshwater production through the up-flow sand filtration system coupled with microfiltration. This research is especially critical and vital to fulfilling the needs of particularly the people in rural and isolated areas who have strict access to fresh, clean water.

Introduction to Membrane and Its Application

Membrane processes unceasing advancement in the separation process application has made them a viable, sustainable, though "unconventional" unit operation for separation. The main advantage in utilizing the membrane separation process is that most membrane processes require comparatively low energy requirements due to no phase transition unlike what can be seen in most conventional separation processes. The merit that membrane processes possess makes them be expected to replace a significant amount of conventional separation processes such as distillation, absorption-stripping, crystallization, adsorption, and extraction.

The principle operation of membrane processes is predominantly driven by pressure, while they may differ on the perm-selectivity and separation means. A common membrane process would have a purified permeate product and a rejected concentrated retentate (Figure 1). Membrane filtration method utilizes a permeable medium with a specific pore size through which a suspension will be passed. The process is strictly physical, and the performance will depend on the membrane pore size and its operating parameters (Anselme et al., 1994). In this process, the need for prior chemical addition can be

Figure 1. Common elements in a membrane process

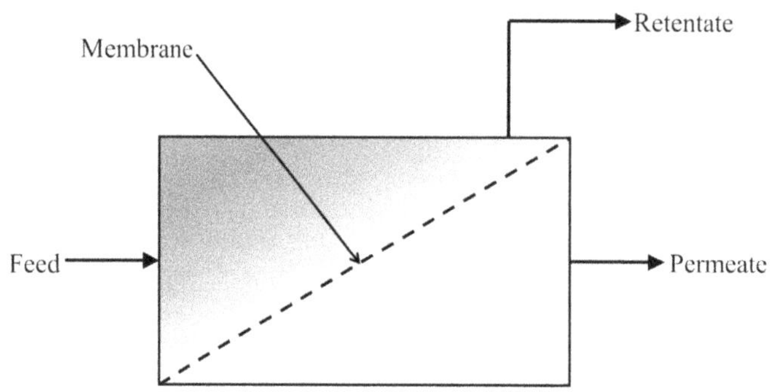

avoided (Wicaksana et al., 2012) and the membranes permeate can be maintained below the specified limit without any further treatment.

The most broadly used membrane filtration processes in water and wastewater industry comprise microfiltration, ultrafiltration, and reverse osmosis. These processes are related processes differing essentially in the standard pore diameter of the membrane filter. The relative size of solutes removable by each membrane process is demonstrated in Figure 2. In addition to that, membrane processes also include but not limited to nanofiltration, membrane distillation, forward osmosis, pervaporation, and electrodialysis. These membrane processes are distinct in terms of their operation driving force, membrane properties, consumption of energy, and costs. In Table 1, some key differences are tabulated to compare between different membrane processes. Some of the most critical factors in choosing the best separation processes are the costs associated with the process and the energy requirement. Costs is particularly an important factor but are difficult to be quantified because the actual number varies widely depending on many factors such as the membrane, membrane module, application, and operating conditions. Generally, high-pressure units are more expensive in both capital and operating costs than low-pressure units. This is also generally true for high and low-temperature units.

With the evolution and advancement of technology, membrane filter which has a smaller footprint and mild operating condition were able to be developed in providing a more efficient separation process (Wicaksana et al., 2012). Membrane filter with great variation to choose from including the type of membrane operation (Figure 3), either in dead-end or cross-flow operation, the membrane pore size, either microfiltration, ultrafiltration, or reverse osmosis which is further discussed on Section 5-2 – 5-4; and the attachment of membrane, either internal or external attachment are available. These membranes also come in a lot of different materials with different preferences of separations. With regards to the cost of membrane filter, improvement of the manufacturing techniques and discoveries of better materials has steadily decreased the cost of membrane filters (Zhang et al., 2010).

However, careful consideration should be made upon application the membrane filter as the well-known drawbacks of membranes filtration is its fouling. This phenomenon is mostly caused by the retentate accumulation at the membrane surface, which either causes less water flows through the membrane or increase the transmembrane pressure for a given flux (i.e. reducing the membrane flux or permeability respectively). Many methods can be applied to combat the issue with fouling, but complete avoidance may not be practical. Thus, membrane cleaning, either through physical or chemical means can be employed to remove the foulants material from the membrane filter.

Figure 2. Relative pore diameter for various membrane processes

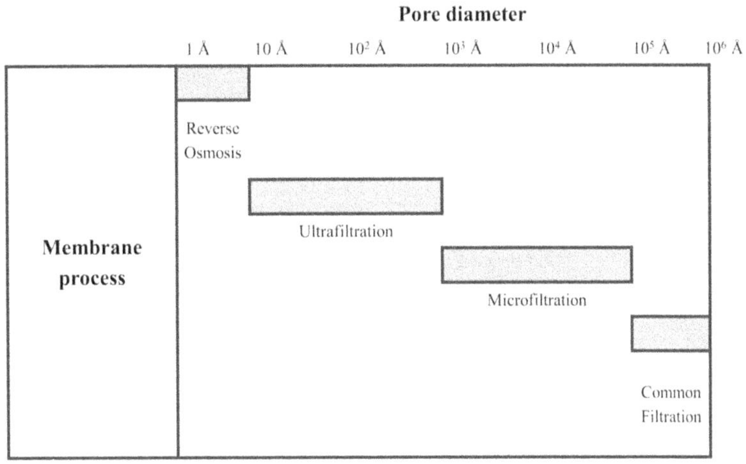

Table 1. General comparison of some membrane processes

Membrane Process	Driving Force	Capital Cost	Operating Cost	Energy Consumption
Microfiltration	Pressure	Low	Low	Low
Ultrafiltration	Pressure	Medium	Medium	Medium
Nanofiltration	Pressure	High	High	High
Reverse osmosis	Pressure	High	High	High
Forward osmosis	Concentration	Medium	Low	Low
Membrane distillation	Temperature	Medium	Low	Low
Pervaporation	Concentration	High	High	High
Electrodialysis	Electric	High	High	High

Figure 3. Type of membrane operation in the filtration process (a) dead-end, and (b) cross-flow

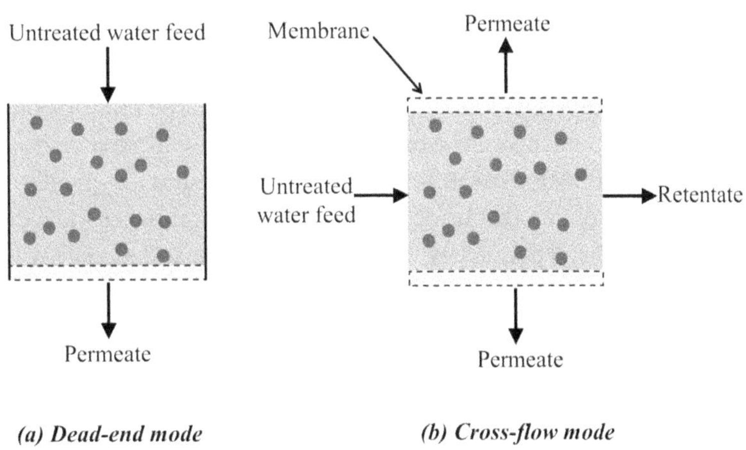

Table 2. Application of various membrane processes in water and wastewater industry

Application	Membrane Process
Desalination	Reverse osmosis Forward osmosis Membrane distillation Nanofiltration
Drinking water production	Reverse osmosis Forward osmosis Membrane distillation Nanofiltration Ultrafiltration (as pretreatment) Microfiltration (as pretreatment)
Industrial wastewater treatment	Microfiltration Ultrafiltration Nanofiltration Reverse osmosis Forward osmosis Membrane distillation Osmotic membrane distillation
Recovery of organic compounds	Pervaporation Membrane distillation

The application of membrane technology in water and wastewater treatment includes desalination, water purification, drinking water production and recovery of organic compounds from water. The application of various membrane processes in water and wastewater industry are presented in Table 2.

MICROFILTRATION

Microfiltration is a membrane filtration process which separates suspended particles from water by utilizing porous membranes medium with a diameter of 0.1 to 10 μm. This makes microfiltration membranes fall between ultrafiltration and a common filter. The early applications of microfiltration were used in laboratory experimentation to a very small size industrial usage. Various materials were used to develop the membrane unit such as cellulosic material, comprising cellulose acetate, and nitrocellulose; and noncellulosic materials, including poly(tetrafluoroethylene), poly(vinylidene fluoride), polyolefins, and polyamides. Nevertheless, a membrane made using the blend between cellulose acetate and nitrocellulose materials is the extensively used microfilter until today.

Symmetric or asymmetric porous membrane structure is used in the microfiltration process. Microfiltration process operating mechanisms can be based upon the dead-end filter or cross-flow. The driving force of the separation process is by pressure. Transmembrane pressure is typically used to describe the membrane operation. It is the pressure that is required to force water through the membrane, or the pressure difference between the feed and permeates. Microfiltration is regarded as a low-pressure membrane filtration process which typically operates at a relatively low pressure of 0.1 – 3.0 bar (transmembrane). Figure 4 illustrated the common setup of a microfiltration process.

Figure 4. Microfiltration process setup

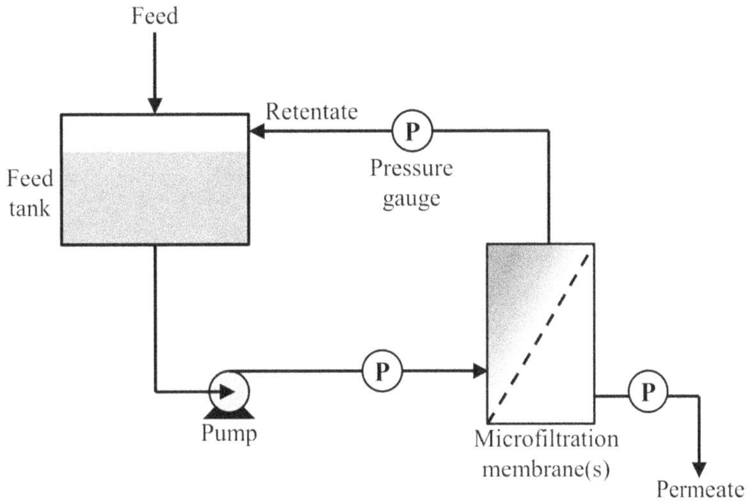

Principle Operation in Microfiltration

The physical separation process involved in microfiltration between the suspended particles and the permeate/filtrate is through sieving mechanism based upon particle size segregation as depicted in Figure 5. Depending on the application, particles larger than the pores sizes of the membrane will be retained at the retentate side and cannot pass through the membrane, while molecules smaller than it will only be partially removed depending on the construction of a refuse layer formed on the membrane surface.

The volume flow across the microfiltration membrane can be depicted by Darcy's law, where flux across the membrane is directly proportional to the transmembrane pressure.

$$J = A.\Delta P \quad (1)$$

where J is the flux, A is the permeability constant containing structural factors such as the porosity and pore size distribution, and ΔP is the transmembrane pressure.

Microfiltration can operate at two different modes of operation, dead-end or cross-flow. In dead-end filtration, the water recovery is very high in most cases. In this mode of operation, the feed is pumped through the membrane, where water molecules flow directly across the membrane, while the suspended particles accumulate on the surface of the membrane interior. The pressure required to sustain the water flow across the membrane increases with the accumulation of particles. The membrane would need to be replaced when the surface is too clogged with the particles accumulation.

In cross-flow mechanism, the feed is pumped across the membrane surface, allowing clear water known as permeate to pass through the membrane, while the refused water remaining in the membrane interior is recycled and mixed with the feed water as a concentrate. Compared to dead-end filtration, the cross-flow operation for membrane gives it a longer lifecycle than dead-end filtration.

The choice of which operating mode to be used is mostly depending on the type of application, particulate characteristics, and feed concentration. Feed having lower than 0.1% solid matters is usually treated using a dead-end filter, while feed with 0.5% or higher amount of solid matters is always treated

Figure 5. Principle of microfiltration

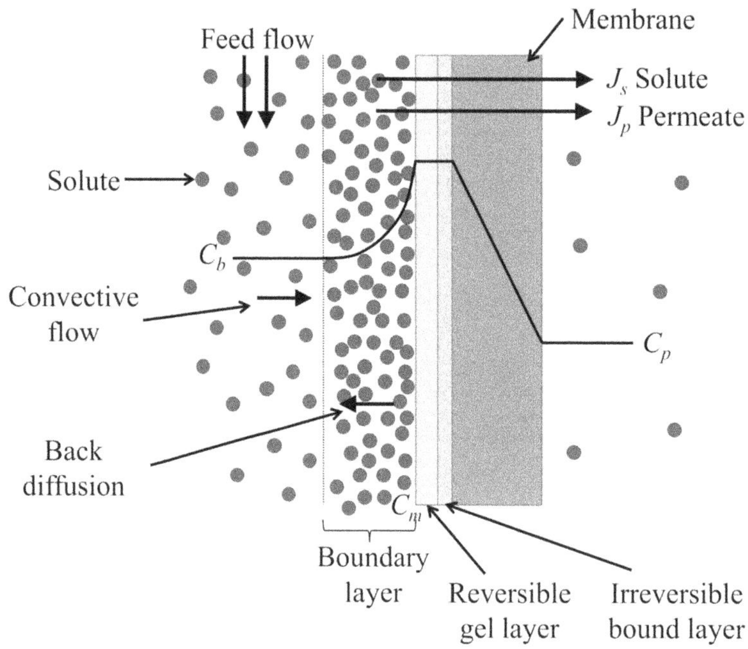

with cross-flow operational mode. Figure 6 shows the illustration for the cross-flow and dead-end filtration operations.

Application of Microfiltration in Water and Wastewater Treatment

The early application of microfiltration was used for biological laboratory experiment to observe the water supply for contamination. In this application, the membrane is used to filter the water supply, trapping the microorganisms on the membrane filter surface. The membrane is then left in a nutrient broth for 24 hours, allowing the bacteria to grow into colonies and subsequently counted under a microscope and analyzed.

In water and wastewater treatment, microfiltration can be applied as final filtration to treat surface water or groundwater after pretreatment using screening and/or sand filtration. Microfilter can be operated as pre-filtration as well in many cases of reverse osmosis system to prolong the lifespan of a downstream membrane due to its relatively lower capital costs. In operating the microfiltration process, it is important to ensure that the operation is properly applied within its operating parameters in order to reduce the risks of membrane fouling. Microfiltration is extensively used to remove contaminants such as suspended solids, large colloids, debris, and bacteria.

ULTRAFILTRATION

Ultrafiltration is the process of removing very small suspended particles, colloids, and dissolved materials ranging in the size of 1 – 100 nm from water using finely porous membrane filter. The fundamental

Figure 6. (a) Cross-flow operation, (b) dead-end filtration operation for microfiltration process

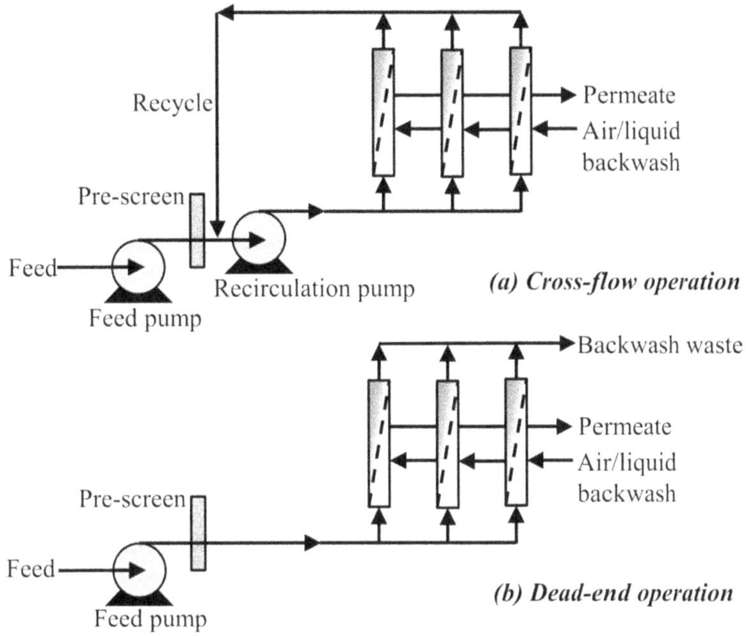

basis of ultrafiltration separation is based on the physical molecular size segregation. In the early 1900s, ultrafilter was first used for laboratory application using membrane made from nitrocellulose material. It wasn't until the mid-1900s before ultrafilter is utilized for industrial application. Cellulosic material including nitrocellulose and cellulose acetate and other types of polymers such as polyacrylonitrile, poly(vinylidene fluoride), polysulfone, and polyamides are used to manufacture ultrafiltration membranes.

Ultrafiltration membrane structure is typically asymmetric with a pore size smaller than those in the microfiltration membrane. Like microfiltration, ultrafiltration also has two types of operational modes, the dead-end filter or cross-flow operation which was discussed in Section 5-2.1. Ultrafiltration membrane operation can be carried out at ambient temperature or at a lower temperature room. The usual operating pressure is at 0.5 – 10 bar (transmembrane). The general setup for the ultrafiltration process is shown in Figure 7.

Principle Operation in Ultrafiltration

The principle operation involved in ultrafiltration is in fact very similar to the operation of the microfiltration membrane process. However, instead of being characterized only by particulate size cut-off, the ultrafiltration membrane is distinguished by solute molecular weight cut-off. This means that matters retainment by the membranes is based on their molecular weight as well as their sizes; heavier and larger materials are retained while allowing water and salts passing through the membrane structure. What's more, the membrane permeability is affected by some other factors such as the shape of the molecules to be retained, and the feed solution pH.

Ultrafiltration membrane can be used to retain or separate particulate bits, tiny colloidal matter, and macromolecules such as protein. The ultrafiltration solutes rejection model as compared to the micro-

Figure 7. Ultrafiltration process set up for clean water production

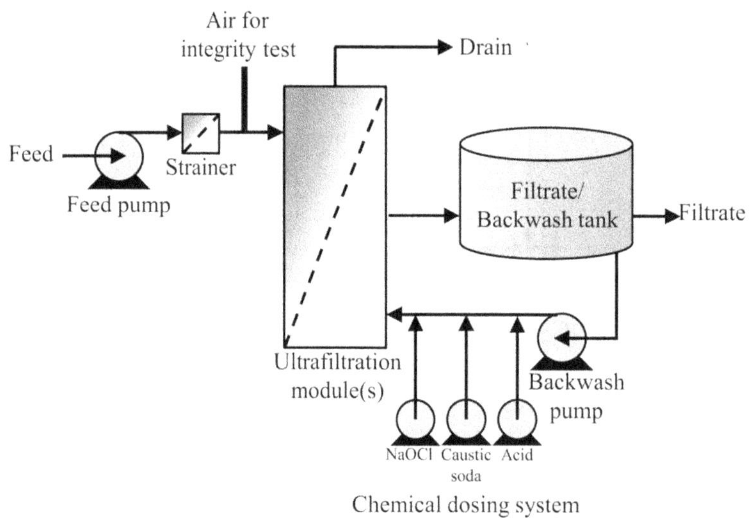

filtration's and reverse osmosis' is shown in Figure 8. Also, the ultrafiltration process can be operated by two modes of operation – dead-end filter and also cross-flow operation as discussed in Section 5-2.1.

Application of Ultrafiltration in Water and Wastewater Treatment

Ultrafiltration can be used to treat industrial and municipal wastewater. The process has high efficiency to remove suspended particles, macromolecules, turbidity, and bacteria contained in wastewater. Pretreatment is crucial when utilizing ultrafiltration process in order to prevent severe fouling of the membrane. Introduction of coagulant as a pretreatment before the ultrafiltration process for highly turbid water can improve the quality of water as permeate and decrease the membrane fouling (Xia, Liu, & Li, 2014). This is due to floc formation having large sizes that are retained by the membrane which prevents fouling of the membrane pores.

REVERSE OSMOSIS

Reverse osmosis is a process that separates salts and tiny molecules from water using membrane at relatively high pressure. This process is typically used in water desalination and purification treatment to produce clean and drinkable water. Reverse osmosis membranes have very small pore diameter which is usually in the range of 1 – 10 Å.

In the third quarter of the twentieth century, the reverse osmosis membrane was made using cellulose acetate material with asymmetric structure as the standard. It wasn't until 1975 before the interfacial composite membrane was introduced. This subsequently brought to the development of a fully aromatic interfacial composite membrane which becomes the industry standard. The continuous advancement in the reverse osmosis process makes it a practical key in increasing water reserve availability for drinking water production.

Figure 8. Solutes rejection model of membrane processes, (a) microfiltration, (b) ultrafiltration, and (c) reverse osmosis

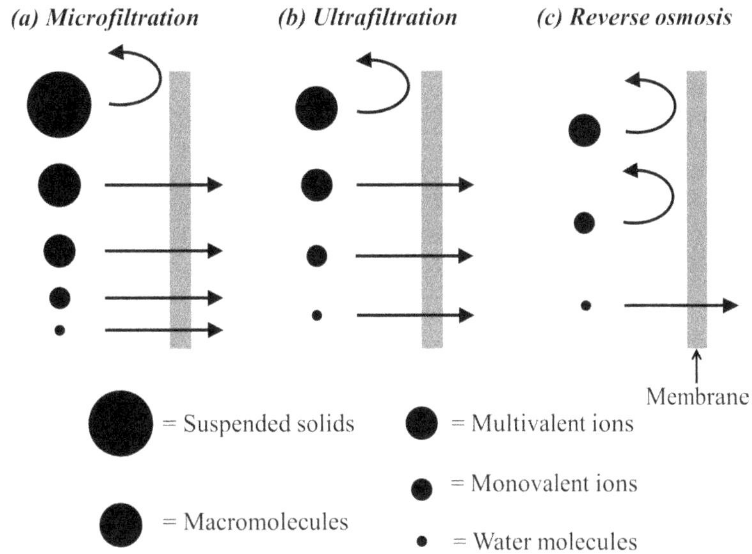

The conventional operating pressure (transmembrane) in reverse osmosis process typically range from 1 – 3 bar for brackish water, and 23 – 35 bar for seawater to overcome the water osmotic pressure. Figure 9 shows the typical setup of the reverse osmosis filtration process in water treatment.

Principle Operation in Reverse Osmosis

The primary principle in reverse osmosis membrane is based upon its capability to retain salt, i.e. sodium chloride. Its ability in allowing only water molecules to pass through the membrane is characterized by the membrane's pores that have a very small diameter. The term "reverse" comes from the fact that water molecules flow against their osmotic pressure gradient in this process, driven by the operating pressure applied. Thus, the saline water in the retentate side becomes more concentrated as water permeates across the membrane.

The pressure required to operate the reverse osmosis process is the opposing pressure needed to completely inhibit the water molecules from the permeate side from passing through via osmosis. This pressure is known as hydrostatic pressure. For the reverse osmosis process to occur, the hydrostatic pres-

Figure 9. Reverse osmosis filtration process setup

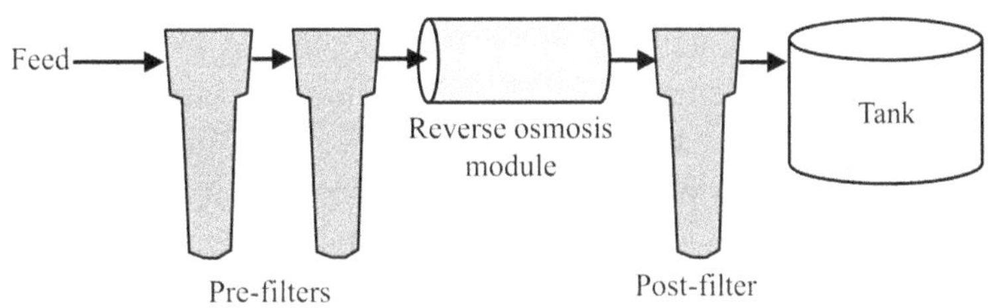

sure must exceed the osmotic pressure for the water molecules to pass through from high concentrate solute to low concentrate solute. The water transports occur through three sequential stages, which are:

1. Water absorbed into the membrane retentate surface;
2. Water diffuses through the membrane thickness;
3. Water discharged from the membrane permeate surface.

The water flux through the membrane during the reverse osmosis process can be calculated using Equation 2, where the flux is directly proportional to the pressure difference.

$$J_w = A_w \left(\Delta P - \Delta \pi\right) \quad (2)$$

where J_w is the water flux, A_w is the water permeability coefficient of water, ΔP is the transmembrane pressure difference, and $\Delta \pi$ is the difference between the osmotic pressure of feed and permeate.

Application of Reverse Osmosis in Water and Wastewater Treatment

Reverse osmosis process is widely used in treating water for drinking water production. In reverse osmosis operation, pretreatment is the utmost crucial stage to reduce fouling tendency in the reverse osmosis membrane system. Conventional pretreatment includes chemical treatment using coagulant, flocculant, and acid, and physical treatment using common filtration such as a sand filter. A rising trend is seen in using a series of larger pore sized membrane pretreatment before being subjected to the reverse osmosis treatment. This includes microfiltration and ultrafiltration. Microfiltration is used to separate large particulate matters, while the ultrafiltration is an apposite choice to remove colloids and macromolecules present in the feed water. Pretreatment is very important as it reduces the potential of membrane fouling and damage as well as ensuring efficient water treatment operation.

COMPARATIVE PROCESS BETWEEN MICROFILTRATION, ULTRAFILTRATION AND REVERSE OSMOSIS

Table 3 shows the general comparison between microfiltration, ultrafiltration and reverse osmosis membranes discussed earlier in this chapter.

ADVANCED MEMBRANE PROCESSES

The progressions in membrane separation technologies are associated with the notable advancement in many aspects of the field such as in materials used, the innovation of membrane modules and fabrication techniques, and the rising technological efficiency and applicability. In this section, several advanced membrane processes are discussed which highlights the emerging membrane technologies for various applications. Table 4 highlights some general comparison between some advance membrane processes in terms of their driving force and application.

Table 3. Characteristic features of microfiltration, ultrafiltration and reverse osmosis membranes

	Microfiltration	Ultrafiltration	Reverse Osmosis
Pore diameter (Å)	1,000 -100,000	10 – 1,000	1 – 10
Mechanism of separation	Based on size exclusion	Based size and molecular weight exclusion	Based on the size and ionic exclusion
Solute rejection	Suspended solid particle, debris, large colloid, bacteria	Macromolecules like proteins, colloids, organic matters	Salts, metals, small organic matters
Membrane structure	Mostly symmetric; asymmetric	Asymmetric	Asymmetric
Operation mode	Dead-end filter and cross-flow operation	Dead-end filter and cross-flow operation	Dead-end filter and cross-flow operation
Operating transmembrane pressure (bar)	0.1 – 3.0	0.5 – 10	1 – 3 for brackish water; 23 – 35 for seawater
Driving force	Pressure gradient	Pressure gradient	Pressure gradient
Pre-treatment requirements	Low	Medium	High
Capital cost	Low	Medium	High

Pressure-retarded Osmosis

Pressure-retarded osmosis is a membrane process which uses osmotic pressure difference between two water sources of different salinity to do work and as a result produces energy. In this process, the saline water pulls freshwater across a semipermeable membrane due to the osmotic pressure provided by the concentration gradient between the two media. This process is illustrated in Figure 10.

The osmosis principle applied in this process is closely related to the reverse osmosis and forward osmosis, whereby the process falls between the two. The schematic differences can be observed in Figure 11. At osmotic equilibrium, the net transfer of water molecules across the semipermeable membrane is zero. In forward osmosis, $\Delta P = 0$, no pressure is applied as water moves freely across the semipermeable membrane along the concentration gradient until equilibrium is reached; while in reverse osmosis, $\Delta P > \Delta\pi$, strong pressure is applied to overcome the osmotic pressure to allow water molecules move against the concentration gradient.

Table 4. Comparison of some advanced membrane separation processes

Membrane Process	Driving Force	Application
Pressure-retarded osmosis	Concentration gradient	Brine water desalination Juice concentration/dehydration Power generation
Membrane distillation	Vapour pressure gradient	Brine water desalination Valuable compound recovery/extraction Ultrapure water production
Osmotic membrane distillation	Concentration gradient	Brine water desalination Juice concentration/dehydration
Pervaporation	Vapour pressure gradient	Organic recovery/removal from water Organic-organic separation

Figure 10. Pressure retarded osmosis process

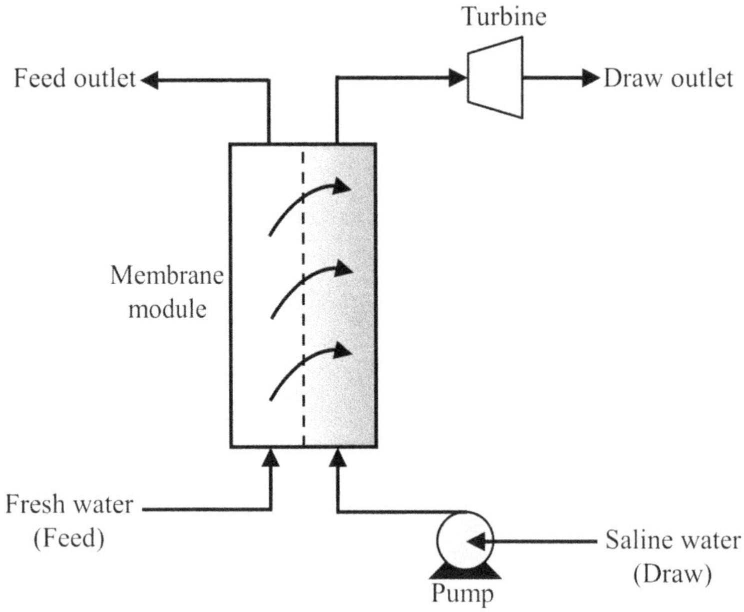

In the case of pressure retarded osmosis process, $\Delta P < \Delta\pi$. Light pressure is applied, and the draw solution of higher concentration is used to generate a net flow of water from the feed solution across the membrane to the permeate region. Typically, a salt or sucrose solution is used as the draw solution due to their inexpensive nature.

Membrane Distillation

Membrane distillation is a membrane separation process for the liquid removal across a porous hydrophobic membrane which is driven by the vapour pressure difference, due to the thermal gradient at the liquid/vapour interface. The process is affected upon the vapour-liquid equilibrium of the feed, concentration, and pressure; whereby mass transfer occurs from the hot side to the cold side as illustrated in Figure 12.

Unlike conventional distillation processes, membrane distillation occurs below the liquid normal boiling point. The lower operating temperature permits for the usage of unconventional energy sources such as solar energy or waste heat. There are four fundamental configurations of membrane distillation system: direct contact, air gap, sweeping gas, and vacuum membrane distillation as illustrated in Figure 13.

In membrane distillation, it is paramount crucial that liquid does not fill the pores of the membrane as it may affect the membrane efficiency in the separation process, which explains the normal usage of the hydrophobic membrane. This phenomenon is coined as wetting which is a major issue in membrane distillation. Liquid entry pressure can be defined as the pressure difference where the liquid is able to diffuse into the membrane pores. To prevent wetting of the membrane, the transmembrane pressure should not exceed the liquid entry pressure as this can hinder the separation activity at the vapour-liquid interface. In any case, the wetted membrane can be effectively regenerated such as by soaking in ethanol and removed through evaporation at high temperature, or by applying vacuum regeneration method. The membrane distillation topic is further discussed in Chapter 8.

Figure 11. Classifications of osmotic pressure (a) at equilibrium, (b) in forward osmosis, (c) in reverse osmosis, and (d) in pressure retarded osmosis

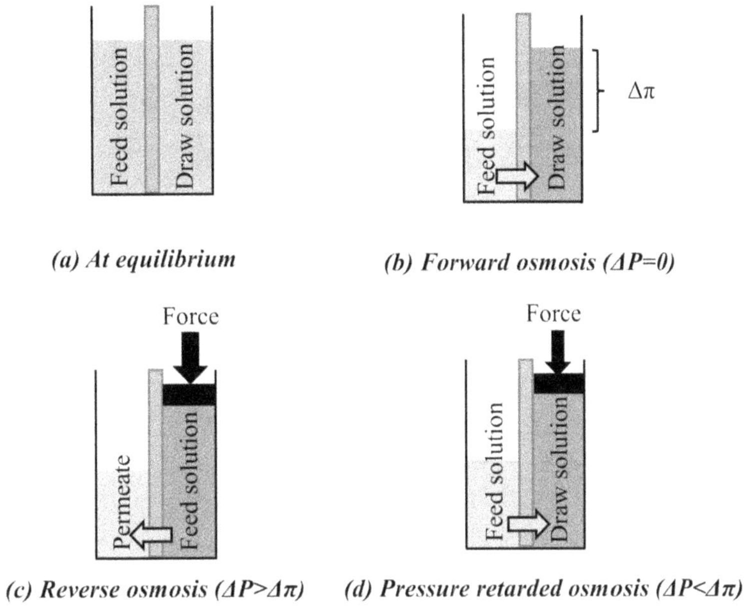

Osmotic Membrane Distillation

Osmotic membrane distillation is a membrane process which employs porous hydrophobic membrane to allow feed solution dehydration between two different concentration media at constant temperature and atmospheric pressure. In this process, the two solutions with different water activity are each on either side of the membrane. This creates a water vapour flux from the feed to the stripping solution.

Figure 12. A schematic showing the mass transfer from hot feed to cold permeate in the membrane distillation process

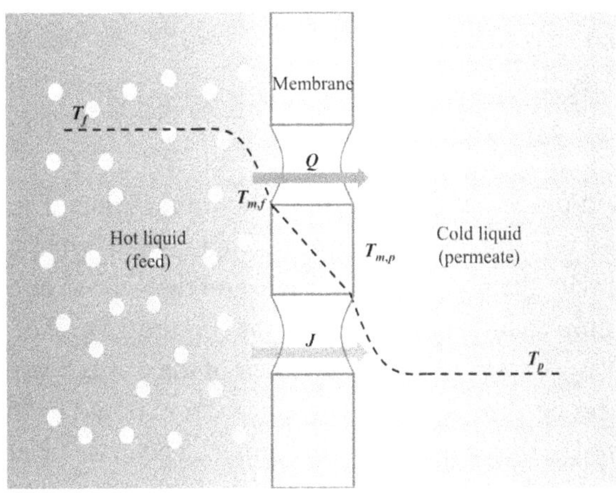

Figure 13. Fundamental configurations of the membrane distillation process, (a) direct contact, (b) air gap, (c) sweeping gas, and (d) vacuum membrane distillations

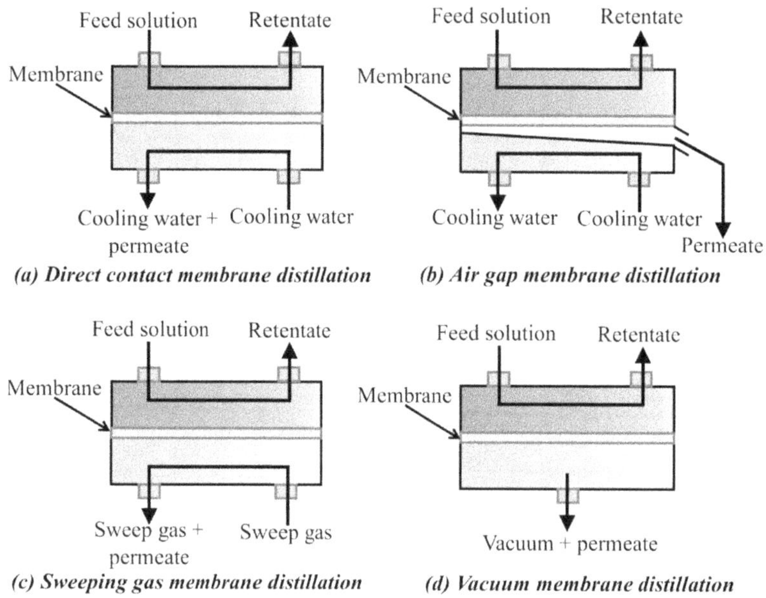

The higher water activity in a feed solution allows the water vapour passing through the semipermeable membrane into the more concentrated stripping solution (Figure 14).

Unlike membrane distillation, osmotic membrane distillation is an athermal process, requiring no heating or change in temperature. The water molecules transport through the membrane occurs through evaporation of water on the feed solution surface-membrane interface, water vapour diffusion across the hydrophobic membrane, and finally condensation of water vapour on the permeate side. A similar case of pores wetting can be encountered in this process like the one found in membrane distillation practice. The wetting is typically induced by membrane fouling or deposition of salt which ultimately hinders the process efficiency. This is because wetting of the membrane will cause the salt (or solute of stripping solution) to leak into the feed side which devalues the quality of the product. Thus, the stringent operation is important in handling the process by making sure that salt deposition can be prevented in any way. Brine solution of 25 – 27% is usually used as the stripping solution to prevent fouling, as at above which salt crystallization can start to be observed.

Pervaporation

Pervaporation is a membrane process used to separate liquid mixtures through vapour pressure difference between the heated feed mixture and the permeate vapour. The separation occurs when the heated feed mixture makes contact with the membrane surface, and the vapour permeates across the other side of the membrane into a vacuum chamber or sweeping gas (Figure 15). The term pervaporation was first introduced by Kober in 1917 which is essentially the fusion of the words "permeation" and "evaporation".

Since the pervaporation process involves phase changes from liquid to vapour for the permeate compound, heat supply is required in this separation mechanism. The main principle in pervaporation is the

Figure 14. Schematic showing the water vapour flux in osmotic membrane distillation

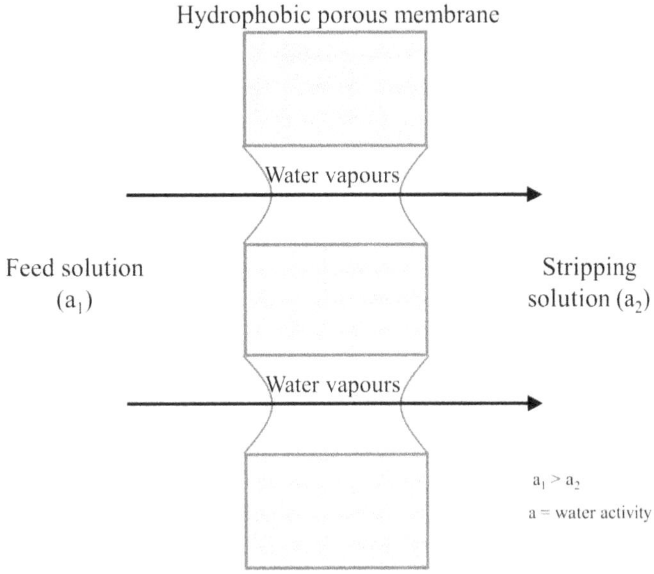

evaporation of the feed when it is heated to the point that saturated vapour is produced. When this vapour comes in contact with the membrane in the feed side, it diffuses through the membrane to the permeate side. Typically, a vacuum pump or a condenser is fitted at the permeate side to create an area of lower pressure. The flux across the membrane is characterized by the partial pressure difference between the vapour pressure of feed and vapour pressure of permeate. Thus, the temperature can greatly affect the process efficiency as at increased temperature, the system can handle higher permeate flux and vice versa.

MEMBRANE BIOREACTOR

Fundamentals of Membrane Bioreactor

Water scarcity is one of the major problems faced by many regions in the world. Almost anything that we do needs clean water such as industries, agriculture, and municipal uses. Water is the source of life, but overwhelming development that we faced in lots of sectors has contributed to the water contamination which leads to the deprivation of clean water. Water polluted from industries, agriculture and municipal usage with suspended and dissolved organic compounds, metallic ions, pesticides, toxic materials, and sewage should be properly treated to ensure a continuous supply of clean water. Conventional methods of treating water and wastewater require many stages in the treatment operation. This means high capital costs.

Membrane bioreactor is a promising technology for water and wastewater treatment. This technology has a lower footprint than a corresponding conventional facility and can produce a higher quality of water to meet stringent water quality standard imposed. However, several drawbacks are associated with membrane bioreactors as they are susceptible to clogging of the pores, resulting in fouling of the

Figure 15. Schematic diagram of pervaporation process

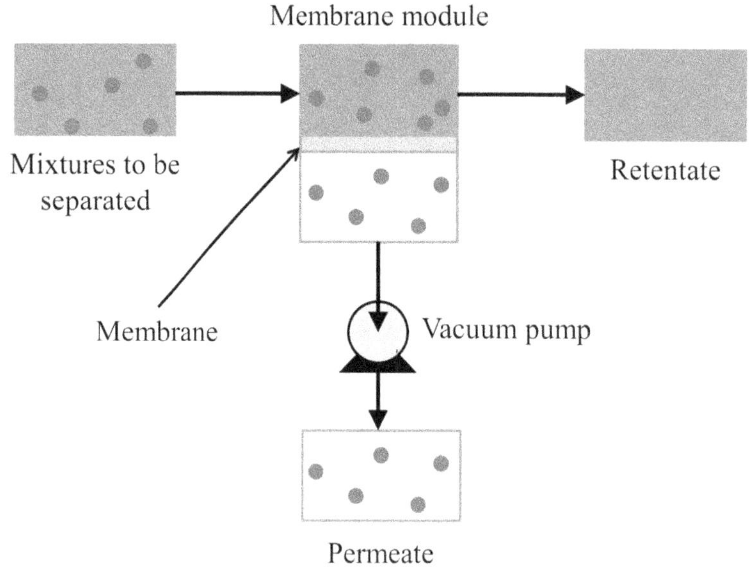

membrane. This will require regular cleaning and substitution which makes them less attractive for maximum scale and long term applications.

Membrane bioreactor technology in wastewater treatment combines two units: a bioreactor for the biological reaction, and a membrane system for the separation process. Typically, the membrane used is either a microfiltration or ultrafiltration membrane system. In the membrane bioreactor system, the bioreactor unit is in charge of biodegradation of the wastes, while the membrane unit is for the physical separation of treated water from the sludge.

Membrane bioreactor is slowly gaining wide application in sewage treatment. The conventional treatment of sewage is through the activated sludge system through which the sewage is treated to produce water effluent that can be reused in many industrial applications. Yet, this conventional method is associated with a debottleneck focused on the settler unit, as sludge settlement takes quite a long time. This problem ultimately hinders the productivity of the treatment plant. The utilization of membrane bioreactor in sewage treatment to replace the conventional method can solve this issue, as it will be able to reduce the operating time as well as the number of unit operations. As conventional activated sludge system uses a settler for the separation of solid-liquid, membrane bioreactor uses membrane system. The membrane bioreactor uses membrane as a filter which separates the solid matters and bacteria from the biological process to produce clear, disinfected effluent. The technical feasibility of membrane bioreactor technology has been evident through wide application in small-scale to large-scale applications. Figure 16 shows the comparison of process flowchart between the activated sludge process and the membrane bioreactor.

Membrane Material and Module

The overall membrane performance is directly affected by its material of construction and type of module configuration. Membrane module can be described as the smallest unit it can pack in. The membrane can

Figure 16. Comparative process flowchart of (a) conventional activated sludge system and (b) membrane bioreactor system for wastewater treatment

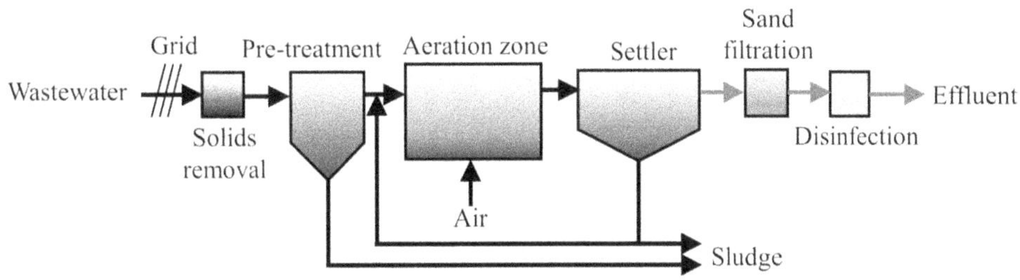

(a) Conventional activated sludge system

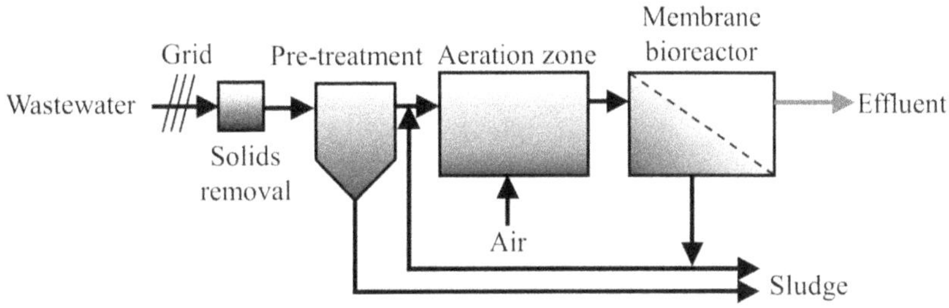

(b) Membrane bioreactor system for wastewater treatment

be installed and positioned in the different geometrical arrangement. There are six primary configurations of the membrane module, which are flat sheet (plate and frame), hollow fibre, tubular, capillary tube, pleated filter cartridge, and spiral wound; though only the former three configurations are suitable for membrane bioreactor application. The chief differences between these configurations are tabulated in Table 5. A good geometric structure of membrane module can be characterized by its capability to reduce fouling during the filtration process, its physical simplicity, and control flexibility.

Membrane fabrication can be accomplished using different types of material, mainly from two distinct types which are polymeric and ceramic based. The selection is carried out so as to get a membrane structure that has high mechanical strength and stability, selective to the target permeate, can resist thermal and chemical strike, and has some degree of resistance to fouling. The most extensively used material in membrane production is the nitrocellulose and cellulose acetate. Besides, other polymers that are commonly used are polyvinylidene difluoride, polyethersulphone, polyethene, and polypropylene.

Membrane Bioreactor Configurations

Membrane bioreactor configuration can be classified into two types, which are side-stream membrane bioreactor, and submerged membrane bioreactor.

In side-stream membrane bioreactor, the membrane module is located outside the membrane bioreactor system (Figure 17(a)). The sludge is pumped from the bioreactor into the membrane module. The membrane filtration process is pressure-driven, requiring high velocity of liquid to effectively prevent membrane fouling. The concentrate is recycled back to the membrane bioreactor using a recirculation pump. This offers high flux, but at the price of regular cleaning and high energy requirement. The pro-

Table 5. Membrane configurations

Membrane Configuration	Cost	Backflushability	Application
Flat sheet	High	No	Ultrafiltration Reverse osmosis
Hollow fibre	Very low	Yes	Microfiltration Ultrafiltration Reverse osmosis
Tubular	Very high	No	Cross-flow microfiltration Ultrafiltration Nanofiltration
Capillary tube	Low	Yes	Ultrafiltration
Pleated filter cartridge	Very low	No	Dead-end microfiltration
Spiral wound	Low	No	Ultrafiltration Nanofiltration Reverse osmosis

cess requires a relatively high transmembrane pressure of 3 – 6 bar to maintain high cross-flow on the membrane surface. However, high cross-flow gives an undesirable impact on biomass activities. The maintenance of side-stream membrane bioreactor is quite easy due to its ease of accessibility owing to the external installation of the membrane module. In term of costs, side-stream membrane bioreactor requires high capital cost and operating cost, which concentrates on the membrane housing for high-pressure operation and energy cost to recirculate some liquid back to the bioreactor respectively. The recirculation is vital as only <5% of the feed can be recovered as permeate in one pass, meaning that large volume of the concentrate needs to recycled back for efficient recovery. The high capital cost and operating cost associated with side-stream membrane bioreactor system make it excessively expensive and less attractive for most wastewater treatment operation. Even so, a slight modification can be employed to reduce the high energy consumption by installing a suction pump to the permeate side. This can significantly reduce the energy consumption associated with this type of membrane bioreactor configuration.

In submerged (or immersed) membrane bioreactor, the membrane module is submerged into the membrane bioreactor system (Figure 17(b)). In this operation, a suction or vacuum pump is required to produce suction pressure or transmembrane pressure of 0.05 – 0.3 bar for permeate production. Aeration of the immersed membrane module is needed to create cross-flow and reduce the membrane fouling. It also functions to provide oxygen for the biological process inside the membrane bioreactor. Since the membrane module is directly submerged inside the membrane bioreactor, the system requires less equipment which contributes to lower capital cost compared to side-stream configuration. However, this makes it difficult for cleaning as it is hard to access the membrane module residing inside the bioreactor. The operating cost is also exceptionally lower (39 times lower) than the side-stream configuration owing to the fact that only aeration is needed without the recirculating pump. In any case, both membrane bioreactor systems can be used as aerobic (with the aeration system) and anaerobic (without the aeration system for the module). Table 6 shows the general comparison between the two configurations of a membrane bioreactor.

Figure 17. Membrane bioreactor configuration; (a) side-stream, and (b) submerged

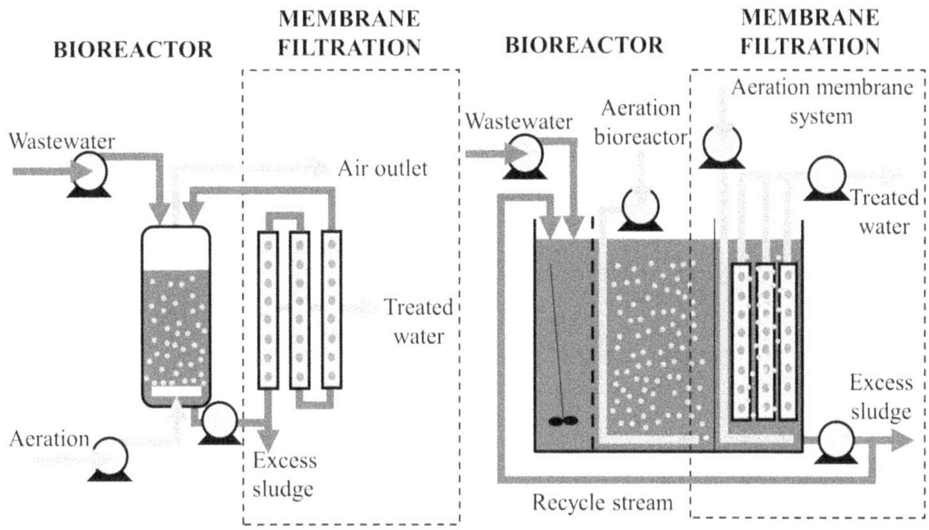

(a) Side-stream membrane bioreactor ***(b) Submerged membrane bioreactor***

Types of Membrane Bioreactor

Aerobic Membrane Bioreactor

Aerobic membrane bioreactor is a type of membrane bioreactor connected to aeration system utilization. Aeration is required for two main purposes: 1) for the bio-treatment, i.e. to supply sufficient oxygen to the microbes for biodegradation of organic materials; and 2) to scour solids from the membrane surface to keep it relatively clean through the cross-flow generated.

The aeration is normally achieved using air diffusers, either for production of coarse bubble aeration or fine bubble aeration. The principal application for each type of aerator system is for membrane scouring and microorganisms aeration respectively. Aerobic type membrane bioreactor is generally operated for low to medium organic load wastewater. Some examples of aerobic membrane bioreactor application in wastewater treatment are:

- Municipal wastewater treatment;
- Industrial wastewater treatment (e.g. textile, food processing industries);
- Slaughterhouse/fisheries wastewater treatment; and
- Agricultural wastewater treatment

Figure 18 illustrates the typical schematic arrangement of the aerobic type membrane bioreactor for both side-stream and submerged configurations.

Table 6. Comparison between side-stream membrane bioreactor and submerged membrane bioreactor

	Side-stream Membrane Bioreactor	Submerged Membrane Bioreactor
Operation mode	Cross-flow	Cross-flow
Typical membrane configuration	Plate and frame Tubular	Flat sheet Hollow fibre
Complexity	Complex	Simple
Cleaning	Easy	Hard
Operating pressure (bar)	3 – 6	0.05 – 0.3 (vacuum)
Flux (LMH)	50 – 100	15 – 35
Permeability[a] (LMH/bar)	7 – 30	50 – 500
Specific energy demand[b] (kWh/m^3 permeate)	4 – 12	0.1 – 0.5
Capital cost	High	Low
Operating cost	High	Low

[a]Permeability at operating state
[b]Include suction energy of permeate but exclude biological aeration
Source: (Yoon, 2016)

Anaerobic Membrane Bioreactor

Anaerobic membrane bioreactor is essentially similar to aerobic membrane bioreactor, except for the absence of an oxygen-supplier unit, i.e. the aerator system. This type of membrane bioreactor has the advantage of coupling water pollution treatment with energy production. The operational mechanism is almost the same with the aerobic-type which combines the biotreatment process with separation of the solid-liquid using a membrane system. The underlying contrasts are on the exclusion of oxygen-supplier, biogas (methane) production, and lesser yield of sludge. The anaerobic membrane bioreactor consists of a digester tank which is responsible for the biogas production and membrane module to separate the sludge from the effluent.

The advantage of anaerobic membrane reactor over a conventional anaerobic digester is on the total biomass retention due to the presence of the membrane filter. In a conventional anaerobic digester, the biomass is easily lost to the effluent as a result of the slow settlement of biomass. Due to no air-supply system used, it cannot produce cross-flow across the membrane as seen in an anaerobic membrane bioreactor. Thus, the generated biogas is typically used to scour the membrane. Anaerobic type membrane bioreactor is used mainly in wastewater treatment for high organic load wastewater, either from municipal or industrial wastewater.

Figure 19 demonstrates the schematic of both side-stream and submerged configurations of the anaerobic membrane bioreactor.

Critical Flux Operation

Field et al. (1995) first introduced the concept of critical flux in the membrane process and its effects of the membrane operation. They hypothesized that for microfiltration membrane, the critical flux "... is that on start-up, there exists a flux below which a decline of flux with time does not occur; above

Figure 18. Schematic of aerobic type membrane bioreactor configurations (a) side-stream, (b) submerged

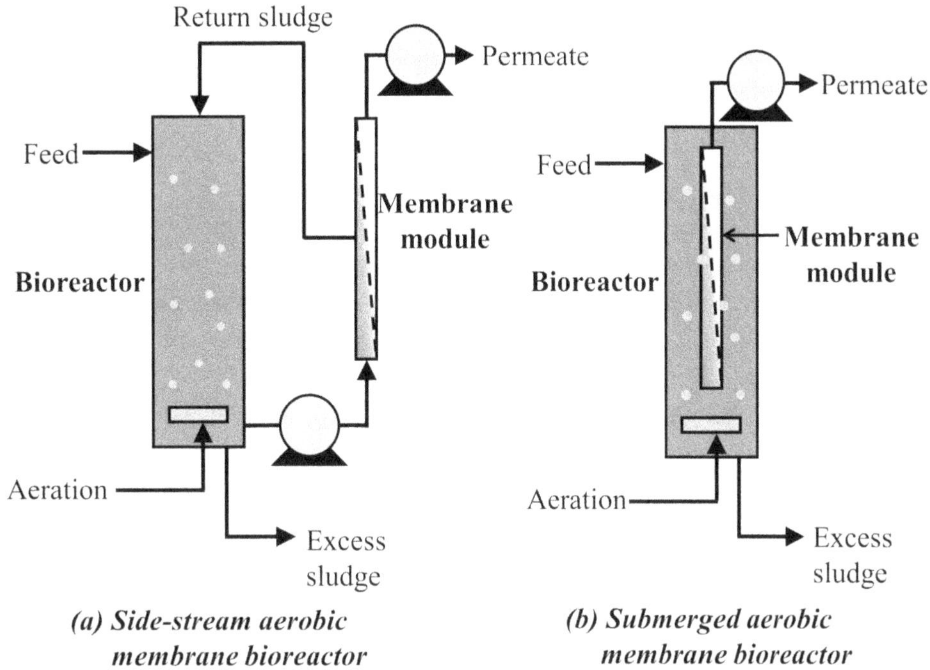

Figure 19. Schematic of anaerobic type membrane bioreactor configurations (a) side-stream, (b) submerged

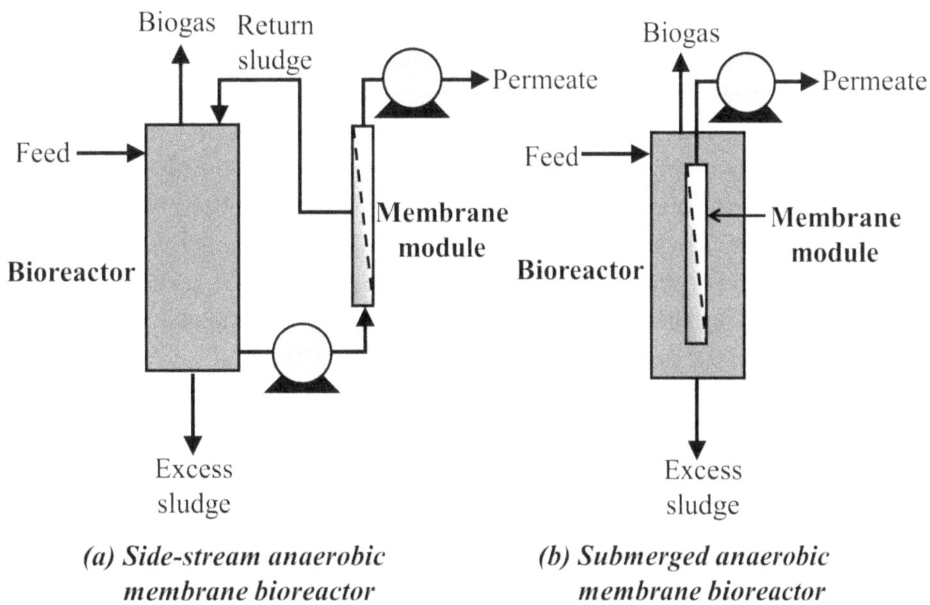

Table 7. Fouling reducing methods for membrane bioreactor

Side-stream Membrane Bioreactor	Submerged Membrane Bioreactor
Cross-flow Backwashing Chemical cleaning	Air bubble scouring Backwashing Chemical cleaning

it fouling is observed". The hypothesis basically describes critical flux as a flux above which fouling occurs. Thus, it is crucial for a membrane system to be operated below this value to prevent fouling which can clog the membrane pores and affects the membrane performance. A study by Benkahla et al. (1993) remarked that flux is directly proportional to the transmembrane pressure, provided that the critical flux is not exceeded.

In membrane bioreactor operation, the critical flux is implied as to the operating flux at which no fouling occurs provided it is inappropriate fouling control operation. Nonetheless, fouling still results in membrane bioreactor operation even if operated under the critical flux condition due to various factors affecting membrane fouling and complexity of the system. Thus, the critical flux in membrane bioreactor is regarded as flux which withstands rigorous fouling with the utilization of suitable fouling control operation, for instance, the air scouring or regular membrane cleaning. Due to this lenient concept of critical flux in membrane bioreactor operation, it is normally viewed as the sustainable flux to distinguish it from the original concept of critical flux. The usual range of critical flux in membrane bioreactor is from 10 – 40 LMH subjected to a number of factors which affect membrane fouling.

Fouling Control and Preventive Mechanism

The standard operation and maintenance of membrane bioreactor require membrane cleaning which is closely related to membrane fouling. To maintain the performance of the membrane unit, a periodical cleaning should be conducted on the membrane. Membrane cleaning can reduce the chance of short membrane lifetime as well as reducing the capital cost for its replacement. There are a number of strategies that can be imposed pertaining to membrane fouling control and preventive methods. To this date, several strategies to operate the membrane in order to avoid fouling have been proposed and researched, including running under the critical flux, enhancing cross velocity and scouring the membrane surface with air bubbles. Table 7 shows the typical fouling reducing methods that are employed for membrane bioreactor operation. However, these preventive strategies could not completely guarantee of the membrane fouling avoidance, as the membrane fouling is inevitable, especially irreversible fouling which could only be eliminated by chemical cleaning (Kimura et al., 2004; Zhang et al., 2011).

Membrane cleaning in either physical or chemical approach is necessary if fouling should occur. The cleaning process is when the material is relieved of a substance, which is not an integral part of the material. Physical cleaning is conducted using mechanical or hydraulic forces in removing foulant from the membrane surface. The cleaning process may include backwashing or forward flushing or both. On the other hand, chemical cleaning of the membrane will depend on the chemical reaction to weaken the cohesion forces between the foulant and the membrane surface.

In the account of physical cleaning, membrane backwashing has been reported to be more effective than forwarding flushing (Liang *et al.,* 2008; Chen *et al.,* 2003). This is due to the mechanical stress on the membrane in a resistance direction provided by the backwashing process (Liang *et al.,* 2008).

However, a single hydraulic cleaning mode will sometimes not satisfy the need for flux recovery. Therefore, a combination of backwashing and forward flushing may be necessary. This case was studied by several researchers (Liang *et al.,* 2008; Shengji *et al.,* 2008; Chen *et al.,* 2003). Based on their findings, a combination of the hydraulic cleaning mode (i.e. backwashing combined with forwarding flushing) will present a higher flux recovery compared to a single mode of cleaning.

Nonetheless, a total recovery of the membrane flux by using hydraulic cleaning procedure will not be successful every time, especially when dealing with irreversible fouling. If this type of fouling occurs, chemical cleaning may be required to restore the initial performance of the membrane (Shengji *et al.,* 2008). Chemical cleaning of the membrane can be achieved (97% flux recovery) by combining the use of NaOH (0.02 N) and NaClO (100 mg/L), being soaked in the solution for 4 hours (Liang *et al.,* 2008). Zhang *et al.,* (2011) find out that 80% of the flux can be recovered by soaking the membrane in NaClO of 100 mg/L for 1 hour. The use of NaClO solution for membrane cleaning was reported elsewhere to be promising as it can eliminate almost all the major organic foulants such as carbohydrate-like and protein-like material on the membrane surface (Bilad *et al.,* 2012; Tzotzi *et al.,* 2007).

Prospect of Membrane Bioreactor Implementation in Sabah Wastewater Treatment

The prospect on the environmental future of Malaysia, particularly in Sabah is highly reliant on the waste management and treatment applied along with the public and authority awareness. Severe consequences on environmental quality deterioration are highly potential if the current trend of waste management is used without any improvement in the technologies utilized. With the accelerated increase in the number of population, an increase in demand for life essentials such as food and energy will follow. The rapid development of the industrial sector will occur in order to cope with the demand. Consequently, a larger amount of wastewater will be produced. On the other hand, smaller land will be available to be utilized for wastewater treatment. Therefore, technologies with higher efficiency and smaller physical and environmental footprint are required to ensure Sabah's environmental quality is not compromised.

Over the past decades, the treatment of wastewater in Sabah has been conducted by using conventional technologies such as biological treatment and chemical treatment. Open ponding and intermittently decanted extended aeration (IDEA) are two conventional sewage treatment technologies in Sabah. These technologies have been proven to have its complication and low treatment efficiency. The major problem associated with conventional treatment is the effect of Sabah's weather on its operating performance.

In open ponding treatment, the technology utilized an open system concept with relatively small loading capacity. The average annual rainfall in Sabah is 2,788 mm with seasonal monsoon between November to February. During the monsoon season, the open system treatment facility is often flooded with influent which subsequently overload and flows to the nearby area and discharge point. This brought detrimental effect to the hygienic condition around the affected region, especially in the urban area of Sabah.

In Malaysian water quality standard, the limit for recreational with body contact for total coliform bacteria count is 5,000 counts/100 mL, while faecal coliform at 400 counts/100 mL. The major constituent of coliform bacteria is *E. coli* which could spread detrimental effects to human health. In Likas, Kota Kinabalu, the point of discharge for sewage treatment effluent would consequently contaminate the nearby shoreline. Improperly treated effluent discharged from a sewage treatment plant contributes to the rise of *E. coli* population along the recreational area around the local shoreline. This consequently hinders the recreation activities if the contaminants exceed the standard limit which is a major social implication.

Optimistic discoveries have been made on new technologies with higher efficiency, low cost and lower footprint. Many of these technologies can be applied in a single system or coupled together with the conventional or new wastewater treatment system. One of the technologies that have recently gained interest in wastewater treatment is membrane technology. Application of membrane technology for wastewater treatment is highly potential in Sabah. Combining the membrane technology with the conventional biological treatment, membrane bioreactor has the ability to produce effluent that is high in quality with shorter treatment time requirement. Modifications and improvements have been made to this technology such as by using the anaerobic membrane bioreactor. This system requires less energy due to the absence of aeration to operate and can compensate its energy usage by producing methane during its operation. Based on its promising efficiency recorded by others, it is highly feasible that these technologies are used in Sabah. These technologies are versatile enough to be operated for Sabah's main contributors of wastewater such as POME, sewage, municipal as well as oil and gas wastewaters. The nonpoint source wastewater can also be treated by utilizing these membrane technologies.

REFERENCES

Anselme, C., Mandra, V., Baudin, I., & Mallevialle, J. (2004). Optimum use of membrane processes in drinking water treatment. *Water Supply*, *12*, 1–2.

Benkahla, Y. K., Ould-Dris, A., & Jaffrin, M. Y. (1993). *Hydrodynamic anti-fouling mechanism in crossflow microfiltration.* Poster Presentation at ICOM '93, Aug. 30-Sept. 3, 1993, Heidelberg.

Bilad, M. R., Vandamme, D., Foubert, I., Muylaert, K., & Vankelecom, I. F. J. (2012). Harvesting microalgal biomass using submerged microfiltration membranes. *Bioresource Technology*, *111*, 343–352. doi:10.1016/j.biortech.2012.02.009 PMID:22391586

Chen, J. P., Kim, S. L., & Ting, Y. P. (2003). Optimization of membrane physical and chemical cleaning by a statistically designed approach. *Journal of Membrane Science*, *219*(1-2), 27–45. doi:10.1016/S0376-7388(03)00174-1

Field, R. W., Wu, D., Howell, J. A., & Gupta, B. B. (1995). Critical flux concept for microfiltration fouling. *Journal of Membrane Science*, *100*(3), 259–272. doi:10.1016/0376-7388(94)00265-Z

Kimura, K., Hane, Y., Watanabe, Y., Amy, G., & Ohkuma, N. (2004). Irreversible membrane fouling during ultrafiltration of surface water. *Water Research*, *38*(14-15), 3431–3441. doi:10.1016/j.watres.2004.05.007 PMID:15276760

Liang, H., Gong, W., Chen, J., & Li, G. (2008). Cleaning of fouled ultrafiltration (UF) membrane by algae during reservoir water treatment. *Desalination*, *220*(1-3), 267–272. doi:10.1016/j.desal.2007.01.033

Shengji, X., Xing, L., Bingzhi, D., & Juanjuan, Y. (2008). Application of membrane techniques to produce drinking water in China. *Desalination*, *222*(1-3), 497–501. doi:10.1016/j.desal.2007.01.142

Tzotzi, C., Pahiadaki, T., Yiantsios, S. G., Karabelas, A. J., & Andritsos, N. (2007). A study of $CaCO_3$ scale formation and inhibition in RO and NF membrane processes. *Journal of Membrane Science*, *296*(1-2), 171–184. doi:10.1016/j.memsci.2007.03.031

Wicaksana, F., Anthony, G. F., Pharima, P., & Robert, F. (2012). Microfiltration of algae (Chlorella sorokiniana): Critical flux, fouling and transmission. *Journal of Membrane Science, 387–388*, 83–92.

Xia, S., Nan, J., Liu, R., & Li, G. (2004). Study of drinking water treatment by ultrafiltration of surface water and its application to China. *Desalination*, *170*(1), 41–47. doi:10.1016/j.desal.2004.03.014

Yoon, S. H. (2016). *Membrane Bioreactor Processes: Principles and Applications*. Boca Raton, FL: CRC Press.

Zhang, X., Qiang, H., Sommerfeld, M., Puruhito, E., & Chen, Y. (2010). Harvesting algal biomass for biofuels using ultrafiltration membranes. *Bioresource Technology*, *101*(14), 5297–5304. doi:10.1016/j.biortech.2010.02.007 PMID:20206502

Zhang, Y., Tian, J., Liang, H., Nan, J., Chen, Z., & Li, G. (2011). Chemical cleaning of fouled PVC membrane during ultrafiltration of algal-rich water. *Journal of Environmental Sciences (China)*, *23*(4), 529–536. doi:10.1016/S1001-0742(10)60444-5 PMID:21793392

ADDITIONAL READING

Bodzek, M., Konieczny, K., & Kwiecinska, A. (2012). Application of membrane processes in drinking water treatment – state of art. *Desalination and Water Treatment*, *35*(1–3), 164–184.

Iorhemen, O. T., Hamza, R. A., & Tay, J. H. (2016). Membrane bioreactor (MBR) technology for wastewater treatment and reclamation: Membrane fouling. *Membranes*, *6*(2), 33–61. doi:10.3390/membranes6020033 PMID:27314394

Li, N. N., Fane, A. G., Ho, W. S. W., & Matsuura, T. (2008). *Advanced Membrane Technology and Applications*. USA: Wiley. doi:10.1002/9780470276280

Raut, N., Jahan, S., & Shinde, G. B. (2008). Membrane bioreactors for wastewater treatment: A review. *Chemical Engineering World*, *43*(3), 74–81.

Rautenbach, R., Vossenkaul, K., Linn, T., & Katz, T. (1997). Wastewater treatment by membrane processes – new development in ultrafiltration, nanofiltration and reverse osmosis. *Desalination*, *108*(1–3), 247–253. doi:10.1016/S0011-9164(97)00032-5

Singh, R. (2015). Design, Energy and Cost Analyses of Membrane Processes. In R. Singh (Ed.), *Membrane Technology and Engineering for Water Purification* (pp. 339–368). Oxford: Butterworth-Heinemann. doi:10.1016/B978-0-444-63362-0.00005-7

Van Der Bruggen, B., Vandecasteele, C., Van Gestel, T., Doyen, W., & Leysen, R. (2004). A review of pressure-driven membrane processes in wastewater treatment and drinking water production. *Environmental Progress & Sustainable Energy*, *22*(1), 46–56.

Van Der Bruggen, B., Verberk, J. Q. J. C., & Verhack, J. (2004). Comparison of pressure-driven membrane processes and traditional processes for drinking water production in Europe based on specific impact criteria. *Water S.A.*, *30*(3), 413–419. doi:10.4314/wsa.v30i3.5091

Warsinger, D. M., Chakraborty, S., Tow, E. W., Plumlee, M. H., Bellona, C., Loutatidou, S., ... Lienhard, J. H. (2016). A review of polymeric membranes and processes for potable water reuse. *Progress in Polymer Science*, *81*, 209–237. doi:10.1016/j.progpolymsci.2018.01.004 PMID:29937599

Zularisam, A. W., Ismail, A. F., & Sakinah, M. (2010). Application and challenges of membrane in surface water treatment. *Journal of Applied Sciences (Faisalabad)*, *10*(5), 380–390. doi:10.3923/jas.2010.380.390

KEY TERMS AND DEFINITIONS

Critical Flux: The flux above which fouling occurs.

Cross-Flow Filtration: A type of membrane operation mode where feed flow is tangential (or parallel) to the membrane surface.

Dead-End Filtration: A type of membrane operation mode where feed flow is perpendicular to the membrane surface.

Fouling: Accumulation of unwanted substances on the membrane surface or inside the membrane pores which reduces the membrane performance.

Membrane Bioreactor: A membrane technology which combines two units: a bioreactor for the biological reaction, and a membrane system for the separation process, typically either microfiltration or ultrafiltration.

Membrane Technology: Collection of the separation processes which utilize membrane.

Microfiltration: Membrane filtration process which separates suspended particles from water by utilizing porous membranes medium with a diameter in the range of 0.1 to 10 µm.

Reverse Osmosis: Membrane process that separates salts and tiny molecules from water using membrane at relatively high pressure, which is generally used in water desalination and purification treatment to produce clean and drinkable water.

Transmembrane Pressure: Pressure that is required to force water through the membrane, or the pressure difference between the feed and permeate.

Ultrafiltration: Membrane process of removing very small suspended particles, colloids, and dissolved materials ranging in the size of 1 – 100 nm from water using finely porous membrane filter.

Section 2

Chapter 6
Conventional Water Treatments

ABSTRACT

Conventional water treatments have several successive processes in series to produce potable water. This chapter talks about the conventional water treatment processes which are mainly used to treat water originated from freshwater sources. Besides, the discussion covers some typical water quality, both raw and treated, as well as the standards of water quality. One of the highlighted topics in this chapter is the common issues that are frequently happening in the conventional water treatment facilities around the rural regions experiencing tropical climate, which is centred on the issues affecting the raw water quality and treatment processes. The major issue during post-treatment which is on sludge management is also discussed by underlining some alternative to the traditional way of using sludge lagoons. Topics in this chapter provide a better perspective to the water treatment operators and students who are interested in this topic of major processes used in conventional water treatment plants as well as the common issues encountered.

INTRODUCTION

Water treatment is one of the most important processes that are crucial and necessary for everyone and everywhere around the world. Water is the most essential element for the continuation of life on Earth. Most importantly, water treatment is the pivotal stage in order to ensure that clean and safe water is adequately treated before it is distributed to consumers. Water treatment has several phases in its operation which divides into the upstream, treatment plant, and downstream. The downstream refers to the input or feed to the water treatment plants, whereby the downstream denotes the output or products of the water treatment plants. One of the many importance of treating water is to eliminate or reduce the pollutants especially the sediments and pathogenic entities present in untreated water in order to avoid contracting any water-borne diseases.

Conventional water treatment is a treatment process consisting of several individual technologies which deal with specific water treatment needs. In most municipals, conventional water treatment plants

DOI: 10.4018/978-1-7998-2645-3.ch006

are operated by either government or private companies to treat raw water originating from rivers or lakes to produce potable water for domestic uses to a certain size of populations. Some industrial companies also possess their own water treatment facilities, whereby the treated water is utilized in the specific industrial process. This practice is usually carried out in industrial plants with large consumption of water to ensure that it can run efficiently and subsequently at cheaper operating costs.

The main problem with most engineering processes is that they are designed based on a fixed range of parameters at which the processes can perform optimally. This means most plants are desired to operate at a steady state, whereby the parameters are within the desired range of values fixed during the design stage. The selection of parameters range is usually determined from engineering calculations, engineers' expertise and experiences, and also from the standard that was devised which works best in most environments. However, in water treatment operations, changing seasons cause dynamic conditions to the plant which could exceed or upset the extreme values. In water treatment, the critical parameters that are important in the process design are the feed (raw water) quality (i.e. the chemical, physical, or biological), and the capacity (flow rate) of the feed water. The capacity is selected based on the maximum demand for treated water for a projected amount of time in a certain location.

Generally, variations on the feed quality are widely encountered which results from various factors both controllable such as human activities, and non-controllable like the weather. Human activities ranging from logging, livestock and crop farming, and some non-controllable factors like rains, drought and changing season lead to an adverse effect on the quality of raw water supply. These conditions often lead to some issues during the conventional water treatment operation.

All municipal water treatment plants are under the jurisdiction of water authorities. They are required to comply with the regulations imposed on the standard of drinking water quality. In addition, the treated water features – taste, colour, odour etc. must be controlled to maintain consumer's satisfaction. In essence, water treatment plants are under the requirements of delivering safe treated water to protect public health through the awareness of water security.

SOURCES OF WATER

Water is a tasteless, odourless, and transparent substance that occurs naturally on earth in all three physical states: solid (ice), liquid (liquid water), and gas (steam). Surface water and groundwater are the main sources of potable water. The surface water includes freshwater from streams and lakes, brackish water, and seawater from the ocean. The groundwater is a type of freshwater that lies under the ground surface and can be recovered by digging a well. Although around 70% of the earth's surface is covered with water, only about 2.5% of its total is drinkable freshwater, while the rest is seawater. The earth holds about 1.4 quintillion (1.4×10^{18}) m^3 of water of which 97.5% are composed of seawater, around 1.7% frozen in glaciers, ice caps, and ice sheets, while the other 0.8% exists as liquid water found in lakes, rivers, swamps, atmospheric water vapour, and underground (Caldecott, 2008).

COMPOSITION OF RAW WATER

Raw water supplies particulate contents varies according to its origin, size, composition, and concentration. The particulate composition can be derived from the land- or atmospheric-based which water has contacted such as ash, clays, silts, pathogens, and bacteria, and also produced as a result of chemical or biological activities in the water source, for examples the dissolved ions and minerals and algal growth. In addition, natural organic constituents such as humic substances derived from organic matters (e.g. cellulose and other polysaccharides, proteins and amino acids, lipids, lignin, etc.) are found in all surface water resources. Besides, synthetic organic constituents from agricultural runoff or industrial, commercial or domestic effluents are also found in some raw water resources depending on the activities surrounding the water supply. Trace amounts of heavy metals and toxic metals such as manganese, iron, lead and arsenic are also associated with the raw water supply contents.

Several studies have been conducted on the physicochemical composition of raw surface water originated from several rivers and lakes throughout the rural areas of Sabah. The physicochemical analysis is typically conducted to assess the quality of the water resources. Table 1 shows some examples of the physicochemical characteristics of Sabah rural raw surface water as compared to the Malaysian Raw Water Standards (RWS).

CONVENTIONAL WATER TREATMENT PROCESS

A typical water treatment plant has the combination of processes needed to treat the contaminants in the source water treated by the facility. The presence of unbeatable organic or mineral substances causes some problems in obtaining drinking water. Understanding these phenomena requires taking into account the physical and chemical natures of the water to be treated. Optimization of conventional drinking water treatment plant means "to attain the most efficient or effective use" of your water treatment plant regarding some principles, there are: the achievement of consistently high quality finished water on a continuous basis and the importance to focus on overall plant performance, instead of focusing too much on individual processes.

Figure 1 shows a conventional water treatment plant which generally consists of a ten-step process to produce potable water to the consumers. However, other steps may be added/removed as required. These steps typically proceed as follows: (1) screening, (2) aeration, (3) chemical dosing, (4) coagulation, (5) flocculation, (6) sedimentation, (7) filtration, (8) fluoridation, (9) disinfection, and (10) storage.

Screening

Screening is the first physical operation of water treatment. In the context of water treatment, there are two objects concerned in the screening process, i.e. the water, and the object to be screened out. The screening process includes the removals of large non-biodegradable and floating solids such as papers, rags, plastics, tins, containers, leaves and wood. The screening process is positioned before the pumping stations in order to protect the treatment facilities from clogging and damage; thus reducing the maintenance costs of the following treatment facilities. The screening operation can be performed by several types of equipment such as trash racks, bar racks and coarse screens, fine screens, very fine screens and micro-screens. The typical opening sizes are between 1 μm and 150 mm.

Table 1. Physicochemical characteristics of Sabah rural raw surface water against Malaysian Raw Water Standards (RWS)

Location	Crocker Range National Park	Kota Marudu	Kota Marudu	Kota Marudu	Sukau	Sukau	Beaufort	RWS [d]
Main River	Ulu Kimanis River [a]	n.s.* [b]			Kinabatangan River [c]		Padas River [c]	
Tributaries / Lake	Kimanis River	Manggaris River	Ragaroh River	Taritipan River	Kalandaun Lake	Kalanapan Lake	Luangan Rompong Lake	
pH	6.62 – 7.75	7.15 – 8.39	6.73 – 7.79	6.83 – 7.16	6.1 – 7.6	6.5 – 9.5	6.0 - 8.5	5.5 - 9.0
DO (mg/L)	7.39 – 8.97	3.80 – 8.39	3.39 – 5.09	3.72 – 6.56	4.6 – 9.0	4.0 – 12.4	3.9 - 11.4	-
Salinity	0.04	0.08 – 0.13	0.07 – 0.10	0.06 – 0.08	n.m	n.m	n.m	-
Conductivity (µS/cm)	79.83 – 92.80	190 - 320	170 - 210	150 - 190	75 - 150	65 - 130	71.8 - 86.2	-
BOD (mg/L)	n.m*	0.15 – 1.12	0.16 – 2.34	0.01 – 1.92	n.m	n.m	n.m	6
COD (mg/L)	n.m	0 – 13.0	6.0 – 20.0	2.0 – 24.0	n.m	n.m	n.m	10
Total Coliform (CFU/100 mL)	n.m	n.m	n.m	n.m	n.m	n.m	n.m	5,000
Ammonia-Nitrogen (mg/L)	0 – 0.24	0 – 0.04	0 – 0.36	0 – 0.07	n.m	n.m	n.m	1.5
TSS (mg/L)	1.87 – 3.73	4.62 – 13.12	9.12 – 13.68	16.96 – 55.92	n.m	n.m	n.m	-

*n.m: not measured; *n.s.: not stated

Source: [a] (Chaw et al., 2018); [b](Aris et al., 2014); [c] (Lee et al., 2006); [d] RWS is based on standard enforced by the Ministry of Health Malaysia from (MMOH, 2010)

Aeration

Aeration is a process whereby water is brought into close contact with air to remove dissolved gases (e.g. carbon dioxide) and oxidize dissolved metals (e.g. iron, manganese), hydrogen sulfide, and volatile organic chemicals (VOC). Oxygen is mixed in the water through the aeration process to remove odour and the unpleasant taste of water. The addition of oxygen is also to remove dissolved metal through oxidation from their soluble to an insoluble state so that they can form precipitates which can be removed by sedimentation or filtration. In the raw water, iron and manganese exist as soluble ferrous and manganous bicarbonates. During aeration, the oxygen saturated in water promotes the following reactions:

$$4\mathrm{Fe}\left(\mathrm{HCO_3}\right)_2 + \mathrm{O_2} + 2\mathrm{H_2O} \rightarrow 2\mathrm{Fe}\left(\mathrm{OH}\right)_3 + 8\mathrm{CO_2} \quad (1)$$

$$\underset{soluble}{Ferrous\,bicarbonate} + oxygen + water \rightarrow \underset{insoluble}{ferric\,hydroxide} + carbon\,dioxide$$

$$2\mathrm{Mn}\left(\mathrm{HCO_3}\right)_2 + \mathrm{O_2} \rightarrow 2\mathrm{MnO_2} + 4\mathrm{CO_2} + 2\mathrm{H_2O} \quad (2)$$

$$\underset{soluble}{Manganese\,bicarbonate} + oxygen \rightarrow \underset{insoluble}{manganese\,dioxide} + carbon\,dioxide + water$$

Figure 1. Conventional water treatment processes

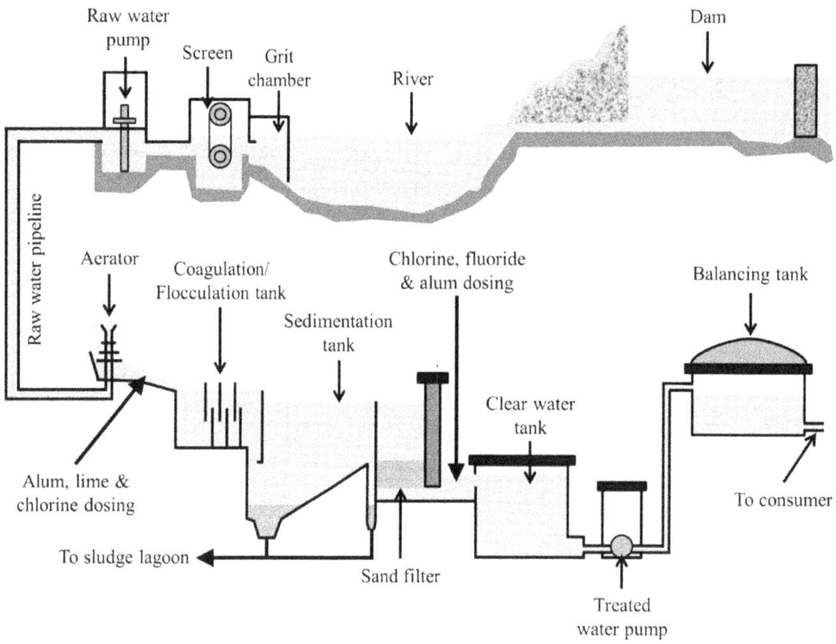

In this process, drops or thin sheets of water is exposed to the air in the aerator. The turbulence produced in the aeration promotes scrubbing (removal) of the undesired dissolved gases from water and allow them to escape into the surrounding air. The efficiency of aeration depends on the number of surface contact between water and air, i.e. the size of water drops or air bubble, contact duration, and type of aerator.

There are several types of aerator used in conventional water treatment plants, such as the cascade aerator, spray nozzle aerator, forced draft aerator, coke tray aerator, cone tray aerator, and air diffusion aerator as shown in Figure 2.

Chemical Dosing

In chemical dosing step, the typical chemicals that are added into the water aerator or mixing tank are lime (calcium hydroxide, $Ca(OH)_2$), and chlorine (pre-chlorine). Lime is added to adjust the pH/alkalinity of water to the desired value. In water treatment, the pH/alkalinity of water throughout the process will affect the dosing and effectiveness of other chemicals used. For example, the coagulation/flocculation process is very dependent on water alkalinity, as the alkalinity in water will be consumed by the formation of flocs. Thus, the right dose of lime must be present in order to offset the loss of alkalinity to flocs formation.

During this step, the addition of lime helps to adjust the pH of water into the alkalinity that is best for the alum coagulant, and also for CO_2 removal and softening process. Lime will react with alum, thus direct mixing is prevented. The amount of dissolved calcium and magnesium in the raw water contributes to the hardness of the water. Hardness in water may cause some negative effects, such as scale build-up in the distribution pipes that can cause clogging, and form deposits on equipment like a pump, heater etc. that can reduce the life of the equipment. When lime is added, the minerals form nearly insoluble

Figure 2. Types of aerator (a) cascade aerator, (b) spray nozzle aerator, (c) forced draft aerator, (d) coke tray aerator, (e) cone tray aerator, and (f) air diffusion aerator

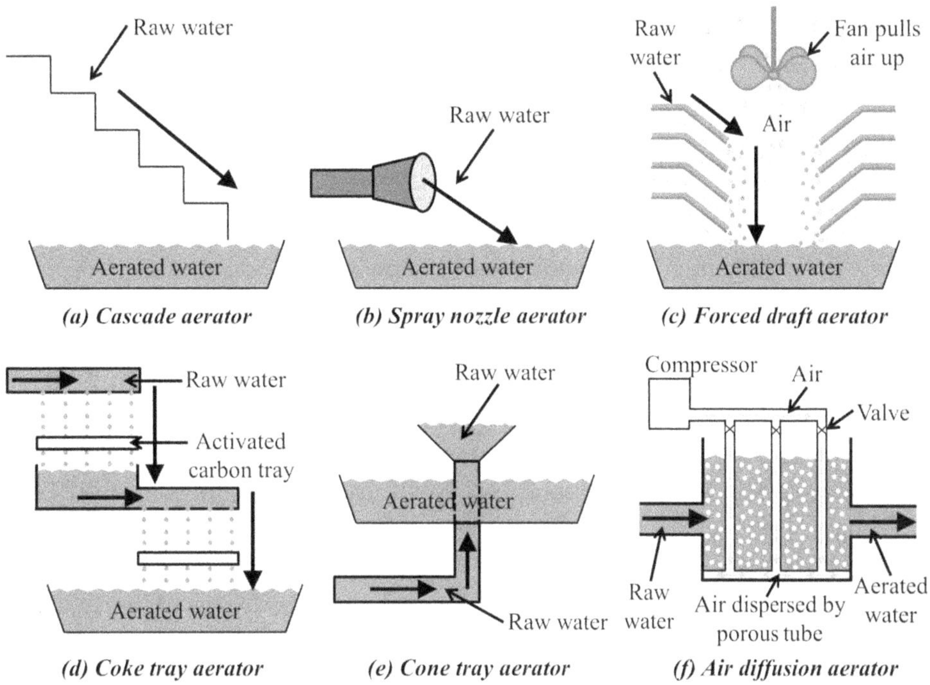

precipitates which can be removed in the subsequent processes and at the same time reduce the number of CO_2. The reactions can be seen as shown in Box 1.

Pre-chlorination is mainly used as a disinfectant to kill potentially harmful microorganisms and pathogens. There are two types of chlorine that are used in most water treatment plants which are sodium hypochlorite (NaOCl) and chlorine gas (Cl_2). NaOCl, commonly known as liquid bleach, is the liquid form of chlorine which is easy to be used and has lower health hazard compared to other chlorine agents. Chlorine gas has a yellowish-green colour and a distinctive odour. It is the cheapest method for chemical disinfection, but is hazardous to the respiratory system and has a risk of explosion when stored as compressed gas. Thus, it must be stored in a separate room with good ventilation, heat and light. Besides being used as a disinfectant, chlorine can be utilized as an oxidizing agent to oxidize undesirable contaminants such as iron, manganese and arsenic and allow them to precipitate which can then be removed by sedimentation or filtration; it also helps to control odour, colour, and prevent the growth of algae in the tank.

Box 1.

Hardness		Lime		Precipitate
CO_2	+	$Ca(OH)_2$	→	$CaCO_3 + H_2O$
$Ca(HCO_3)_2$	+	$Ca(OH)_2$	→	$2CaCO_3 + 2H_2O$
$Mg(HCO_3)_2$	+	$Ca(OH)_2$	→	$Mg(OH)_2 + 2CaCO_3 + 2H_2O$

Coagulation and Flocculation

Coagulation is the clumping together of suspended matter in the water due to either the physical texture of the chemical or through the electrical charges of the coagulant and the colloidal particles. Colloids are very fine particles of several kinds ranging from 1 nm to 100 µm. They are composed of inorganic and organic material derived from soils and rocks such as clays, fibrous particles of asbestos minerals, and organic material from the decomposition of plant/animal debris in the soil. In addition, bacteria which cannot be removed by chlorine disinfection such as Cryptosporidium and the beaver fever causing organism, Giardia sp. can be eliminated by alum during coagulation. Initially, a coagulant is added into turbid water and rapidly mixed (typically for 1 – 3 minutes) to ensure uniform dispersion of alum throughout the raw water and promotes particles collision.

Coagulants are nontoxic at the working dosage, relatively high charge density and insoluble in the neutral pH range. The commonly used coagulants are aluminium sulphate (alum), sodium aluminate, aluminium chloride, poly-aluminium chloride, poly-aluminium sulphate, polyiron chloride, ferric chloride and ferric sulphate. Alum is used as a coagulant to clump suspended/colloidal particles together and settle out of water. Alum is one of the most commonly used coagulants and is relatively cheap than other types of coagulants. In order for alum to be effective, the pH of raw water must be between 6 and 8. Moreover, there must be enough alkalinity present, as alum consumes alkalinity during the coagulation/flocculation reactions. Thus, there might be some adjustment needed for the pH and alkalinity of water during the treatment. When alum is added into raw water, it reacts with natural alkalinity (as calcium bicarbonate) forming aluminium hydroxide precipitate.

$$Al_2\left(SO_4\right)_3.14H_2O + 3Ca(HCO_3)_2 \rightarrow 2Al(OH)_3 + 3CaSO_4 + 6CO_2 + 14H_2O \quad (3)$$

When there is not enough alkalinity in water to react with alum, the addition of lime provides the alkalinity needed:

$$Al_2\left(SO_4\right)_3.14H_2O + 3Ca(OH)_2 \rightarrow 2Al(OH)_3 + 3CaSO_4 + 14H_2O \quad (4)$$

The principle coagulation mechanisms are the bonding of $Al(OH)_3$ with colloids. The $Al(OH)_3$ reacts with the surface of charged colloids (small suspended particles) in raw water by neutralizing its charges (known as destabilization). Once the charges are neutralized, the small suspended particles are capable of sticking together to form micro-flocs. The overall effect of the addition of lime will decrease in the pH of water because CO_2 is produced in the reaction.

Jar Test

A jar test is a pilot-scale batch test to simulate the coagulation/flocculation and sedimentation processes in a water treatment plant. In this test, raw water samples are mixed rapidly and then slowly and then allowed to settle. It is important to conduct this test to help operators in understanding: the way a treatment chemicals such as alum will behave and react with a particular type of raw water, determine which chemical will work best with the type of raw water used, and determine whether the operators are using the right amount of chemical and thus, improves the WTP's performance.

According to David Pask (1993) on his article "Jar Testing: Getting Started on a Low Budget", "By performing jar tests, you can try alternative treatment doses and strategies without altering the performance of the full-scale treatment plant and easily compare the results of several different chemical treatments for time of formation, floc size, settle ability, and, perhaps, filtration characteristics. One cannot make such comparisons with the full plant's treatment.".

Besides, doing jar test also helps water department in saving coagulant expenses in a way that it helps to determine the optimum dose of coagulant, thus eliminating the problem of underfeeding or overdosing of coagulant into the water treatment plants.

However, several disadvantages associated with this test is that it is a batch test, which can be time-consuming, and the results obtained from a series of jar tests might not correspond to the results obtained on a full-scale plant. This is because, jar test procedures and the technical limitations of the equipment used only shows the approximate assessment on the required intensity and time of rapid mixing, flocculation and sedimentation processes.

The amount of coagulants required during a certain period of operation varies depending on the condition of feed water. Important parameters which need to be measured prior to jar testing is the feed water turbidity, flow rate, and pH in the water treatment plant which are periodically measured in the laboratory in hourly basis or in several hours gap. Figure 3 illustrates the experimental setup of a jar test in the laboratory. The test can be carried out using the steps as follows:

Step 1: Check the pH, turbidity, aluminium content and inflow rate of raw water to the treatment plant and record the data.

Step 2: Label 6 500 mL beakers from 1 – 6 and fill with 500 mL of raw water into each beaker. Place the beakers into the flocculator machine.

Step 3: To prepare the alum 'stock solution' (1% alum), calculate the concentration of 'mother solution' and determine the amount needed to be diluted into 1% alum.

Step 4: Calculate the amount of 1% stock solution needed to be added into each beaker by referring to Table 2, and add the amounts calculated into each respective beaker.

Step 5: Turn ON the flocculator machine agitator at 200 rpm for 3 minutes.

Step 6: Lower the agitation speed to 50 rpm for a period of 10 minutes.

Step 7: Turn OFF the agitator and let the flocs settle at the bottom of the jars for 10 minutes.

Step 8: Check the turbidity of the water samples in all 6 beakers. Choose the optimum alum dosage which corresponds to the lowest turbidity among the 6 samples.

EXAMPLE 1: Jar Test

Problem. A water treatment plant operates at a feed water flow rate of 500 m^3/h which contain turbidity of 170 NTU, pH of 8, and 0 mg/L aluminium content. The plant's laboratory is required to calculate the flow rate of alum dose needed (in 20 mL pump stroke) to treat the raw water feed in the water treatment plant by using the jar test. The plant uses alum with 6.35% alum concentration, while the jar test is carried out with a stock solution of 1.00% alum. The experiment is carried out using the apparatus and material as follows:

Table 2. Recommended alum dosage for jar test based on water samples turbidity

Turbidity (NTU)	Alum Dosage (mg/L)					
	Jar 1	Jar 2	Jar 3	Jar 4	Jar 5	Jar 6
1 – 99	5	10	15	20	25	30
100 – 199	10	20	30	40	50	60
200 – 299	20	30	40	50	60	70
300 – 399	30	40	50	60	70	80
400 – 499	40	50	60	70	80	90
500 – 599	50	60	70	80	90	100
600 – 699	60	70	80	90	100	110
700 – 799	70	80	90	100	110	120
800 – 899	80	90	100	110	120	130
900 – 999	90	100	110	120	130	140
≥ 1000	100	110	120	130	140	150

Apparatus and material:

- 6-station Jar Tester (Flocculator machine)
- Beakers (500 mL)
- Syringes
- Untreated raw water
- Alum 'mother solution'
- Distilled water

Figure 3. Schematic of the jar test setup

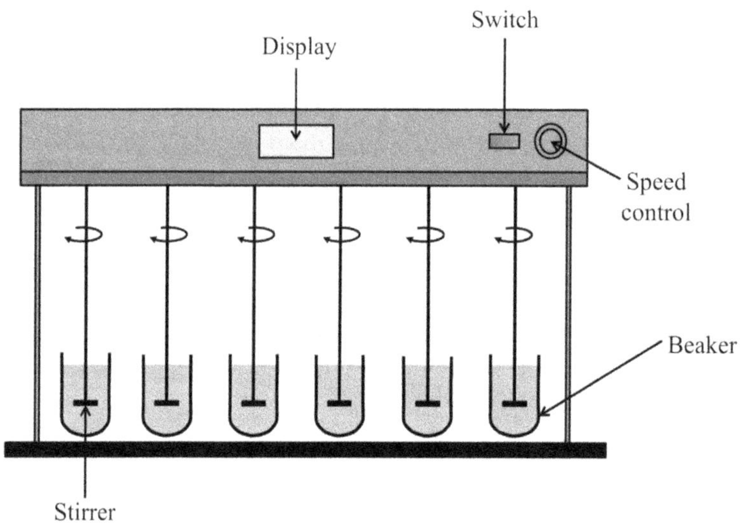

Table 3. Jar test result

	Jar 1	Jar 2	Jar 3	Jar 4	Jar 5	Jar 6
[Alum], mg/L	10	20	30	40	50	60
$V_{j,alum}$, mL	0.50	1.00	1.50	2.00	2.50	3.00
Turbidity, NTU	26	25	13	9	5	3

Since the pH is within the working range of alum, no chemical modification is required on the raw water. By referring to Table 2, the recommended alum concentration for each jar test is used to calculate the volume of stock solution to be added in each jar. The jar test yields the result as shown in Table 3.

Jar test result:

Solution. Based on the problem statement, the data that can be extracted are:

1. Raw water flow rate, $Q_{r.w}$: 500 m^3/hr
2. pH: 8
3. Turbidity: 170 NTU
4. Aluminum content: 0.0 mg/L

Since the maximum acceptable value for turbidity is 5 NTU (Ministry of Health of Malaysia), the minimum and maximum alum dosage that is optimum for the plant based on the jar test result are:

Minimum alum dosage:

50 mg/L (Jar 5) which produces water with a turbidity of 5 NTU

Maximum alum dosage:

60 mg/L (Jar 6) which produces water with a turbidity of 3 NTU

The optimum alum dosage flow rate can be calculated as:

$$C_{alum,m} Q_{alum} = C_{alum,opt} Q_{w,raw} \tag{5}$$

where

$C_{alum,m}$ = Concentration of 'mother solution'
Q_{alum} = Optimum alum dosage flow rate
$C_{alum,opt}$ = Concentration of optimum alum dosage
$Q_{w,raw}$ = Raw water flow rate

Minimum alum dosage flow rate:

$$Q_{alum} = \frac{C_{alum,opt} Q_{w,raw}}{C_{alum,m}} = \frac{50mg / L \times 500m^3 / hr}{63500mg / L}$$

$$= 0.394m^3 / hr = 109.44mL / s$$

Maximum alum dosage flow rate:

$$Q_{alum} = \frac{C_{alum,opt} Q_{w,raw}}{C_{alum,m}} = \frac{60mg / L \times 500m^3 / hr}{63500mg / L}$$

$$= 0.472m^3 / hr = 131.11mL / s$$

Minimum alum pump stroke:

$$= 109.44mL / s \times \frac{1stroke}{20\frac{mL}{s}} = 5.5stroke$$

Maximum alum pump stroke:

$$= 131.11mL / s \times \frac{1stroke}{20\frac{mL}{s}} = 6.6stroke$$

Flocculation

Flocculation is a slow mixing stage to increase the particle size from microscopic flocs (microflocs) to larger particles (flocs). During the slow mixing, the micro-flocs are brought into contact and collide, causing them to bond, producing larger, heavier, and visible flocs enough for it to settle at the bottom of the sedimentation tank, or large enough to be trapped in the filter. The particles are held together in weak intermolecular Van Der Waals forces which are encouraged through longer detention time and lower mixing energy, hence low-velocity water flow is highly preferable. Thus, if the water is agitated at high mixing energy, the weak Van Der Waals forces holding the flocs together can be sheered or broken easily.

A mechanical process of flocculation brings the coagulated particles to join and form a mass of flocs and then a precipitate. Slow agitation of the particles suspended water encourages the particles to clump together and agglomerate. Mixing and sedimentation are applied to remove the precipitates from clear water. The commonly used flocculation tank in most conventional water treatment plants in here is the hydraulic flocculation tank. There are two types of hydraulic flocculation, which are the horizontally baffled tank and the vertically baffled tank as shown in Figure 4. The baffle channels are designed to cause the water flow in a "zig-zag" (or direction changes) motions. The baffles are placed with increasing distant from inlet to an outlet such that the channel sizes are increasing. The purpose of the baffles is to produce a longer flow path to enhance the agglomeration of flocs. This gradual incre-

Figure 4. Types of hydraulic flocculation, (a) horizontal flow, and (b) vertical flow

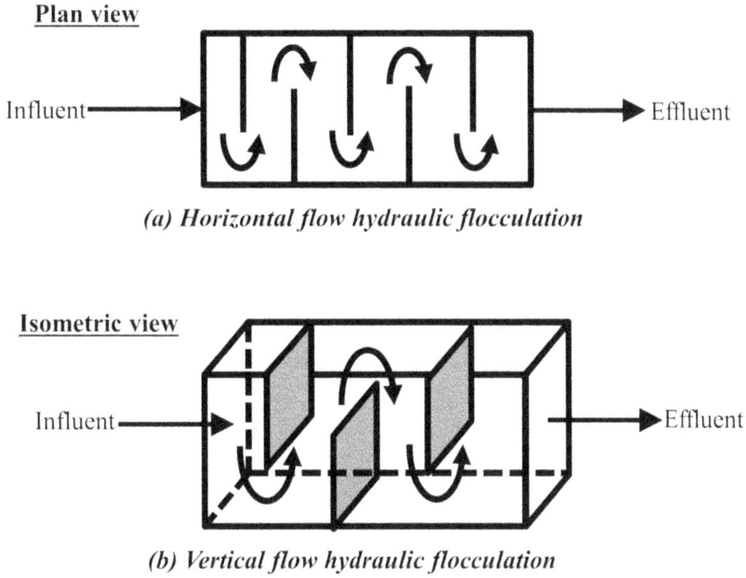

ment in baffles spacing will result in decreasing mixing speed and create velocity gradient as the water flows through the tank.

Sedimentation

Sedimentation, also called clarification allows suspended solids to settle by gravity to form sludge. In the physical water treatment, the sedimentation tanks can be used as pre-treatment or primary treatment. The pre-sedimentation tank is employed to the raw water supply before it enters the treatment facility to facilitate large objects that can clog the pumps to settle; while the primary sedimentation tanks are used in the treatment facility for settling down of suspended material or floc by gravity. It can be accomplished by decreasing the water velocity to a point where the flocs will no longer remain in suspension. When the driving force of velocity no longer supports the flocs, gravity will remove them from the water flow.

Some important parameters to be considered in the sedimentation process are detention time and water flow through velocity. Longer detention time permits more suspended solids to be removed in the water treatment. In practice, the common detention time for the primary sedimentation tank in water treatment is in the range of 1.5 – 2.5 hours. If the inlet flow rate is very high, the water will have low detention time, and all of the flocs may not be removed. On the other hand, if the incoming flow rate is very low, the water is going to have high detention time. While this might be good as all of the flocs will likely settle at the bottom of the tank, the water treatment plant may not be operating to its fully optimum capacity. Nonetheless, the settling time would largely depend on the size and type of particles to be settled. Table 4 shows the typical settling time for some particles found in water treatment plants. It also goes to show the importance of coagulation and flocculation to agglomerate fine objects so as to increase its size and subsequently reduce the settling time. Also, the sedimentation operation needs to maintain the water flow through velocity so as to prevent sweeping and carrying off of the solid materi-

Table 4. Settling time for particles of various sizes

Diameter of Particle	Type of Particle	Settling Time Through 1 Meter of Water
10 mm	Gravel	1 second
1 mm	Sand	10 seconds
0.1 mm	Fine sand	2 minutes
10 μm	Protozoa, algae, clay	2 hour
1 μm	Bacteria, algae	8 days
0.1 μm	Viruses, colloids	2 years
10 nm	Viruses, colloids	20 years
1 nm	Viruses, colloids	200 years

Source: (Petterson, 2001)

als already settled at the bottom of the sedimentation tank. Thus, the water flow through velocity should be retained at less or equal to 9 m/s at typical operation.

Factors that affect the operation of the sedimentation tank are particle size (large particle, easy removal), water temperature (high temperature, high settling rate), density and shape of particles (high density, high settling rate), water velocity (low velocity, more time for flocs to settle), type of sedimentation tank, and chemical dosage.

Most sedimentation tanks are separated into 4 zones: inlet, settling, sludge, and outlet zones. The inlet zone should distribute uniform flow across the inlet to the tank. The inlet should be designed so that the flow velocity near the bottom is minimized in order to prevent the sludge from being swept up and out of the tank. The settling zone is the largest portion of the tank which provides a calm area that is necessary for the flocs to settle. The sludge zone is a storage area for the sludge before they are treated or disposed of, which is located at the bottom of the tank. The sludge disposal should be well controlled so that the zone will not be overloaded and cause the sludge to float around and not settled. The settled water will flow out through the outlet zone into the filtration tanks. This zone controls the depth of water in the tank. Weir is set at the outlet to control the overflow rate and prevent the solids from rising and leaving the tank before they settle out.

Filtration

Filtration process aims to remove fine suspended particles and reduce the number of bacteria. Water from sedimentation tank flows into the gravity filter, where it will be filtered through filters composed of layers of sand and gravels. In the tank, there are two physical mechanisms for the suspended particles removal from the tank: mechanical and adsorption processes. In mechanical removal, large particles are trapped and become embedded and fixed in the voids between the filtration media (sand, gravel) and cannot continue down through the media. Adsorption is a process of particles sticking on the surfaces of each filter grains or onto the previously deposited particles, thus cannot continue down through the media. The visual difference between these two physical mechanisms can be seen in Figure 5.

Rapid sand filters or rapid gravity filters are employed when slow sand filters are not applicable because of land areas are limited. The sand used in the rapid sand filters can be coarser to remove the flocs which

Figure 5. Physical mechanisms used for particles removal in a filtration tank, (a) mechanical, and (b) adsorption mechanisms

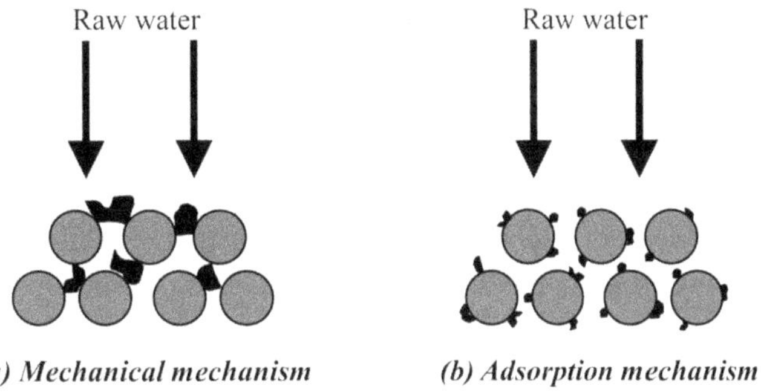

are produced in the coagulation-flocculation process. The flocs flow along with water through the sand media by gravity action or pressure created from pumps. The flocs are trapped within the sand matrix.

Slow sand filters are typically 1 to 2 m in depth. The designs of sand filters can be rectangular or cylindrical in cross-section and basically used to treat the surface water. The loading rates range from 0.1 to 0.2 m^3/m^2 h. The sand inside the filters does not perform any filtration function but simply works as a substrate. From a biological perspective, a gelatinous layer also called biofilm which consists of bacteria, fungi, protozoa, etc. formed on the top few millimetres of the fine sand layer. The microbes metabolise the organic materials contained in the raw water. The slow sand filters can reduce the bacteria by more than 90%. The development of biofilm with time may reduce the flow rate of the sand filters. The top few millimetres of the fine sand is scraped off and new fine sand surface emerges. Backwashing is then performed to clean the sand filters.

Fluoridation

The purposes of adding fluoride in water are to prevent tooth decay and cavities. Sodium fluorosilicate (Na_2SiF_6) is a white, dry crystalline fluoridation compound which is both tasteless and odourless. This chemical is added to the filtered water as one of the fluoridation agents. It can be toxic when inhaled or ingested, thus proper handling procedures should be executed at all times. This compound also has a low rate of solubility in water; hence it requires a longer time to dissolve before it can be added to water.

Disinfection

Disinfection ensures the water is safe to drink. Disinfection destroys the microbes which can cause disease in humans. The primary disinfection methods include chlorination, chloramines, ozone and ultraviolet (UV) light. Chlorine is a very effective agent for removing almost all microbial pathogens. However, chlorine is a dangerous gas because it is lethal at concentrations as low as 0.1% in the air by volume. Sodium hypochlorite is easier in handling than gaseous chlorine or calcium hypochlorite. Compared to the calcium hypochlorite, sodium hypochlorite is less stable and it should not be stored for more than one month, while the calcium hypochlorite is very stable and it can be stored for about one year. Chloramines are effective to eliminate the bacteria in water and generate fewer disinfection by-products. The most disadvantages of using chloramines are weak disinfectants and much less effective against viruses and protozoa. Ozonation requires shorter contact time and dosage than chlorination. Cryptosporidium which is resistant to chlorination or UV radiation can be effectively destroyed by the combination of ozonation and UV light. Nevertheless, ozone gas is unstable and must be generated onsite while the UV radiation application is not suitable for water containing high levels of suspended solids, turbidity, colour or soluble organic matter.

The filtered water needs are disinfected before it is being stored in the reservoir. Disinfection (post-chlorine) process aims to eliminate bacteria and pathogens in the water. The reaction of chlorine with water produces two types of acid: hypochlorous acid (HClO) and hydrochloric acid (HCl). HCl has high dissociation in water, forming H^+ and Cl^-, while HClO will partially dissociate to form H^+ and ClO^-. HClO and ClO^- are oxidizers and primary disinfection agent, and also known as the Free Available Chlorine.

Storage

The common practice of water storage system in conventional water treatment plants involves the use of balancing tank and reservoir. A balancing tank is a vessel that keeps water at a constant level above a pump inlet. The purpose of a balancing tank is to equalize or balance the fluctuating demand for water supply. Then, water from the balancing tank will be pumped through the booster pump house into the reservoir and the service tank. The service tank refers to the tank used in the plant to supply water into the office, backwashing and other plant usages, while the water from the reservoir flows into the distribution network to be channelled to the consumers.

COST ESTIMATE FOR CONVENTIONAL WATER TREATMENT PLANT

Estimation of costs consideration for the construction and operation of any treatment plant is usually conducted in the manners calculation using cost equations. The reliability of the cost equations is thoroughly constructed from data that are derived from a study designed that takes into consideration the overall required unit processes, etc. In a study conducted by Clark (1982), these topics have been completely discussed pertaining to the costs equations related to estimating the costs of conventional water treatment plants. Two cost equations were developed for both capital cost and operating and maintenance costs, as shown in Equation 6 and 7.

$$CC = K_1 USRT^c CCI^d UNTS^e f^{TDH} g^{G} \tag{6}$$

where CC is the annual capital cost in \$/yr, $USRT$ is the design parameter (e.g. size of filter area etc.), CCI is the Engineering News Record Construction Cost Index divided by 100 (This is the three-digit ENR Index), $UNTS$ is the number of process units, TDH is the total dynamic head, G is the energy gradient per second, and K_1, c, d, e, f, and g are the constants determined by regression.

$$OC = K_2 USRT^h PR^i PPI^j DHR^k NGS^l DSL^m UNTS^n o^{TDH} p^{G} q^{MI} \tag{7}$$

where OC is the annual operating and maintenance cost in \$/yr, PR is the power cost in \$/kWh, PPI is the producers price index divided by 100, DHR is the direct hourly wage rate in \$/h, NGS is the natural gas cost in \$/cu ft, DSL is the price of diesel fuel in &/gal, MI is the miles (one way), and K_2, h, i, j, k, l, m, n, o, p, and q are the constants determined by regression.

As shown in Table 5, the cost analysis for a 378.5 MLD water treatment plant has been calculated for both the capital cost and operation and maintenance cost using the cost equations presented in the study. These can, of course, be adjusted to calculate other capacities of water treatment plants. The constants can be determined for each process before the total cost is calculated.

DRINKING WATER STANDARDS

Malaysian Raw Water and Drinking Water Quality Standards

Untreated water may pose threats and detrimental effects on human health and well-being as it is often contaminated with solid suspensions, microorganisms, chemical constituents, and metal content. Impurities and salts contained in the surface water and groundwater must be reduced before the water is suitable for daily uses and drinking purposes. The drinking water quality standards imposed by the Ministry of Health of Malaysia are tabulated in Table 6.

EXAMPLES OF CONVENTIONAL WATER TREATMENT PLANTS IN SABAH

There are currently 58 conventional water treatment plants operating throughout Sabah. These plants produced approximately more than one billion litres of treated water per day in 25 districts with a population count of around 3.2 million censuses 2010 (Department of Statistics Malaysia, 2017). Table 7 shows some examples of conventional water treatment plants in Sabah which lists the type of processes involved, the capacity of treated water production, and the source of raw water for the plant. In these conventional water treatment plants, there are several issues that can surface during the operation. These issues will be discussed in Section 6-8 from the perspective of different sections of the water treatment plants, which are the upstream, treatment processes, and downstream.

COMMON ISSUES IN CONVENTIONAL WATER TREATMENT PLANTS

Water Treatment Plants: Upstream and Treatment Processes

The water treatment plants upstream refer to the influent, the raw water intake, the feed water etc., which are water that has yet to be treated. It contains varieties of minerals, ions, contaminant particles, and bacteria. Water treatment plants are usually developed in areas with nearby water bodies as the main source of feed water, or man-made reservoirs to collect and hold a large amount of water. The water can be sourced from groundwater, rainwater, surface water from water bodies like lakes and rivers, and also dams and water reservoirs. Large scale water treatment plants typically utilize a dam or reservoir

Table 5. Cost estimate for a conventional water treatment plant

Process	Flow (MLD)		Process Parameter		Costs (Dollars)		Cost (Cents/MLD)		
	Design	Actual	Design	Operating	Construction (Dollars/yr)	Capital (Dollars/ yr)	Operation & Maintenance	Debt	Total
Alum (Feed–liquid stock, 30 mg/L)	378.5	264.95	473.4 kg /hr	331.4 kg/hr	120,487	160,650	235.40	16.91	252.31
Polymer (Feed system, 0.20 mg/L)	378.5	264.95	75.3 kg/day	52.7 kg/day	37,680	50,240	92.21	5.28	97.49
Rapid mix (G=600)	378.5	264.95	262.8 m^3	262.8 m^3	217,147	289,529	61.29	30.38	91.94
Flocculation (Horizontal, G=50)	378.5	264.95	5,258.4 m^3	5,258.4 m^3	704,116	938, 822	24.31	98.81	123.12
Rectangle clarifier (Units=3)	378.5	264.95	3,093.7 m^2	3,093.7 m^2	3,807,421	5,076,561	50.46	534.74	584.94
Gravity filtration (81.4 LPM/ m^2)	378.5	264.95	1,300.6 m^2	1,300.6 m^2	3,941,906	5,255,874	141.61	553.50	695.11
Filtration media (Mixed)	378.5	264.95	1,300.6 m^2	1,300.6 m^2	415,922	554,563	0.00	58.39	58.39
Hydraulic surface wash	378.5	264.95	1,300.6 m^2	1,300.6 m^2	423,958	565,277	13.47	59.45	73.18
Backwash pumping facilities	378.5	264.95	238,480 LPM	4,097 m^2	467,281	623,041	49.93	65.52	115.72
Wash water surge basin	378.5	264.95	45,400,000 L	45,400,000 L	764,577	1,019,436	0.00	107.27	107.27
Cl_2 feed cylinder storage (2 mg/L)	378.5	264.95	762 kg/day	533.4 kg/ day	165,418	220,558	99.60	23.25	122.85
Clearwell storage (Below)	378.5	264.95	30,300,000 L	30,300,000 L	2,592,880	3,457,174	0.00	364.07	364.07
Filter press	378.5	264.95	10.7 m^3	10.7 m^3	1,403,159	1,870,879	215.32	197.09	412.42
Dewatered sludge hauling (Miles=20)	378.5	264.95	17,000 cy/yr	11,900 mg/yr	84,015	112,020	13.74	11.89	25.63
Administration building & lab	378.5	264.95	378.5 MLD	378.5 MLD	391,521	522,028	167.77	54.95	222.72
Total					15,537,488	20,716,650	1,165.13	2,181.51	3,347.16

to store a large quantity of water to fulfil the large demand for treated water which is commonly seen in urban and developed areas. On the other hand, small scale water treatment plants usually receive raw water resource directly from water bodies such as rivers and lakes.

In recent years, our freshwater resources are under mounting pressure primarily on both quality and quantity. Many factors contribute to the deterioration of both attributes which are of global concern. In general, the quality of feed water is affected by human activities and also natural constraints such as variations of rainfall and seasonal change. Because of these contributing factors, the feed water will experience changes in composition which makes it difficult for water treatment operation. In extreme situations, the water treatment plants may encounter difficulties during treatment as a result of the current technologies are not built-up for certain feed water conditions. Several examples are high microalgae contents, a high heavy metal such as manganese contents, high sediments contents, high pesticides and fertilizers contents which are commonly encountered in tropical countries such in Sabah, Malaysia. Thus, water operators must always ready and are prepared to overcome the many uncertainties of the raw water supply conditions into the water treatment plant.

Table 6. Malaysian raw water and drinking water quality standard

Parameter	Group	Recommended Raw Water Quality	Drinking Water Quality Standards
		Acceptable Value (mg/litre (unless Otherwise Stated))	Maximum Acceptable Value (mg/litre (Unless Otherwise Stated))
Total Coliform	1	5000 MPN / 100 ml	0 in 100 ml
E. coli	1	5000 MPN / 100 ml	0 in 100 ml
Turbidity	1	1000 NTU	5 NTU
Colour	1	300 TCU	15 TCU
pH	1	5.5 – 9.0	6.5 – 9.0
Free Residual Chlorine	1	–	0.2 – 5.0
Combined Chlorine	1	–	Not less than 1.0
Temperature	1	–	–
Clostridium perfringens (including spores)	1	–	Absent
Coliform bacteria	1	–	–
Colony count 22°	1	–	–
Conductivity	1	–	–
Enterococci	1	–	–
Odour	1	–	–
Taste	1	–	–
Oxidisability	1	–	–
Total Dissolved Solids	2	1500	1000
Chloride	2	250	250
Ammonia	2	1.5	1.5
Nitrate	2	10	10
Iron	2	1.0	0.3
Fluoride	2	1.5	0.4 – 0.6
Hardness	2	500	500
Aluminium	2	–	0.2
Manganese	2	0.2	0.1
Chemical Oxygen Demand	2	10	–
Anionic Detergent MBAS	2	1.0	1.0
Biological Oxygen Demand	2	6	–
Nitrite	2	–	–
Total Organic Carbon (TOC)	2	–	–
Mercury	3	0.001	0.001
Cadmium	3	0.003	0.003
Arsenic	3	0.01	0.01
Cyanide	3	0.07	0.07
Lead	3	0.05	0.01
Chromium	3	0.05	0.05
Copper	3	1.0	1.0
Zinc	3	3	3
Sodium	3	200	200
Sulphate	3	250	250
Selenium	3	0.01	0.01
Argentum	3	0.05	0.05
Magnesium	3	150	150
Mineral oil	3	0.3	0.3
Chloroform	3	–	0.2
Bromoform	3	–	0.1
Dibromochloromethane	3	–	0.1
Bromodichloromethane			
Phenol	3	0.002	0.002
Antimony	3	–	0.005
Nickel	3	–	0.02
Dibromoacetonitrile	3	–	0.1
Dichloroacetic acid	3	–	0.05

continued on following page

Table 6 continued

Parameter	Group	Recommended Raw Water Quality	Drinking Water Quality Standards
		Acceptable Value (mg/litre (unless Otherwise Stated))	Maximum Acceptable Value (mg/litre (Unless Otherwise Stated))
Dichloroacetonitrile	3	–	0.09
Trichloroacetic acid	3	–	0.1
Trichloroacetonitrile	3	–	0.001
Trihalomethanes – total	3	–	1.00
Aldrin/Dealdrin	4	0.00003	0.00003
DDT	4	0.002	0.002
Heptachlor & Heptachlor Epoxide	4	0.00003	0.00003
Methoxychlor	4	0.02	0.02
Lindane	4	0.002	0.002
Chlordane	4	0.0002	0.0002
Endosulfan	4	0.03	0.03
Hexachlorobenzene	4	0.001	0.001
1,2-dichloroethane	4	–	0.03
2,4,5-T	4	–	0.009
2,4,6-trichlorophenol	4	–	0.2
2,4-D	4	0.03	0.03
2,4-DB	4	–	0.09
2,4-dichlorophenol	4	–	0.09
Acrylamide	4	–	0.0005
Alachlor	4	–	0.02
Aldicarb	4	–	0.01
Benzene	4	–	0.01
Carbofuran	4	–	0.007
MCPA	4	–	0.002
Pendimethalin	4	–	0.02
Pentachlorophenol	4	–	0.009
Permethrin	4	–	0.02
Pesticides	4	–	–
Pesticides – total	4	–	–
Polycyclic aromatic hydrocarbons	4	–	–
Propanil	4	–	0.02
Tetrachloroethene and Trichloroethene	4	–	–
Vinyl chloride	4	–	0.005
Gross alpha (α)	5	0.1 Bq/l	0.1 Bq/l
Gross beta (β)	5	1.0 Bq/l	1.0 Bq/l
Tritium	5	–	–
Total indicative dose	5	–	–

Source: (Ministry of Health of Malaysia)

Microalgae

The microalgae contents spike during the algal bloom season where rapid growth in the population of microalgae in water causing discolouration of water due to a high density of pigmented algae species, typically in the tints of bright green, green, yellowish-brown, or red. The rising growth is induced by an excess of nutrients into waters particularly phosphorus and nitrogen content which is a source of food to the algal cells, sunlight, water, and carbon dioxide which are naturally occurring in water. During the drought season, the increase in weather and water temperature leads to the exponential growth of algae

Table 7. Examples of conventional water treatment plants in Sabah

Water Treatment Plant	Processes Involved	Capacity	Raw Water Source
Telibong II Water Treatment Plant (Tuaran, Sabah)	Screening, aeration, coagulation, flocculation, clarification, filtration, disinfection, fluoridation, storage.	80 MLD	Telibong II Dam
Moyog Water Treatment Plant (Penampang, Sabah)		170 MLD	Babagon Dam
Lawa Gadong Water Treatment Plant (Beaufort, Sabah)		76 MLD	Padas River
Kogopon Water Treatment Plant (Papar, Sabah)		40 MLD	Papar River
Bambangan & Kimolohing Water Treatment Plant (Ranau, Sabah)		13 MLD	Liwagu River

as compared to other days. Since the water in dams and reservoirs are fairly stagnant, the low convective flows of water give the algae an edge for competition with other organisms. Increasing concentration of these nutrients led to an increasing number of algae, which cause blockage of the sunlight penetration through the water surface. Decomposition of algae requires the bacteria present in the water to use up the dissolved oxygen content in the water, creating an anoxic environment which subsequently dismisses survivability of other aquatic plants and animals. At this stage, the eutrophication has produced the phenomenon of algal bloom.

A particular concern for algal bloom in water reservoirs is the growth of blue-green algae, of which some species produce toxins (cyanotoxins) which are harmful for human consumption. Thus, even though an only small percentage of harmful algae species, an early indication of algal bloom intended for water treatment plants should be treated as toxic until it is tested. Algal bloom causes negative consequences to the water treatment plants in several ways, which are deterioration of water quality, disturbing the water supply, and increasing costs of operation. On a more larger scale, uncontrollable algal bloom can lead to the creation of a dead zone which destroys the aquatic system, toxic contaminations of both surface water and groundwater supplies, and finally sickness or death from consumption of contaminated water.

Some algae species are difficult to be removed by water treatment processes as they do not settle easily. Microalgae suspension can be flocked together by coagulation and flocculation (Lee et al., 2015). The effectiveness of the coagulation process is affected by the alum dosage, initial concentration of algae, and the shape and size of the microalgae species found in the water (Lin et al., 1971). Despite that, the high content of microalgae can clog the sand filters, which lead to frequent backwashing requirements. As a result, this consumes a large volume of water and consequently lowers the production of drinking water and monetary loss. Nonetheless, conventional water treatment is not effective in removing the toxins released by the microalgae which are soluble in water. Traditional processes such as aeration are not effective to remove soluble toxins. It would require a much more sophisticated operation to remove the toxins, such as activated carbon adsorption which can incur high purchase cost for this mere seasonal problem.

The accumulation of nutrient concentrations in our water supply is primarily caused by the agricultural activities through the use of plants fertilizer and discharge of livestock manure which contain a high amount of phosphorus and nitrogen. Some of the farms and plantations, especially the palm oil plantation in Sabah carry out fertilization of oil palm trees manually by workers. Some workers who failed to strictly follow the standard operating procedures can cause the fertilizers to be washed out to the water bodies. The water run-off from these activities can seep through the soils and subsequently causes intrusion to the water catchment areas, increasing the nutrient contents.

The best way to prevent algal bloom in water reservoirs is to prevent the growth of algal bloom in the first place. This can be accomplished by keeping the nutrient contents in the water at bay. The number of nutrients entering the water bodies should be prevented and this starts by addressing the seriousness of this issue and getting the state regulatory body to protect the drinking water supply, perform holistic and sustainable management of the water resources, and taking strict action at the root causes. By decreasing and controlling the discharge of plant fertilizers and livestock manure into the environment, the occurrence of algal bloom can be regulated. Another way is through the forecast of algal bloom phenomenon. This can be accomplished by monitoring devices which can predict the seasonal blooms and prepare the water treatment plants for prompt actions.

Sediments

The surface water quality is closely related to the characteristics of water catchment area which are influenced by the surrounding human activities. In Sabah, over 50% or 3.9 million hectares of Sabah's total land which have been gazetted as tropical rainforest reserves are important natural water catchment areas. Forestry logging is a controlled activity by law around the world. In Sabah, a lot of illegal logging occurrences have happened these past few decades which threatens our valuable Borneo tropical rainforest. However, uncontrolled illegal logging activities and land-clearing and conversions for plantations present threats to the water catchment areas. Trees harvesting and land clearings involve the construction of a logging pathway, tress cutting, and trails to transport the logs which result in mass clearance of land areas left bare. Logging activities left the soils exposed to rain, whereby heavy rainfall leads to erosion of soils in the catchment areas for large river systems. The eroded soils carry a high load of sediments to the downstream rivers, increasing the level of the sediment by two to fifty times the normal load (Douglas et al., 1993).

Sedimentation (geology) is a process whereby eroded soils are carried by flowing water or other media to water bodies like rivers, lakes, and reservoirs which are then deposited in the form of layers of solids (Tundu, Tumbare, & Onema, 2018). Sediment particles are composed of a mixture of silt, clay, and sand in the range size of 4 – 6 μm in diameter. Sedimentation is one of the major threat to water bodies ecosystem all over the world, as it decreases the water quality and quantity, increases water treatment cost, lowers the availability of water for other uses, as well as threatens the life span of the water bodies. Sediments can be divided into two types, namely, are the suspended sediments and bedload sediments. Suspended sediments refer to components which remains in suspension for a significant amount of time due to the flowing current of the water stream. Bedload sediments are the sediments that are in saltating (movement due to the flow of fluid) or rolling motion (Kasran, 1996).

High sediments content in raw water supply makes it difficult for water treatment plants to perform its treatment operations since the sediments loading need to be reduced before it can be treated regularly. Besides, sediments loading which sinks into the lakes and rivers could fill and clog the waterways. High sediments loading can also cause an increase in the water level of rivers, ponds, and lakes which increase the risk of flooding, while simultaneously reduce their storage capacity. In water treatment plants, sediments are closely related to the total suspended solids and turbidity in water. Sediments carried from the result of logging activities contain muds, silts, branches and twigs, nutrients from the soil, clay, etc. Comprehensive pre-treatment is required before the raw water supply can be pumped to the water treatment plants. The pre-treatment can include the processes of macro-screening of large solid objects, pre-sedimentation of the sediments and silts, and micro-screening of fine particles. Without pre-treatment, the sediments transported through the pumps and piping system can damage the system, which will only incur additional costs to the treatment plants.

One way to regulate the number of sediments in water bodies is through soil conservation practice. In this practice, several methods can be practised to reduce the erosion of soils especially on slope-lands near the water catchment areas. However, this practice may only work if illegal activities which cause soil erosion can be controlled and prevented. Some practices that can be implemented are slope shaping to reduce slope inclination thereby controlling the running water acceleration down the area, and also planting cover plants on exposed soil so as to reduce soil losses due to the interaction between plant roots with the soil.

Droughts and Floods

Most tropical countries typically experience only two seasons: the dry and wet seasons. In Sabah, the climate is largely influenced by the Southwest and Northeast monsoons which occur in the months of April to September and October to March respectively. The dry season generally occurs during the months of May until September, whereby the wet season mainly happens from November to March. This is attributed by the Southwest monsoon which carries drier weather and less rainfall, compared to the Northeast monsoon which brings about more frequent rainfalls.

During these seasons, there are significant differences observed in terms of overall raw water quality, quantity, and composition. In the wet season, heavy rainfall caused floods especially in lower ground areas which span from days or prolonged to weeks. The water quality during this time experiences high sediments and silts. Fertilizers and leachate leaching into the water supply can also be observed. For the period of drought, most water resources experiences drop in water turbidity and supposedly lower water discharge. However, the water supply in Sabah is rarely seen to reach the minimum threshold alert due to the copious amount of available water.

Based on a study by Hrdinka et al. (2012), there is a significant correlation between the flood and drought on the water quality for certain parameters. In this research, the water quality during the 2006 flood in Luznice River at Bechyne, Czech Republic is analyzed at the beginning of the flood, and 4 and 9 days after the flood. These data were then compared to the pre- and post- water quality that was obtained approximately 1 month prior and after the flood. The parameters that were tested in this study are the river discharge, chemical oxygen demand (COD_{Cr}), nitrates (NO_3^-), specific conductivity, organically bound halogens (AOX), lead (Pb^{2+}), suspended solids, faecal coliform bacteria, and zinc (Zn^{2+}) contents. The trends of the analysis results for the selected parameters are shown in Figure 6. Based on the results of the analysis, the river discharge peaked during the flood periods by manifolds. Whilst the dissolved solids concentration (specific conductivity) is gradually decreased as a result of dilution as the flood progressed. It shows that the concentration of dissolved solids per volume of water is decreased, yet the large discharge amount is exceptionally higher than usual. Besides, the concentration of suspended solids, chemical oxygen demand, organically bound halogens, nitrates, lead and zinc is also increased during the beginning of flood and gradually decreased. This is attributed by the surface runoff and other sources of pollution that is carried to the river system. Microbiological indicators suggest that the faecal coliform bacteria due to washout from swage point source or agricultural areas.

On the drought impact assessment, the study was conducted on the drought impact in 2003 in Skalice River at Varvazov, Czech Republic. The result is compared with the result of analysis for 2 years before and 2 years after the drought. The data was taken from June to September during each of these 5 years periods and then represented as the arithmetic averages of the results. The selected parameters for this study are the river discharge, nitrates, biological oxygen demand, (BOD_5), water temperature, orthophosphates, suspended solids, dissolved oxygen, ammonium nitrates, and dissolved solids. Figure 7 illustrates the results obtained for these specific parameters during the research period. The results indicate that during the 2003 drought, the river discharge is very much decreased due to low water velocity. On top of that, the water temperature rose by several degree Celsius as a result of the warming weather. Other parameters that are significantly affected are a decrease of dissolved oxygen in the water which is closely related to the higher water temperature and accelerated phyto- and zooplankton growth, decrease in nitrates and suspended solids concentrations, and increased in dissolved solids concentration as a result of the undiluted factor due to small volume of available water. Furthermore, the orthophosphate concentration is increased since the water resource experiences increased evaporation and point pollution, yet this is not observed for the ammonium nitrates concentration.

Direct solutions for flood and drought prevention are difficult to be devised as these conditions are non-controllable factors that give direct effect to the water supply quality and quantity. Nonetheless, mitigation steps can be taken to lessen the adverse impacts of floods and droughts. One of the effective steps is to develop the drought and flood control master plan to the water management. The master plan should cover both short-term and long-term plans which aim to prevent and minimize the effects of floods and droughts for both media to large scale scenarios. Besides, the plan should devise new or improve existing mitigation and warning systems that deal with these issues, while continuously improving the sustainable management of water, forest, and land.

Water Treatment Plants: Downstream

Sludge Accumulation

All water treatment plants produce the main product of treated water as well as the by-product of residues known as the water treatment sludge. One major problem encountered by many conventional water treatment plants in Sabah is the enormous amount of sludge accumulation from the coagulation and sedimentation processes. Water treatment sludge chemically contains silica, alum, lime, ferric oxide, and trace heavy metals precipitation and moisture. Besides, the water treatment sludge contains a mixture of 60% of fine sand (150 – 75 µm), 24% of silt, and 16% of clay (Ahmad et al., 2016). The sludge resulting from the conventional treatment must be properly treated and disposed of since it contains high chemical contents from the treatments. However, the current practices of sludge management are through sludge treatment using lagoon, or directly discharged into water bodies or disposed of in the landfill. This improper sludge management is not an environmentally friendly disposal option and should be addressed accordingly.

Sludge lagoon is a non-mechanical treatment used to store and thicken the water treatment sludge. Generally, there are two types of sludge lagoon which are classified according to its mode of operation, namely the permanent lagoon and dewatering lagoon. The permanent lagoon is used as a final disposal site, whereas the dewatering lagoon is cleaned and reused continuously. The average lagoon is designed

Figure 6. Selected water parameters of Luznice River, Bechyne in 2006 for comparison during a flood (28 March represents the beginning of the flood, 1 April and 6 April represent 4 and 9 days after the flood respectively) with the pre- and post-flood reference periods (20 February and 25 April respectively)
Source: Hrdinka et al., 2012

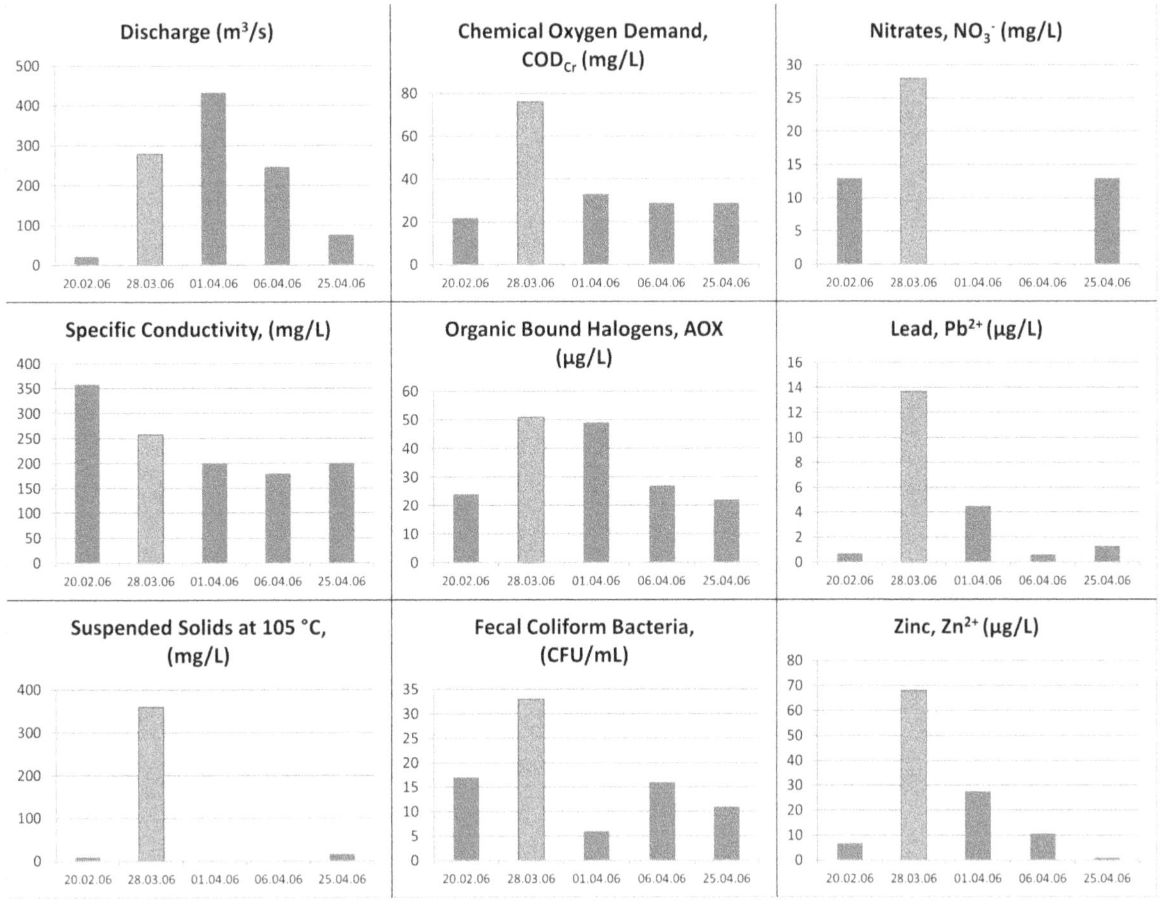

to fit the pre-determined parameters of 3 months filling time and 3 months average drying time, but could vary depending on the plant's operation and sludge loadings.

Water treatment sludge can be classified into lime-softening sludge and alum sludge. The major difference can be seen on the effectiveness of dewatering, at which lime-softening sludge can be dewatered from 30 to 50% (in terms of solid concentration), whereas alum sludge is typically only around 10%. The sludge moisture contents from an alum-based treatment have the properties of viscous, gelatinous and high water content. It is reported that the sludge is difficult to be handled and dewatered since the water is chemically bound to the alum floc. Because of that, most conventional water treatment opt for lagoon treatment.

In the lagoon treatment process, the sludge is discharged into the lagoon until it is filled. Then, another lagoon is employed for the subsequent sludge discharge, whereby the filled lagoon is left to dry until it can be get rid of for final disposal which can take several months, a year, or more. Even after that long retention time, the alum sludge will still hold over 90% of its original water content, which pres-

Figure 7. Selected water parameters of Skalice River, Varvazov for comparison during the 2003 drought with the pre- (2001 and 2002) and post- (2004 and 2005) drought reference periods
Source: Hrdinka et al., 2012

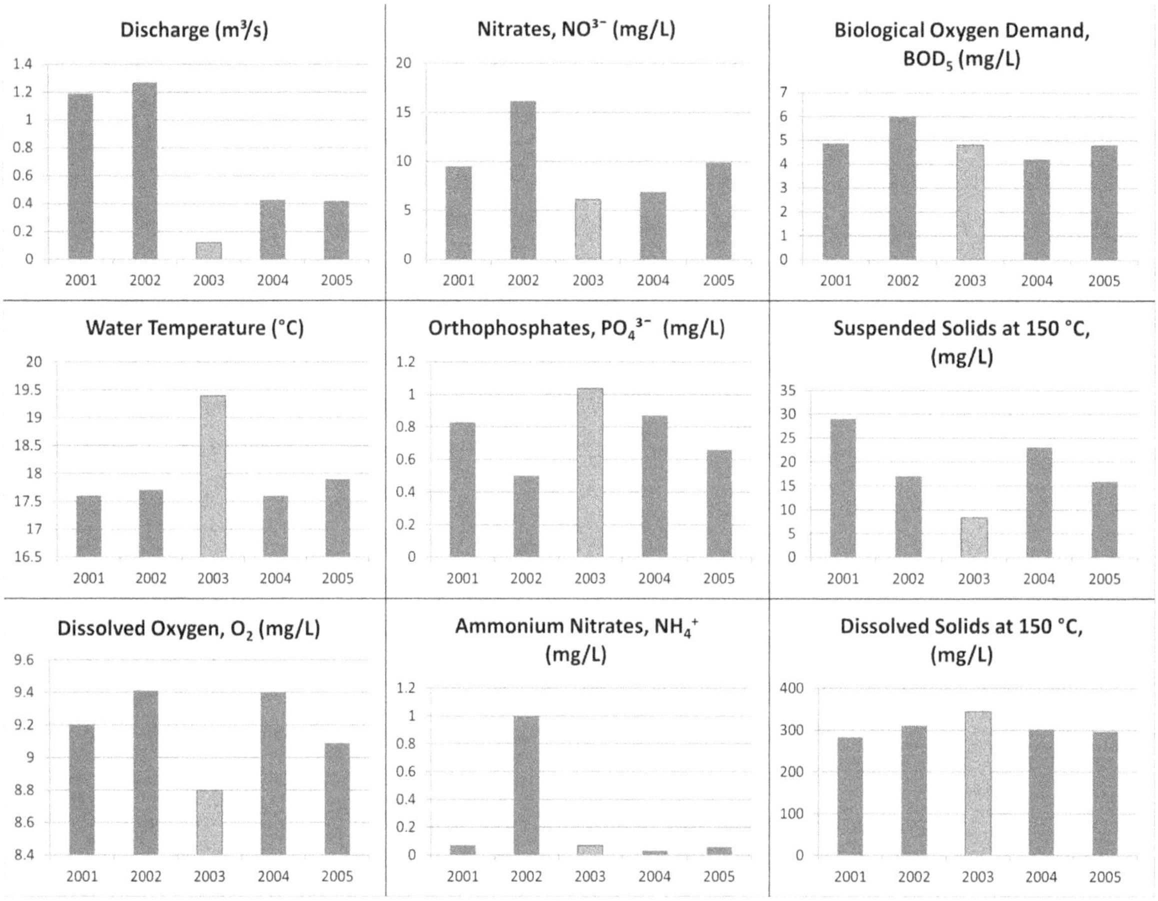

ent difficulties during handling and unsuitable to be disposed of in a landfill (American Water Works Association, 2003). This type of process requires very large land areas to accommodate the high sludge content for the high retention time period. Since there is large land available in Sabah, this practice has become a major preference for its simplicity and cost-effectiveness.

Sludge Management using Dewatering Membrane Technology

Growing environmental concern on the safe disposal of sludge has put significant pressure on the water authorities. Conventional water treatment plants produce huge amounts of sludge which needs to be safely administered. Dewatering is a typical process used to thicken the sludge by removing some of the water content to about 20% of solids before it is disposed of. Currently, the acceptable way of dewatered sludge disposal is through sanitary landfill. Dewatering can be performed mechanically using the principal methods of filter press or centrifuge, or also through non-mechanical means such as drying

Table 8. Advantages and disadvantages of mechanical and non-mechanical dewatering processes

Mechanical Dewatering		Non-mechanical Dewatering	
Advantages	**Disadvantages**	**Advantages**	**Disadvantages**
• High capacity • Smaller footprint • Shorter process duration • Can be fully automated	• High capital and operational costs • Requires expertise for design, handling and maintenance • Produced effluent water and dewatered sludge that is septic and requires further treatment • Requires constant power supply	• High capacity • Low capital and operational costs • Low energy consumption • Low operator skill • Low attention requirement	• Higher footprint • Higher process time • The requirement for prior sludge stabilization • Effect of climate

bed. Table 8 describes some advantages and disadvantages of both the mechanical and non-mechanical dewatering processes.

An alternative and efficient dewatering process, as opposed to these traditional methods, are preferred which eliminates the needs of high pressure, centrifuging, or drying beds. This can be achieved by incorporating membrane technology into the process since the membrane is an excellent separation tool. Membranes in dewatering units act as a medium which retains the solids at one side while allowing water to move across it and drained away. The horizontal membrane works best with regards to the dewatering principle, as water is allowed to be removed by gravity while solids are retained on top of the membrane. Horizontal configuration can also reduce the operating pressure required to dewater the sludge as gravitational force is employed in the operation. The resulting cake can be further pressed to remove excess moisture before it is being disposed of in a landfill.

REFERENCES

Ahmad, T., Ahmad, K., & Alam, M. (2016). Characterization of water treatment plant's sludge and its safe disposal options. *Procedia Environmental Sciences*, *35*, 950–955. doi:10.1016/j.proenv.2016.07.088

American Water Works Association. (2003). *Water Treatment*. USA: American Water Works Association.

Aris, A. Z., Wan, Y. L., Praveena, S. M., Yusoff, M. K., Ramli, M. F., & Juahir, H. (2014). Water quality status of selected rivers in Kota Marudu, Sabah, Malaysia and its suitability for usage. *Sains Malaysiana*, *43*(3), 377–388.

Caldecott, J. (2008). *Water: Life in Every Drop*. London, UK: Virgin Books.

Chaw, V. V., Harun, S., Kueh, B. H., Wong, B. H. A., & Fikri, A. H. (2018). Aquatic insects and water quality study at Kimanis River, Crocker Range National Park, Sabah, Malaysia. *Journal of Tropical Biology & Conservation*, *15*, 223–245.

Clark, R. M. (1982). Cost Estimating for Conventional Water Treatment. *Journal of Environmental Engineering*, *180*(5), 819–834.

Douglas, I., Greer, T., Bidin, K., & Spilsbury, M. (1993). Impacts of rainforest logging on river systems and communities in Malaysia and Kalimantan. *Global Ecology and Biogeography Letters*, *3*(4/6), 245–252. doi:10.2307/2997773

Hrdinka, T., Novicky, O., Hanslik, E., & Rieder, M. (2012). Possible impacts of floods and droughts on water quality. *Journal of Hydro-environment Research*, *6*(2), 145–150. doi:10.1016/j.jher.2012.01.008

Kasran, B. (1996). Bedload sediment yield in a logged-over tropical rainforest. *Journal of Tropical Forest Science*, *8*(3), 427–429.

Lee, R., Jessop, P. G., & Champagne, P. (2015). Carbon dioxide pressure-induced coagulation of microalgae. *Philosophical Transactions of the Royal Society A: Mathematical, Physical, and Engineering Sciences*, *373*(2057), 1–8. doi:10.1098/rsta.2015.0016 PMID:26574522

Lee, Y. H., Lee, N. C., Stuebing, R. B., & Omar, M. (2006). The water quality of several oxbow lakes in Sabah, Malaysia and its relation to fish fauna distribution. *The Journal of Biological Sciences*, *6*(2), 365–369. doi:10.3923/jbs.2006.365.369

Lin, S. D., Evans, R. L., & Beuscher, D. B. (1971). *Algal Removal by Alum Coagulation* (Report Investigation 68). Urbana, IL: Illinois State Water Survey.

Tundu, C., Tumbare, M. J., & Onema, J. M. K. (2018). Sedimentation and its impacts/effects on river system and reservoir water quality: Case study of Mazowe Catchment, Zimbabwe. *Proceedings of the International Association of Hydrological Sciences*, *377*, 57–66. doi:10.5194/piahs-377-57-2018

ADDITIONAL READING

Atari, L., Esmaeili, S., Zahedi, A., Mohammadi, M. J., Zahedi, A., & Babaei, A. A. (2017). Removal of heavy metals by conventional water treatment plants using poly aluminium chloride. *Toxin Reviews*, *2018*, 1–8.

Benitez, J., Rodriguez, A., & Malaver, R. (1995). Stabilization and dewatering of Wastewater using hollow fibre membranes. *Water Research*, *29*(10), 2281–2286. doi:10.1016/0043-1354(95)00068-V

Edzwald, J. K. (Ed.). (2011). *Water Quality & Treatment: A Handbook on Drinking Water*. New York, USA: McGraw-Hill.

Mostafe, K. S., Bahareh, G., Elahe, D., & Pegah, D. (2013). Optimization of conventional water treatment plant using dynamic programming. *Toxicology and Industrial Health*, *31*(12), 1078–1086. doi:10.1177/0748233713485891 PMID:23625909

O'Connor, J., O'Connor, T., & Twait, R. (2009). *Water Treatment: Plant Evaluations and Operations*. New Jersey, USA: Wiley. doi:10.1002/9780470431474

Pal, P. (2017). *Industrial Water Treatment Process Technology*. Oxford, UK: Butterworth-Heinemann.

Thomas, G. (1998). *Advances in Water Treatment and Environmental Management*. England: Elsevier.

Tuckwell, S. B., Carroll, B. A., & Joseph, J. B. (1991). A fresh approach to sludge dewatering using tubular membrane pressing technology. In M. K. Turner (Ed.), *Effective Industrial Membrane Processes: Benefits and Opportunities* (pp. 129–141). England: Elsevier. doi:10.1007/978-94-011-3682-2_9

KEY TERMS AND DEFINITIONS

Conventional Water Treatment Plant: Water treatment facility which consists of several unit processes, typically seven to ten-step process to deliver safe drinking water to consumers.

Dewatering: Process of removing water typically from a mass of sludge.

Drinking Water: Water having the quality that is suitable and safe for drinking.

Microalgae: Microscopic-size algae which can be frequently found in freshwater and marine ecosystems.

Potable Water: *See Drinking Water.*

Raw Water Supply: Water resource found in the environment that has not been treated which contains impurities and microbes making it unsuitable for direct consumption, such as rainwater, groundwater, and surface water of river and lake.

Sludge: Viscous mixture of solid and liquid compounds discharged from the water treatment process.

Tropical Country: Country experiencing tropical climate which is situated within the region known as the tropics.

Water Quality: Variables or parameters which limit water use, for examples the chemical, physical, biological, and radiological properties of water.

Chapter 7
Clean Water Production for Isolated Areas

ABSTRACT

The provision of clean water is a global concern that must be critically addressed by all government bodies. However, lack of attention to this matter presents difficulties to many rural regions, especially in the developing countries. Most rural regions around the world experience lack of clean water due to various factors such as shortage of water resources and lack of water treatment facilities due to geographical constraint and scatter of population. Thus, small-scale water treatment system is an attractive technology for clean water production in isolated areas as it only requires a small footprint and it is more cost-effective than conventional water treatment plants. This chapter focuses on the small-scale water treatment systems for each of surface water, groundwater, rainwater, and brackish water resources. Considerations are taken based on the type of water resources and geographical conditions of the rural areas which include flat grounds, hills, and island areas. Besides, rainwater harvesting and treatment is reviewed for individual house application.

INTRODUCTION

With a land area of 73,619 km^2, which is divided into 23 districts, Sabah is sparsely populated. The rural and remote areas in Sabah are focused at the central region of the state, such as at the districts of Keningau, Tambunan, Ranau, Beluran, Nabawan, Tenom, and Tongod, and also at the islands, for example, the Banggi Island, and Mabul Island. The major economic activities in these areas are agriculture and eco-tourism. Due to this, a large quantity of water is required to fuel the socio-economic activities which are the backbones of trade and industry in these regions. Table 1 shows some general comparison of the major water sources in Sabah rural areas. The major issue identified is the needs of a water treatment facility that is crucial to ensure that the water is free from contaminants and bacteria so that it is safe for consumption.

DOI: 10.4018/978-1-7998-2645-3.ch007

Table 1. A general comparison between the water sources in Sabah rural areas

Surface Water	Groundwater	Rainwater
Water scarcity during drought; contains suspended solids and microbes	The necessity of a deep-dug well for groundwater collecting; contains odour, colour, and microbes	Inconsistent supply especially during drought; water may be contaminated from atmospheric pollutants
Only accounts for ~10% of freshwater availability	[a] Accounts for ~90% of freshwater availability	The average rainfall of 3,000 mm annually

Source: [a] (Mishra & Dubey, 2015)

The rural areas of Sabah are highly associated with scattered low-density population due to its geographical condition that is mountainous and hilly, densely forested and steep hilly terrain. Large distance between populated areas combined with hilly terrains makes the development of a water supply network system a highly costly project. Hilly terrain requires longer pipelines and even greater pumping energy requirement. Even though the areas are rich in water resources, without a proper water treatment system, the populations in the majority of the rural regions have a lack or no access to treated water. The rural communities are mainly dependent on the rainwater harvesting and surface water supplied via gravity pipe as their major water resources. Part of the communities also uses groundwater from hand-dug well for their everyday usage without proper treatment. Thus, there is a need for the building of groundwork for water treatment that can provide the locals with sufficiently clean water in order to allow them to obtain the basic necessity that is potable water.

Hence, the suitable treatments will be discussed in the subsequent sections which cover the rainwater, surface water and groundwater which has been initially identified as the major water resources available. Yet, several systems are discussed that is practical for each type of water resources since not all rural areas will have all three resources in abundance and a single treatment system may not be suitable for all type of water resources. The main focus of the water system for rural areas would be small scale water treatment system that can combat the issues associated as previously discussed. Below are some typical examples of areas in Sabah associated with clean water problems.

Banggi Island

Banggi Island is the largest island in Malaysia with approximately 20,000 population. It is situated at the northeast of Kudat, whereby most of its population work as fisherman or farmers. Located at a remote region of Sabah, it is far from receiving development, inclusive of the water supply system. The local residents have to rely on surface water from rivers, rainwater, and groundwater from a hand-dug well for their everyday usage. Almost 70% of the locals have lack access to supplied clean water. The lack of access to piped water gives a negative impact on the general hygiene and sanitation system like toilet and solid waste disposal. If the usage of contaminated water (from the surface and/or groundwater) continues, sickness outburst among the locals could occur. Though a water treatment plant is recently constructed at the island, not all of the locals are able to experience and benefit from the infrastructure because of high costs for meter installation and pipeline taping to the existing main pipeline distribution.

Mabul Island

Mabul Island is a small island located at Semporna, Sabah, and is a densely populated region as a popular tourist attraction. The residents of the island have improved water supply through the reverse osmosis treatment process, yet the treatment is not satisfactory enough as it yields significantly brackish water. Thus, the locals are forced to depend on the conventional water supply, i.e. the rainwater for their everyday usage.

Ranau

Ranau is located in the west coast division of Sabah. Due to its hilly geographic and cold yearly average temperature, Ranau is the main tourist attraction in Sabah and supplier for agriculture products. Thus, there is a high water demand in the district, yet the supplied water is of low quality with significant turbidity. The inefficiency in operation of the water treatment plants in Ranau using water intake from river branches contaminated with chemical residue from closed Copper Mine in Mamut may lead to severe health risk (Jopony & Tongkul, 2009; Mohammad Ali et al., 2015). Besides, the substantial agriculture activities which use a lot of fertilizers, herbicides and pesticides contribute to the organic and inorganic pollutions, aside from contaminating the water supply through surface runoff and soil seepage.

The water supply in Ranau experienced more complication especially during the rainy season, as heavy water flows through surface runoff into the rivers, carrying mud and debris. This causes an increase in sedimentation loading of the river water. At the intake point, the Liwagu River, which is the main water supply in Ranau will be too flooded by mud, exceeding the allowable level the water treatment plant able to cater. This has led to a water treatment plant shutting down, and no clean pipe water to be distributed to the locals. Furthermore, debris from upstream such as leaves, fallen branches and trees will block water inlet at the intake point further worsen the condition.

Beaufort

Beaufort has experienced frequent water inconsistency despite being one of the districts in Sabah with the highest annual rainfall and having stable water level at the intake from Padas River. The root cause is due to mismanagement of water supply treatment and distribution. With a population of 66,000 censuses 2010, the district has perceived a rapid annual increase of population which consequently generated vast demand for clean water supply. The current water treatment plant is unable to accommodate the continually increasing demand which is presently at 40 million litres per day. The rapid development at upstream of Padas River as eco-tourism has brought to the increase in soil erosion and sedimentation, particularly during heavy rainfall. Highly turbid water influences the efficiency of the water treatment plant, resulting in poor water being produced. This contributes to the dissatisfaction among the locals who are obliged to pay for the charges of water utility bill without receiving any clean water.

SMALL SCALE WATER TREATMENT SYSTEM

A small scale water treatment system treats and distributes water near its points of generation or need and the piping system involved are comparatively shorter. Over the years, the advantages and disadvantages of both centralized and small scale water service system have been discussed by many researchers in terms of technical application and economics. Points highlighted are as the following: (1) small scale system allows flexibility for providing solutions alongside changing condition in rapidly growing capitals, (2) provision of long-distance pipelines to rural and suburban areas in a centralized system are prohibitively expensive (3) Small scale system remove the requirement of pumping station in challenging topographical condition, (4) Localized nature of a small scale system reduce piping infrastructure and decrease pipe length where leakage occurs, (5) cost of operation and maintenance of decentralized system is marginally reduced due to smaller plant scale.

Due to the overwhelming benefits of the small scale system for water service, researchers have shown great interest for a small scale water supply. However, most studies focus not on substituting the existing water supply with a small scale system but rather to complement it. An on-site greywater reuse small scale system by Piratla & Goverdhanam (2015) has found to improve the reliability of water supply up to 17% and able to cater water demand upsurge of 25%. System analysis was conducted by putting the system under various probable issues such as system age, pipeline roughness, treatment efficiency and allowable use of reclaimed water. However, the study was limited by the lack of consideration for the failure of on-site greywater reuse plumbing, life cycle economic, and energy considerations. A life-cycle analysis of a small scale rainwater harvesting (RWH) from a roof runoff by Yan et al. (2017) later found that the system performs poorly compared to a centralized system in terms of environmental aspect, contributed to the high consumption of energy. The outcome was somewhat expected due to the difference in throughput between a small scale and centralized system. They suggested that the application of a renewable source of energy will improve the reliability of the system.

Small scale water service seems to have a positive outcome in developing countries. A small-scale small scale water system implemented in a rural area of Brazil that was intended to reduce the dependency to centralized government-supplied water and run by the local communities. Access rate increase averaged at 6% yearly in ten years of implementation of the system. The involvement of locals is highly important in providing water access in a rural area which also engages a sense of responsibility and ownership among the communities. The management of small scale water treatment was recorded to be 20% more efficient than the centralized ones. This could be attributed to the independence of the management without financial assistance from the government.

In that note, a shift from centralized to a small scale system for potable water supply in Sabah's rural areas are seen to be highly beneficial. The centralized system has many advantages but it may be impractical to be implemented in rural areas due to uneconomical aspects when perceived from the prospects of consumer demands and system installation. This system usually does not reach the poor and remote populations. Thus, the best method that can be used here is the small scale water system which refers to a small-scale treatment and distribution to improve access to drinking water. It would mean the elimination of issues arising from the unique topographic and geographic distance of populated areas. Smaller water treatment plants that are closer to point of needs are viable due to abundant resources of water in the form of groundwater, rainwater and surface water. Energy requirement would be solved by the application of renewable energy options such as solar energy which has proved to be promising due to the equatorial climate of the country.

Sections 7-3 – 7-6 demonstrates small scale water treatment plant that can be implemented for the rural areas in Sabah. These treatment systems can be applied in small remote villages or clustered housing areas situated on flat grounds, hilly areas, or islands that are far from receiving clean piped water. The small scale treatment system can be managed and operated by the locals due to its ease of operation and believed to be able to supply potable water for everyday usage. The system encompassed several unit operations and shorter piping network and pumping requirements as compared to a centralized system.

SMALL SCALE WATER TREATMENT PLANT FOR FLAT GROUND

Main Source of Water 1: Surface Water

The main water intake in Sabah principally comes from river water, but a small percentage of the population in rural areas depends on the lake to assist their day-to-day water necessity. Comparison of the surface water quality against the Malaysian Raw Water Standard (RWS) for potable water is imperative to ensure the safety of consumers. The RWS enforced is comparable to the standard imposed by the World Health Organization (WHO). Table 2 shows the quality of surface water in the rural location around Sabah. Compared to the RWS, the most parameter is within the allowable limit with exception to the COD, Ammonia-Nitrogen and total suspended solids (TSS).

Evaluation of surface water quality is challenging due to the various pollution sources affecting the water quality may it be from a point source and non-point source. This is especially pressing due to increasing socioeconomic activities that generate polluting by-products. A common example of point source pollution is sewage discharge. Untreated or inadequately treated sewage is a major contributor to surface water pollution. Sewage contamination in surface water will lead to nutrient loading that causes eutrophication which in turn will reduce dissolved oxygen. Sewage polluted water is a risk to human health since it contains high in bacteria and pathogenic content.

In suburban and rural areas where agricultural activities are more prominent, pesticide and fertilizer usage will eventually leach into the water bodies via soil leaching or surface run-off which is considered as a non-point pollution source. Apart from the natural occurrence, fertilizer leaching is the main source of ammonia and nitrogen in the water. This pollutant leaches into the water bodies as a non-point pollution source. Furthermore, the quality of surface water varies depending on the precipitation rate. Although higher precipitation causes dilution of contaminant through higher water flow, it increases the suspended solids in water through disturbance of river deposits and accumulation of sediments from erosion of soil surface.

Sabah water scarcity occurs due to two main reason; (i) polluted waters and (ii) high development cost. Surface water is highly affected by anthropogenic emission and weathering. Water pollution in Sabah is categorized based on the location of rural, urban and semi-urban area. In a rural area, the main pollutants are sediments from soil erosion that comes from cleared land for logging, plantation or mining, agricultural residue such as fertilizer nutrients and heavy metals from mining activities. In urban and semi-urban areas, main pollutant contributors are industrial and domestic wastewater discharge. Cleared land for construction sites and landfill area contributes to the erosion of sediments to the nearest water body (NRO, 1994). No proper waste management by housing areas also contributes to heavy pollution of waters causing drifted garbage and sewage discharge to water bodies.

Quality of surface water particularly rivers should conform to the Malaysian Interim National Water Quality Standard (INWQS). General indicator for water bodies quality index is the level of biological oxygen demand, chemical oxygen demand, dissolved oxygen, pH, suspended solids, and ammoniacal nitrogen.

Table 2. Physicochemical characteristics of Sabah rural surface water against Malaysian Raw Water Standards (RWS)

Location	Beluran	Beluran	Beluran	Beluran	Sukau	Sukau	Sukau	Tambunan	Tambunan	Beaufort	RWS[e]
Main River	Sugud River[a]				Lower Kinabatangan Catchment[b]			Liwagu River[c]		Padas River[d]	
Tributaries / Lake	Suguta River	Sabang River	Wansayan River	Kelipatan River	Resang River	Lumun River	Kalinanap River	Nukakatan River	Mesangoh River	Luangan Rompong Lake	
pH	7.01	7.08	6.82	7.02	5.3 - 6.5	5.2 - 6.7	6.6 - 7.0	6.34 - 8.3	7.79 - 8.18	6.0 - 8.5	5.5 - 9.0
DO (mg/L)	4.93	1.93	3.66	3.2	4.3 - 6.2	3.5 - 5.6	1.9 - 4.1	6.19 - 6.81	6.9 - 7.79	3.9 - 11.4	-
Salinity	0.07	27.29	0.11	2.22	0.05 - 0.1	0.01-0.03	0.03-0.04				-
Conductivity (μS/cm)	155.28	45,574.65	180.3	4,373.45	110.8 - 236.5	52.9 - 175.3	76.9 - 180..9	0.05 - 0.09	0.05 - 0..06	71.8 - 86.2	-
BOD (mg/L)	3.19	1.27	4.01	2.49	1.3 - 2.1	2.9 - 3.2	3.1 - 3.6	0.38 - 0.59	0.52 - 0.65	n.m*	6
COD (mg/L)	18.08	751.15	21.03	24.38	53.8 - 100.0	36.8 - 51.4	35.1 - 45.8	0.67 - 3.4	3.5 - 4.8	n.m	10
Total Coliform (CFU/100 mL)	3,594.88	1,108.28	1,753.25	3,029.53	n.m	n.m	n.m	17.0 - 169	33 - 137	n.m	5,000
Ammonia-Nitrogen (mg/L)	0.52	98.44	0.7	4.75	n.m	n.m	n.m	0.09 - 0.27	0.03 - 0.32	n.m	1.5
TSS (mg/L)	n.m	n.m	n.m	n.m	33.0 - 42.0	49 - 75	49 - 96	0.1 - 4.0	1.0 - 3.4	n.m	-

*n.m: not measured

Source: [a](Harun and Fikri, 2016); [b](Harun et al., 2014); [c](Cleophas et al., 2013); [d](Heng et al., 2006); [e]RWS is based on standard enforced by the Ministry of Health Malaysia from (MMOH, 2010)

Water Treatment System 1: Surface Water

The proposed scheme of the small scale water treatment system for surface water supply in Figure 1 is driven by the necessity for an effective solution of the water treatment system at areas with an abundant supply of source water.

In this system, water source from the intake point is first screened from large solid objects such as rocks, tree barks, leaves, and plastic wastes. Then, the screened water is pumped to an overhead tank before it is being treated in the up-flow sand filtration system. For this proposed scheme, the water source is taken from a river or lake which is abundant in Sabah. The main purpose of the overhead tank is to create water gravity flow in order to ensure a natural downward flow which can be designed to give enough pressure for the filtration operation without using pumps. From the overhead tank, water is channelled to an up-flow sand filtration system to filter out the water from any suspended and colloidal micro-solids present in the surface water.

Prior to up-flow sand filtration, chemical dosing is added to the water depending on the characteristic of raw water such as turbidity, pH etc. The typical kinds of chemicals used are lime, alum, and chlorine. Lime, or calcium hydroxide, $Ca(OH)_2$, is added to adjust pH/alkalinity of water to the desired value. In water treatment, the pH/alkalinity of water throughout the process will affect the dosing and effectiveness of other chemicals used. For example, the coagulation/flocculation process is very dependent on water alkalinity, as the alkalinity in water will be consumed by the formation of flocs. Thus, the right dose of lime must be present in order to offset the loss of alkalinity to flocs formation. Lime will react with alum, thus direct mixing should be prevented.

Figure 1. Small scale water treatment plant for raw water from river/lake

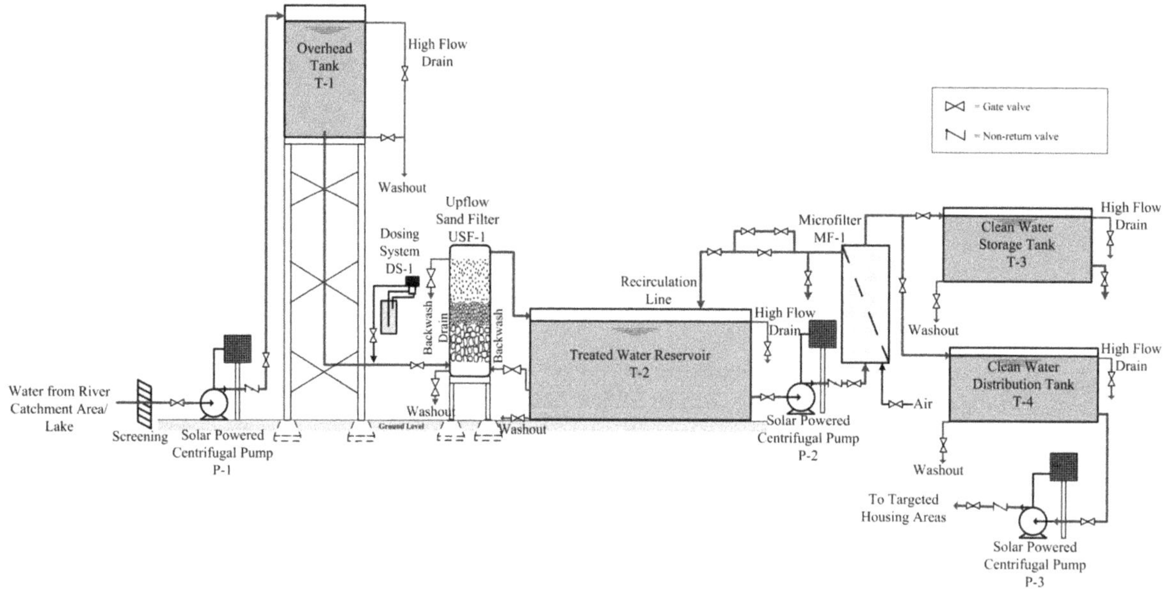

Alum, aluminium sulfate, $Al_2(SO_4)_3$, is used as a coagulant to clump suspended/colloidal particles together and settle out of water. Alum is one of the most commonly used coagulants and is relatively cheap than other types of coagulants. In order for alum to be effective, the pH of raw water must be between 6 and 8. Moreover, there must be enough alkalinity present, as alum consumes alkalinity during the coagulation/flocculation reactions. Thus, there might be some adjustment needed for the pH and alkalinity of water during the treatment.

Chlorine (typically in the form of sodium hypochlorite, NaOCl or chlorine gas, Cl_2) is used as a disinfectant, to kill potentially harmful microorganisms and pathogens in water. NaOCl, commonly known as liquid bleach, is the liquid form of chlorine. It is easy to use and has a lower health hazard compared to other chlorine agents. Chlorine gas has a yellowish-green colour and a distinctive odour. It is the cheapest method for chemical disinfection, but is hazardous to the respiratory system and has a risk of explosion when stored as compressed gas. Besides being used as a disinfectant, chlorine can be used as an oxidizing agent, to oxidize undesirable contaminants such as iron, manganese and arsenic and allow them to precipitate which can then be removed by the filtration process; it also helps to control odour, colour, and prevent the growth of algae in the tank.

Up-flow sand filtration is a method involving granular media filtration which consists of layers of pebbles, gravels and sands which has been extensively studied by the membrane research group, Universiti Malaysia Sabah. The combination of up-flow water and deep filter bed can give high filtration performance. The quality of water filtered is influenced by the amount of solid loading in the influent. Nonetheless, the solids filtered out can be easily cleaned and removed to sustain continuous operation without jeopardizing the quality of effluent. The up-flow sand filter has minimum costs for operation and maintenance as it does not involve with any moving parts, screening, and level controllers. It requires simple construction comprising of the tank and filter media, whereby the tank can be made from fibreglass, concrete or stainless steel, with the utilization of deep bed of up to 3 m.

The filtered water is then collected and stored in the treated water reservoir. From this point, the water is further treated in a microfilter system using the principle of membrane separation. The microfiltration system can ensure that the micro-particles and bacteria are screened out from the water so as to ensure its cleanliness and safety before used as potable water. The clean water is then transferred to the clean water storage tank which served as the clean water intake point that is accessible to the rural communities to accommodate their basic necessities. For the direct distribution to individual housing areas, water from the clean water distribution tank is directly routed to individual houses. This can be implemented to the targeted houses but will require extra installation of pumps and piping at added costs which need to be absorbed by the locals.

Main Source of Water 2: Groundwater

The global freshwater consist of 90% of groundwater and only a fraction of it is from surface water which totals up to 442,000 km^3 of water (Tarbuck & Lutgens, 2015). Utilization of groundwater is known to have its advantages compared to other sources of freshwater due to its lower vulnerability to the prolonged dry season. It requires minimum treatment as it is not directly affected by surface pollution and generally closer to point of demand, therefore extraction, remediation and distribution will require lower capital and operational cost. Regular disruption of water supply and rationing could be avoided as dependence in piped water is reduced, and in turn, reducing the water bill.

In Malaysia, freshwater supply from groundwater accounts for less than 10% of the total water usage (Manap et al., 2013). The main application of groundwater is for domestic, industrial and irrigation which is limited to certain areas with surface water limitation such as Kelantan, Klang Valley in Selangor, Kedah, Perlis and Melaka (Chu, 2004). The state of Kelantan has predominantly utilizing groundwater since the early 1990s as the main water supply source with approximately 70% of its freshwater uses (Abd Razak & Abd Karim, 2009; Suratman, 2004). In Sabah, uses of groundwater are mainly in the coastal region such as Sandakan, Kota Belud and Kuala Penyu and some isolated villages where the supply of piped water is limited or in existence.

Malaysia has several water basins that are highly potential for groundwater extraction yet comprehensive study still to be conducted. Understanding the mineral condition in the soil is important as it controls the type of groundwater lay beneath the soil layers. Generally, groundwater resources in Malaysia are from five types of aquifer: (i) shallow alluvial, (ii) deep alluvial aquifer, (iii) hard rock aquifer, (iv) peat aquifer and (v) island aquifer which are extracted through either dug well, driven well or drilled well. The aquifer is a layer of rocks situated underground that has water-bearing properties due to its permeable characteristics that let fluids pass through, such as sand, gravel, and sandstones. Alluvial aquifer with sand and gravel type are generally the most productive and can yield 50 to 100 m^3 per hour per well. In the hard rock aquifer category, limestone can produce 50 m^3 per hour per well is the most productive, followed by sandstone and volcanic rock that can produce 30 m^3 per hour per well. The least productive aquifers are igneous rock with m^3 per hour per well (Abd Razak & Abd Karim, 2009).

Sabah utilization of groundwater is far low compared to Peninsular region and Sarawak state. Studies conducted on the evaluation of geological formation in Sabah for groundwater potential has found that confined and unconfined aquifer can be found in several areas in Sabah. Several potential aquifers within the sandstone layers have been discovered in three different formations located in Lahad Datu i.e: Lower Ganduman, Togopi and Upper Ganduman Formation and Sandakan Formation in Sandakan area (Abd Razak & Abd Karim, 2009; Saleh & Samsudin, 2013). On the western part of Sabah, the Crocker

Formation provides sandstone layers and Quartenary Alluvium that is a significant groundwater reservoir (Faisal et al., 1995). In coastal areas, quaternary and recent alluvium give a small yield but sufficient for supplies in isolated villages (Abd Razak & Abd Karim, 2009). Islands in Sabah have high dependencies on groundwater due to lack of surface water resource, which is applicable because most of these islands have a high potential for groundwater extraction. Manukan Island as an example is underlain by Quaternary alluvium (Jasin & Tating, 1991), which is unlithified loose sediments and have great water storing capacity, depending on the thickness (Aris et al., 2007).

Study on groundwater quality is limited for Sabah and existing study are more focused towards island areas mainly due to the high dependence of groundwater in the islands. Based on studies conducted in Sabah, most physicochemical parameters for groundwater (Table 3) samples fall within the allowable limit in the INWQS, with a few parameters exceeding the limit such as Total Dissolved Solids (TDS), SO_4^{2-}, Na^+ ion and Cl^- cation. Turbidity and TDS increased could be affected by seepage of from industrial, agricultural and residential runoff and chloride content may be due to contamination from seawater which sometimes due to overpumping. Other pollution of groundwater is from leakage of landfill leachate and sewerage system. The geological characteristic of specific location may contribute to these constituents, for example, SO_4^{2-} which arises from Gypsum originated from rocks and soils. Heavy metals also occur from the geological formation or may be derived from anthropogenic pollution.

Previous studies have shown that groundwater as a main or supplementary source for potable water in Sabah is highly potential not only in islands but also in the mainland. However, groundwater in Malaysia, in general, is underdeveloped mainly due to the failure to recognize the potential of groundwater resources and under-appreciation of groundwater especially because Malaysia is blessed with an abundance of surface water. Limited technical knowledge on groundwater in general as well as local expertise in the field hinders its development. There is also no current applied specific policy on groundwater resources and utilization (NRO, 1994).

Water Treatment System 2: Groundwater

The proposed scheme of the small scale water treatment system in Figure 2 is designed to satisfy the necessity for an efficient, and energy-saving water treatment system specifically for the rural regions. The system poses a similar design and working principle with the small scale water treatment in Section 7-3.2 (Figure 1). It displays the concepts of small scale water treatment plant that can cater for up to 20 house clusters in a day which operates on day-time using the solar-powered system. Thus, this section would mainly focus on the construction of tube well which will become the raw water resources for the water treatment system.

This scheme looks for the utilization of groundwater as the main water resource. The digging of a tube well is particularly vital as the first step for the construction of the system. Typically, a well is constructed to the depth of range of 30 – 100 ft. Placement of screen functions to block out stones, pebbles and soils, while water flows across the screen and collected inside the wall cavity. A submersible pump is utilized for pumping the water out of the well and into the overhead tank, which is powered by energy derived from solar. Generally, the tube well can be drilled at diameter 3 – 4 inch before pipes are inserted and cemented which serves as the tube well cast. Additionally, the well needs to be sterilized before it can be used which can be accomplished in several steps as follow: (1) Pour liquid chlorine (disinfectant) to the groundwater source inside the tube hole and leave for 48 hours; (2) Switch on pump and withdraw groundwater for 12 hours.

The purpose of disinfection is to kill bacteria that contribute to unpleasant water odour and taste. Well, water is generally known to naturally possess distinct rotten-egg-like smell that comes from hydrogen

Table 3. Physicochemical characteristics of groundwater around Sabah against Malaysian Raw Water Standards (RWS)

Location	Tiga Island [a]	Seligaan Island [b]	Bakkugan Kechil Island [b]	Manukan Island [c]	Sahabat Plantation, Lahad Datu [d]	RWS
Underlying Rock	Mudstone, Sandstone and Conglomerate	Carbonate Rock, Mudstone and Quartenary Alluvium		Quaternary Alluvium, Sandstone and Shale	Sandstones and Claystone Interbeds	
Temperature (°C)	26.7 - 29.8	n.m	n.m	n.m	n.m	-
pH	6.84 - 7.35	7.2 - 7.7	7.4 - 7.7	7.2 - 7.96	7.1	5.5 -9.0
DO (mg/L)	1.08 - 6.12	1.5 - 3.6	1.4 - 1.9	-	0.08	-
Salinity (ppt)	0.1 - 0.5	n.m	n.m	0.38 - 1.13	n.m	-
Conductivity (μS/cm)	329- 1005	1.3 - 4.9	0.3 - 0.5	1663 - 8703	n.m	-
TDS (mg/L)	164 – 502	n.m*	n.m	336.67 - 6564	674.5	1500
Total Hardness	n.m	n.m	n.m	-	351.5	500
K^+	5.81 - 8.89	n.m	n.m	10.0 - 71.0	1.72	-
Na^+	0.90 - 7.53	n.m	n.m	45 - 2393	20.62	200
Mg^{2+}	1.70 - 3.43	n.m	n.m	6.67 - 194	29.59	150
Ca^{2+}	21.30 - 51.87	n.m	n.m	197 - 415	72.63	-
HCO_3^-	119.0 – 330	318 - 322	367 - 408	284 - 512	-	-
Cl^-	50.48 - 489.85	53 - 194	43374	359 - 2674	15.09	-
SO_4^{2-}	3.0 - 82.0	n.m	n.m	80 - 490	178	250
Fe Total	n.m	n.m	n.m	n.m	5.87	1
Mn	n.m	n.m	n.m	n.m	0.22	0.2
Si	n.m	n.m	n.m	n.m	56.87	-
Cu	n.m	n.m	n.m	n.m	0.01	1
Pb	n.m	n.m	n.m	n.m	n.d.	0.05
As	n.m	n.m	n.m	n.m	n.d.	0.01
CO_3^{2-}	n.m	n.m	n.m	n.m	2.79	-
PO_4^{3-}	n.m	n.m	n.m	n.m	1.35	-
NH_3-N	n.m	n.m	n.m	n.m	0.7	1.5
NO_2	n.m	n.m	n.m	n.m	0.03	-
NO_3^-	0.8 - 3.70	n.m	n.m	n.m	2.33	10
CN	n.m	n.m	n.m	n.m	n.d**	0.07
F	n.m	n.m	n.m	n.m	0.49	-

*n.m: not measure, **n.d: not detected

Source: [a] (Lin et al., 2009); [b] (Abdullah & Musta, 1999); [c](Aris et al., 2007; Praveena et al., 2010); [d] (Hing, 1994); [e] RWS is based on standard enforced by the Ministry of Health Malaysia from (MMOH, 2010)

Figure 2. Small scale water treatment plant for raw water from groundwater

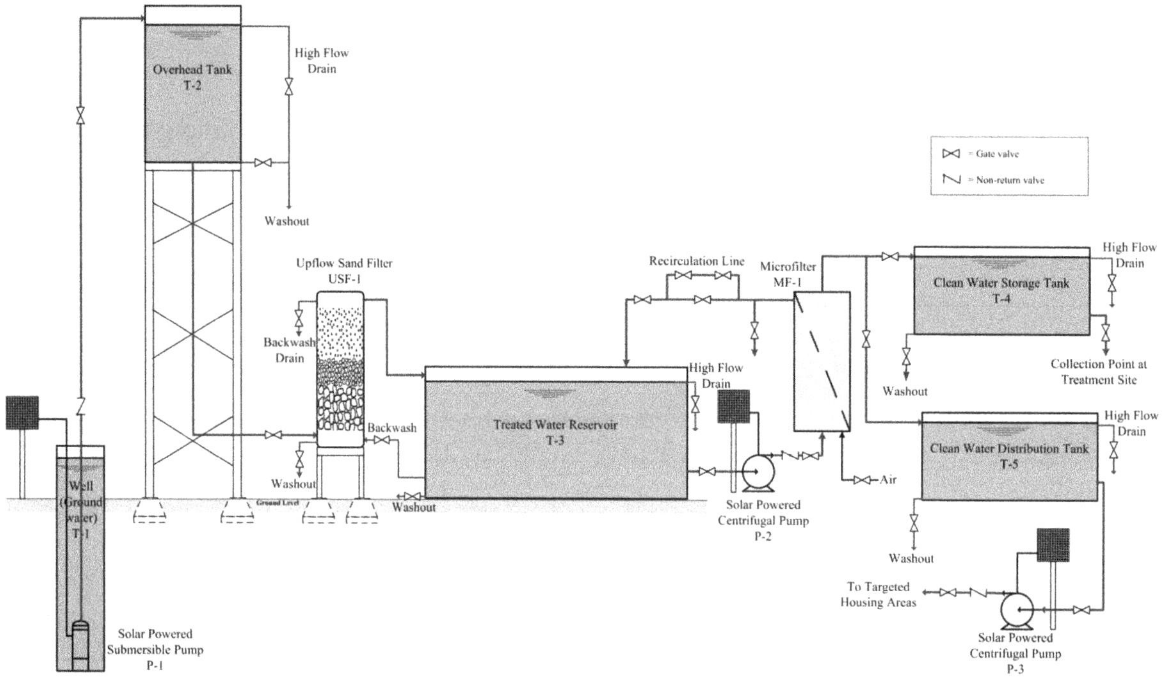

sulfide produced by sulfur bacteria in well water. Some groundwater may also have an unpleasant smell of ferrous like material as a result of the presence of iron bacteria in the groundwater. Iron bacteria can be found naturally living in soil, shallow groundwater, and surface water. These bacteria combine iron/manganese and oxygen which then form rust deposits, bacterial cell, and slime that glue them to good tubes, pumps, and pipe connectors. Although the bacteria are identified as non-disease-causing, they produce unwanted stain, taste and odour to the water.

SMALL SCALE WATER TREATMENT PLANT FOR HILLY AREA

Main Source of Water: Gravity Water

Gravity-fed water is essentially sourced from surface water found in hilly areas or mountainous region. This water source is fed to treatment facilities and/or distributed to consumers (treated water) by utilizing the potential energy (gravity) for delivery. Gravity-fed water offers an advantage in term of energy efficiency, since no pumps are required, and even electricity for that matter. However, this water source can become difficult to be utilized under some circumstances such as (1) distance of water travel to reach destination location; (2) slope of incline which determine how fast and easy the water flow; (3) diameter of pipes and pipe junctions; and (4) topography of the hills.

Figure 3. Small scale water treatment plant for gravity water

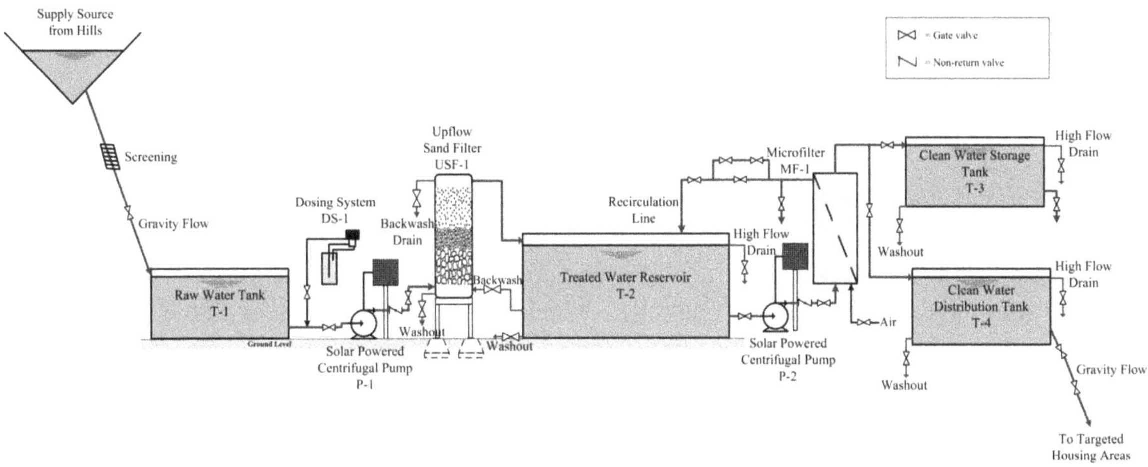

Water Treatment System

The schematic of a small scale water treatment system in Figure 3 is designed for rural communities situated in the hilly area. The system adopts a similar design and working principle with the small scale water treatment in Section 7-3.2 (Figure 1). This system utilizes surface water sources found on hilly regions as these sources can be tapped for their ease-of delivery properties using gravity flow. This contributes to the benefit of energy-saving since no additional energy (electricity for pumping) is required to deliver raw water resources to the water treatment facility. Not only that, but treated water distribution to consumers can also be achieved by gravity flow. This is aided by the difference of elevation (raw water resources from top of the hill, water treatment facility in the middle, and consumer housing area at the lower hill area), allowing potential energy doing the work for water transportation. The concept of pipework construction will be the focus of this section which can be utilized for both raw water and treated water delivery.

Naturally, water flows downhill due to gravity which ensures water delivery from water reservoir to destination point on condition that the reservoir is higher than the destination. The gravity-fed water system can be constructed by the following: (1) water intake with screen for suspended solid/debris removal, the intake should be wholly submerged to make sure that no air can enter the pipelines; (2) pipe network by considering pipe size, pipeline length, difference in elevation, and desired water flow; (3) placement of the air release valve to remove any air trapped inside the pipeline. The pipeline design should consider the factors affecting water delivery performance such as friction loss, pipe sizes, type of fittings, etc.

The pipework construction design typically depends on the geography of the water source to the targeted location of the water treatment system. For a consistent hill terrain, one pipe size is sufficient to be used. On the other hand, inconsistent terrain with downhill, slight hill, and then uphill path before water can flow down toward the treatment system, different sizes of pipes would be required which can positively affect the water flow desired. Using smaller pipe sizes can increase the velocity of the water, permitting the up-flow required. Nevertheless, the utilization of different pipe sizes would bring about the problem of friction and affect water pressure. Careful consideration and calculation should be made when designing the pipe network of gravity-flow water.

ONLINE CONTROL AND MONITORING SYSTEM

Introduction

Many currently established water and wastewater treatment plants and facilities are equipped with automatic control and monitoring system in order to ensure continuous output and standard quality of water. The use of automated control and monitoring is crucial in a big scale facility as it is an effective tool to ensure consistency, increase productivity, improve the quality of output, reduce downtime and operating costs. The advancement in instrumentation allows almost all water variables to be checked and regulated with the use of sensors and controllers which can be placed anywhere on the system. Based on these key advantages, the automation of the control and monitoring system can also be employed in the small scale water treatment system.

Control and monitoring of the water treatment processes in the small scale water treatment system is the main issue which is mainly contributed by the geographical factor i.e. remote and isolated area. Remote and isolated areas are typically difficult to reach due to the geographical conditions such as long-distance or lack of transportation from the urban or more populated areas. A high cost is expected to conduct water quality tests which are essentially done frequently on a daily basis in order to monitor and ensure that the water quality is within the safe drinking water quality. Furthermore, maintaining a small scale water treatment system using workers is not cost-effective. The tasks of control, monitoring, and supervision of the water treatment system can all be done with the implementation of the Industrial Internet of Things (IIoT). IIoT is the use of devices and machines that are connected by the internet network to enhance the industrial process. Nowadays, the IIoT concept can be applied in small scale processes at a low cost but can work as good as the sophisticated industrial technologies.

Open Platform Communication (OPC) is an example of IIoT that can be integrated into the small scale water treatment system. The OPC system can be installed and used for real-time monitoring and gathering the data remotely. The OPC can be used to monitor the water quality for any parameters such as pH, turbidity, temperature, flow rates, dissolved oxygen, and conductivity, which need to be monitored regularly. With the installation of the OPC system, the costs for gathering data would be highly reduced as it eliminates the need to visit the site frequently. The application of OPC is the key to implementing the Industrial Revolution 4.0 concept with the focus of communication and control using the IIoT application.

Open Platform Communication (OPC) also known as Object Linking and Embedding (OLE) for Process Control (OPC) is a platform that allows communication in automation through networked devices. The OPC basically uses the server/client pair. The OPC server is a software program that converts the hardware communication protocol used by a Programmable Logic Controller (PLC) into the OPC protocol. Then, the OPC client software program such as a Human Machine Interface (HMI) and Supervisory Control and Data Acquisition (SCADA) connected to the OPC server would gather the data received from or send commands to the hardware. In this modular water treatment system, the hardware is basically the sensors which would measure the three main parameters. The signal received from the sensor would be converted by an OPC server to a digital data to the HMI or SCADA and can be read from a computer or mobile phone they are installed in.

Monitoring of the water treatment system performance can be viewed online with the installation of the OPC server system like the system shown in Figure 4. The online monitoring system allows remote monitoring which is suitable and practical for rural application. The OPC system typically uses OPC-Unified Architecture (OPC-UA) server to enable connection to any operating system such as android,

Figure 4. Schematic of the OPC online monitoring system

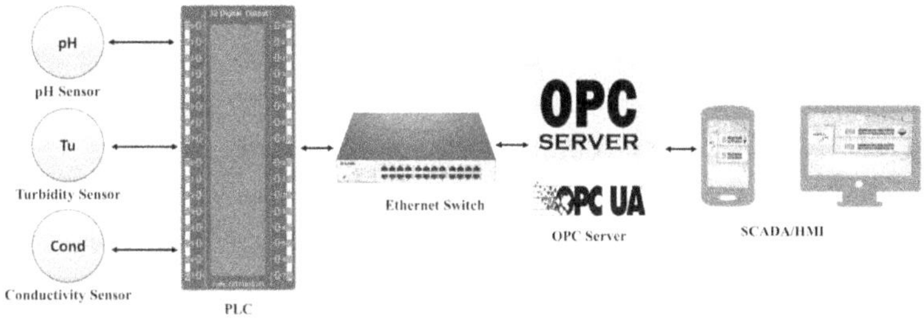

iOS, Windows, macOS, and Linux. This means that the data gathered can be transmitted and read on any devices such as computer and mobile phone. The OPC system is installed in the arrangement of sensors-PLC-OPC-UA-HMI/SCADA as shown in Figure 2. Sensors to check the water parameters such as the pH, turbidity, and conductivity can be mounted on the system. These sensors are connected to a PLC which will continuously monitor the input from sensors and generates the output obtained from the sensors. The PLC is connected to the OPC-UA server via Ethernet switch. The data transmitted from the PLC are received by the OPC-UA server and conveyed to the HMI, where the parameters reading can be read from a computer or mobile phone they are installed in.

Components in Online Control and Monitoring System

Sensors

Sensor is a device that measures the physical quantity of a variable and provides responses which can be read. Sensor converts the physical quantity into a signal that is measurable electrically. Sensors in water treatment application can be used for various parameters detection and monitoring. Wireless sensors or smart sensors allow water quality monitoring to be conducted remotely in real-time. Currently, a smart sensor is able to monitor multiple water quality parameters that can range from pH, dissolved oxygen (DO), conductivity (salinity), turbidity, temperature and dissolved ions (e.g. Fluoride (F^-), Calcium (Ca^{2+}), Nitrate (NO_3^-), Chloride (Cl^-), Iodide (I^-), Ammonia (NH_4), Magnesium (Mg^{2+}), Potassium (K^+), Sodium (Na^+), etc.).

Programmable Logic Controller (PLC)

Programmable Logic Controller (PLC) is a microcomputer device that monitors the inputs and outputs received from the connected sensors or other devices and make decisions based on the pre-programmed parameters for automated operations. The PLC makes logic-based decisions and conducts operations depending on the inputs and outputs. It can be used to monitor and record the process-time data such as the operating pressure and temperature, automatic start-up and shutdown of machines, and create alarms during operation failure.

Online Platform Communication (OPC)

Online Platform Communication (OPC) or also known as the Object Linking and Embedding (OLE) for Process Control (OPC) is a communication standard in the industrial automation of control and monitoring fields which is achieved in Windows operating system. It serves as an interface between Server/Client or Server/Server to interact, access, and share historical data and other applications. The OPC interface standard allows software programs to communicate with hardware devices and exchange information. OPC is typically implemented in server/client pair. OPC Server is a software that converts the communication protocol received from industrial devices such as PLC, SCADA, etc. into OPC protocol. OPC Client, on the other hand, is a software program that needs to be connected to the hardware, such as an HMI. The OPC Client utilizes the OPC Server to acquire data from or to transmit commands to the hardware. In brief, the OPC standard allows intercommunication between hardware devices and controllers and perform control and monitoring from PLC, SCADA, and HMI. OPC-Unified Architecture (OPC-UA) is later developed to enable connection to any operating system such as android, iOS, Windows, macOS, and Linux.

Supervisory Control and Data Acquisition (SCADA)

Supervisory control and data acquisition (SCADA) is a centralized system of software and hardware elements that allows monitoring, control, acquisition, and recording of real-time data from field devices locally or remotely. SCADA system can directly interact with devices such as sensors, actuators, PLCs, motors and valves through HMI software.

The SCADA system starts with the PLCs or Remote Terminal Units (RTUs), which are microcomputers that connect with input devices (e.g. sensors, and switches) and output devices (e.g. motor, valves, relays, HMIs, and light indicators). The information gathered by the PLCs is then transferred to computer devices with the SCADA software. The SCADA software program then processes, and displays the data, allowing the users to analyse the data and perform control and monitoring.

Human Machine Interface (HMI)

Human Machine Interface (HMI) is a device that provides the processed data from SCADA software to the user or human operator. The HMIs are generally screens that allow users to interact with the SCADA system, such as computer monitors, tablets, mobile devices, and any screens that are built on machines. HMI allows the user to perform control processes as it is directly linked to the SCADA system's database. The data provided by the HMI is schematically represented in the form of diagrams that mimics the overall operation that is being controlled.

Control, Monitoring, and Supervision in Small Scale Water Treatment System

The advancement in the field of Information Technology (IT) allows machines to perform many processes that are limited to some level to human capability. Many industrial processes utilize IT particularly to perform control, monitoring, and supervision of their processes. IT has many advantages such as increasing productivity, improve communication, improve data storage and management, improve detection of errors, reduce operation costs, and saves times. The application of IT in the rural and remote areas is

Table 4. Water treatment process with its principle applications

Process	Principle Applications
Chemical dosing	• Chemicals are added to raw water to change the characteristics of water such as pH and turbidity. • Lime is added to adjust the pH to the desired value. • Alum is added to turbid raw water to coagulate small suspended solids into larger particles that are easier to be removed. • Chlorine is added to kill the pathogens and microorganisms presents in the water.
Sand filtration	• Removes particulate matters by filtration through layers of granular sand media. The water flow is by the up-flow mechanism.
Membrane filtration	• Membrane filtration allows further removal of fine particles and bacteria by using porous microfiltration membrane.

Table 5. Possible control parameters in the water treatment processes

Process	Equipment	Control Parameters
Chemical dosing	Dosing system, water sampling and analysis apparatus	Turbidity, pH, bacteria
Sand filtration	Up-flow sand filter, valves, pump	Turbidity
Membrane filtration	Microfilter, valves, pump	Turbidity, pH, bacteria

achievable with the rapid expansion of the internet network. This concept can be implemented to small scale water treatment systems for isolated areas to perform online control, monitoring, and supervision. The integration of this system can eliminate the issue with difficulty in control, monitoring and supervising of the water quality in isolated regions.

Integration of the Online Platform Communication (OPC) system discussed in Section 7-5.1 and 7-5.2 allows remote control and monitoring via wireless communication that is connected between the control centre and local devices installed at the small scale water treatment facility. The control centre can be anywhere as long as the internet connection is present at the treatment facility and control centre since the data are shared and transmitted through internet connectivity. Using OPC Unified Architecture (OPC-UA) Server allows any operating systems to access the OPC-UA Client software which allows monitoring and control of the water quality. This means that any HMI devices such as computer desktop, laptops, tablets, and mobile phones which uses the operating systems such as android, iOS, Windows, macOS, and Linux can be used to access the Client software. Remote supervision can also be conducted with the application of closed-circuit television (CCTV) or Internet Protocol (IP) surveillance camera. The use of supervision camera enables the surveillance of the surrounding water treatment facility to enhance security and help in keeping the facility safe. Moreover, IP camera can be used to control, send and receive the data images via the internet which allows remote supervision at the rural areas.

The small scale water treatment system in Figure 1 - Figure 3 can be integrated with online control and monitoring system which can be managed remotely. The process of small scale water treatment system shown in Figure 1 - Figure 3 generally consists of several stages as shown in Table 4 as follows: (1) chemical dosing (coagulant and disinfectant), (2) sand filtration, and (3) membrane filtration. Each process consists of several types of equipment which have distinct control functions. The process equipment and control parameters for each process are shown in Table 5.

Figure 5. Small scale water treatment system integrated with online control, monitoring, and supervision

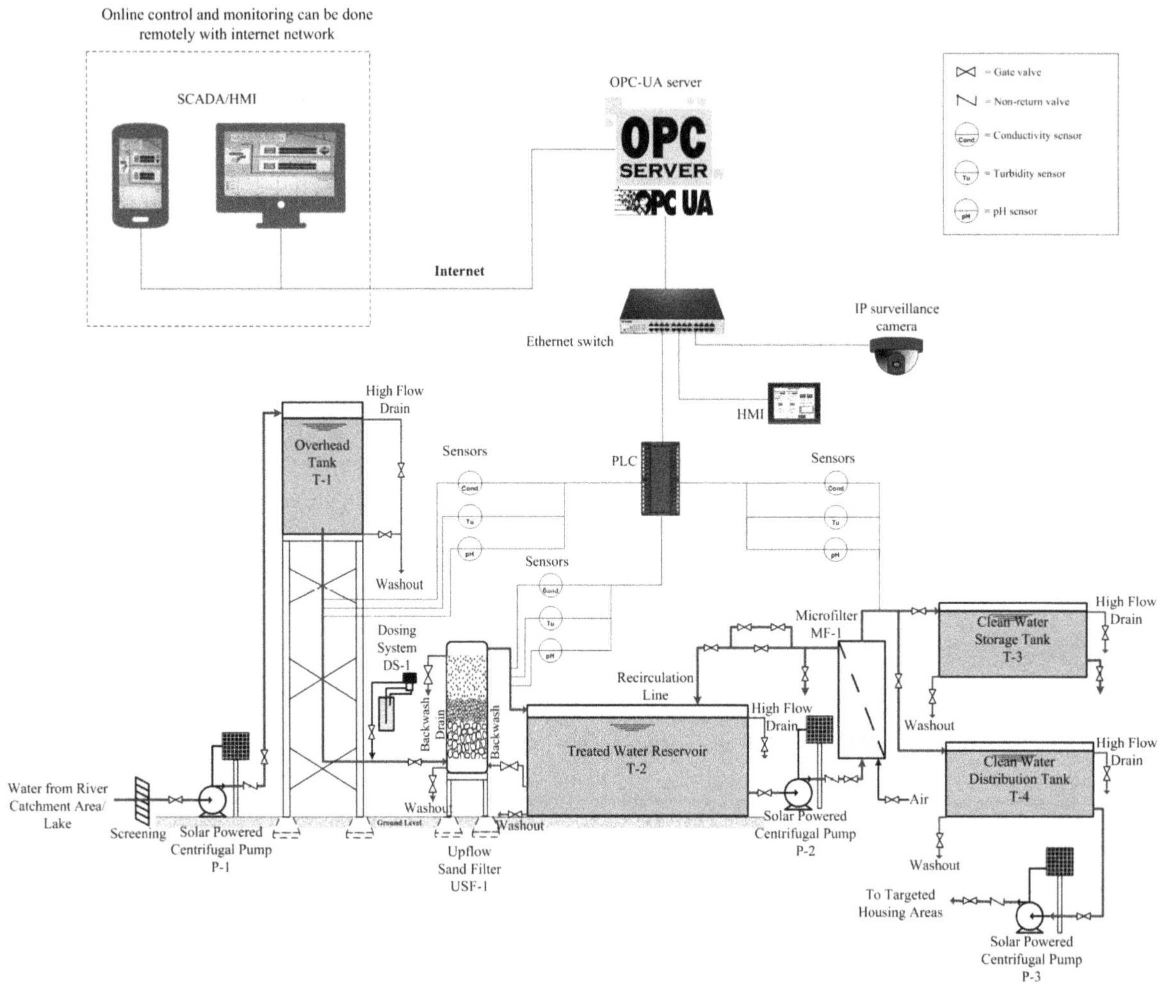

Figure 5 shows an example of the small scale water treatment system from Figure 1 integrated with the online control, monitoring, and supervision system. The system consists of a series of instrumentations of hardware and software elements arranged as sensors-PLC-OPC server-OPC client/SCADA/HMI. The sensors generally measure the magnitude of the control parameters before sending the data to the PLC. The PLC device is a controller that is installed on the site to automatically control the process based on a pre-programmed setting. This also allows it to make decisions in control. With the availability of internet connection, data from the PLC and surveillance camera can be transmitted to an OPC Server that can store and exchange the information with the OPC Client/SCADA software. OPC-UA Server enables universal connections as it can be accessed by various operating systems such as Windows, macOS, Linux, Android, and iOS. These operating systems are accessible by a wide range of HMI devices like computer desktops, laptops, tablets, and smartphones. The OPC Client/SCADA software allows anyone with access to it to monitor the parameters and control the system remotely.

Chemical dosing involves the process of feeding a certain amount of chemicals to the raw water that is measured depending on the characteristics of the raw water. For example, highly turbid water requires more alum dosage, while acidic water requires the addition of lime to adjust the pH. Manual handling of the chemical dosing process requires an operator that will be responsible to frequently conduct water quality analysis as the water characteristics may change from time to time, before adjusting the required dosage accordingly. Manually handling the dosing process will take much time given the dynamic nature of water quality that is dependent on the surrounding activity such as the weather. Furthermore, the dosage of chemicals is directly connected to the quality of treated water and the cost of operation. The precision of dosing is very important to reduce chemical waste and costs. Thus, this process in particular needs online controls and monitoring to work more efficiently.

The chemical dosing control system can be set up particularly on the coagulation (alum) control, pH (lime) control, and disinfection (chlorine) control. This operation can be controlled in fully automatic by the PLC and OPC Client software. Sensors can be applied to monitor water parameters such as the turbidity and pH, which are controlled by the PLC and communicates with the software. The data transmitted from the PLC will be stored in the server and subsequently used by the software for decision making and data report. In this system, monitoring of several parameters (i.e. turbidity, pH, and conductivity) are conducted at several points in the small scale water treatment system. These parameters are important particularly to determine the dosing quantity required at the inlet of the sand filter (e.g. pH affect the amount of lime needed, turbidity affect the quantity of alum required, etc.).

Monitoring the control parameters at the outlet of the up-flow sand filter and microfilter is important to check the efficiency of the treatment processes in removing contaminants. Furthermore, these parameters require regular monitoring to ensure that the treated water quality is safely within the drinking water quality standard. The control system can be programmed to send alerts to the users if the water quality falls outside of the range of drinking water quality standard. The system can also be programmed to make automatic control decisions to bring the water quality within the pre-set values.

The instrumentation devices used to assemble the system can be obtained at relatively low costs as these devices are commercially made and can be bought in retails. Table 6 shows the hardware and software elements that can be used in each process automation and the associated devices/brands. Most of the parameters involved in water treatment can be measured by sensors that are made specifically to measure the parameter. Nowadays, several water parameters can be collectively measured by a single sensor called the smart sensor.

To conclude, successful implementation of the simple process control requires periodic maintenance and knowledge person. Automation of process control and monitoring leads to less critical decisions to be made by the operating personnel. To ensure competency at small scale system, the implementation of the online control, monitoring and supervision system can be accomplished by using the service of a private consulting company for the installation and routine maintenance of the system.

SMALL SCALE WATER TREATMENT PLANT FOR ISLAND

Main Source of Water: Brackish Water

Brackish water is water that contains salinity between that of freshwater and seawater. Brackish water can be formed from the mixing of freshwater and seawater which can occur at surface water sources such as estuaries, or at groundwater sources in aquifers. This type of water is not suitable for livestock or human consumption due to higher salt content than freshwater. Besides, it cannot be used for industrial

Table 6. Process automation and the associated hardware/software elements and devices/brands

Process Automation	Hardware/Software Elements	Devices/Brands
Control	Sensors	• pH sensor • Turbidity sensor • Conductivity sensor • Dissolved oxygen sensor • Temperature sensor • Dissolved ions sensors
	PLC	• Rockwell PLC • Mitsubishi PLC • Siemens PLC • Hitachi PLC • Honeywell PLC • Arduino board (Microcontroller, an alternative to conventional PLC)
	Ethernet switch	• D-Link Ethernet Switch • NETGEAR Ethernet Switch • TP-Link Ethernet Switch • Cisco Ethernet Switch • Arduino Ethernet
	OPC-UA server	• Open source software available
	OPC-UA client/SCADA	• Open source software available
Monitor	HMI	• Computer desktop • Laptop • Tablet • Mobile phone/Smartphone • Built-in screen on the hardware device
Supervise	IP surveillance camera	• Bosch IP Camera • Panasonic IP Camera • Sony IP Camera • Axis IP Camera

usage since the crystallization of salts can damage machinery and equipment. Usually, brackish water is used for the purpose of irrigation.

Currently, there is no universal standard to determine how much salt brackish water contains. Salinity refers to how much salt is dissolved in water and is typically measured in parts per million (ppm). One ppm means that there is 1 gram of salt for every 1,000,000 gram of water. Hillel (2010) has classified the water salinity range into different water type as shown in Table 7, which states that brackish water contains 1,000 – 2,000 ppm of total dissolved salts.

Water Treatment System

Brackish/saline water is typically treated using the reverse osmosis technology which is capable of separating or filtering out salts from water. However, this technology is a high pressure-driven membrane separation process. High energy is required to operate a reverse osmosis filtration system which means high operational costs. A cheaper alternative is available with good performance of brackish water treatment which is the thermally-driven membrane distillation system. Membrane distillation is a relatively new process that utilizes microporous hydrophobic membrane to separate two solutions having different temperatures. This process possesses several advantages over conventional separation

Table 7. Classification of water salinity

Water Type	Salinity (Total Dissolved Salts, ppm)
Freshwater	<500
Slightly brackish	500 – 1,000
Brackish	1,000 – 2,000
Moderately saline	2,000 – 5,000
Saline	5,000 – 10,000
Highly saline	10,000 – 35000
Brine	>35,000

Source: (Hillel, 2000)

processes in ways that it requires low cost and low energy usage. Membrane distillation operation has the characteristics of: (1) having porous membrane; (2) the membrane should not be wetted by liquid; (3) no capillary condensation occurs inside the membrane pores; (4) the membrane should not change the vapour-liquid equilibrium of the liquid components; (5) at least one side of the membrane surface has direct contact with liquid; and (6) partial pressure gradient in vapour phase of each component is the driving force of the process.

The membrane distillation can be performed by four types of configurations; they are (1) direct contact membrane distillation, with downstream side of the membrane in contact with cold water; (2) air gap membrane distillation, with the downstream side of the membrane in contact with stagnant air associated with a cold plate; (3) sweeping gas membrane distillation, with the downstream side of the membrane swept by an inert gas, and (4) vacuum membrane distillation, with the downstream side of the membrane maintained under vacuum conditions or under low pressure. The four types of the membrane distillation have a similar way of feeding the feed solution to the upstream side of the membrane, but they have different ways of condensing the vapour on the downstream side of the membrane.

Figure 6 shows a small scale water treatment system for brackish water which employs an air gap membrane distillation as the main unit operation. Initially, the cold feed (brackish water) flows into the first (cold) chamber which will act as the cooling liquid to induce the evaporation process from the second (hotter) chamber. The feed then exits and being heated up before it enters the second chamber. In this system, the solar cell is used as the solar thermal collector which will provide the heating energy required to increase the temperature of the feed. In the second chamber, the hot water molecules will start to evaporate and pass through the membrane pores into the air gap chamber due to the partial vapour pressure gradient. The cooling plate will assist in the condensation of water vapour into liquid water which is then collected and stored in the freshwater tank.

APPLICATION OF RAINWATER HARVESTING FOR INDIVIDUAL HOUSE

Main Source of Water: Rainwater

Sabah specifically receives rainfall 1,500 to 3,000 mm throughout the year, with distinct difference between regions of northwest, southern and central part the state due to the variance in terrain. In general,

Figure 6. Small scale water treatment system for brackish water

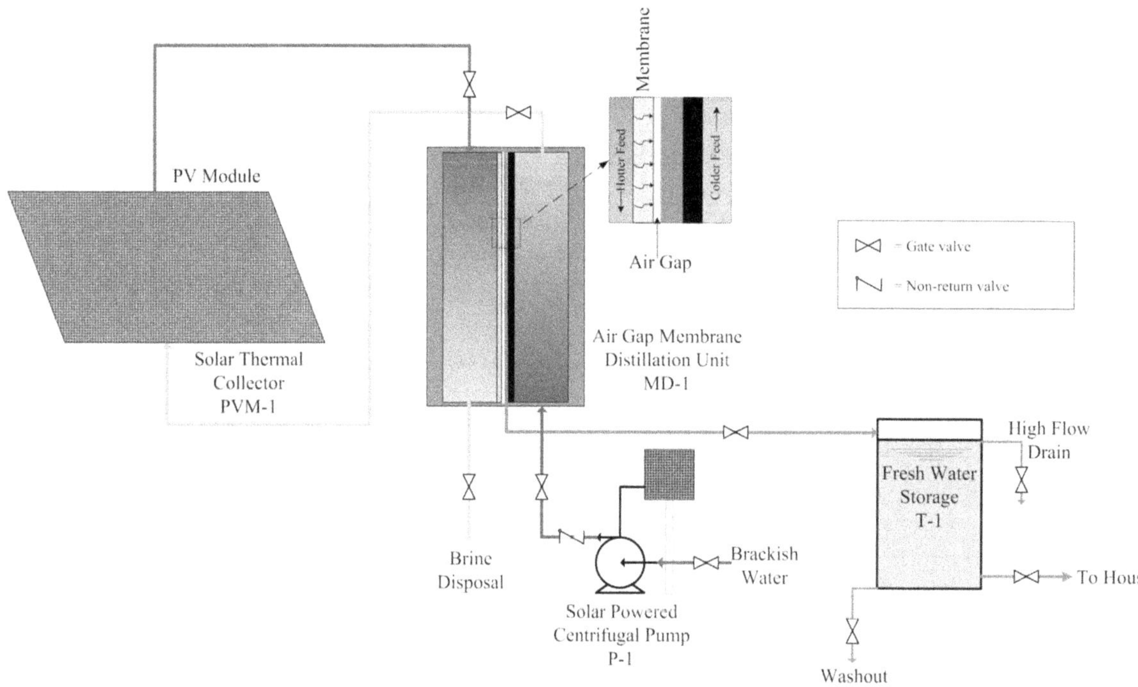

the northwest region receives heavier rainfall in October and June. Drier season is experienced around February and August. The central part of Sabah receives evenly distributed rainfall, however lower compared to other regions due to being sheltered by mountainous terrain. Southern region also has an evenly distributed rainfall at around 150 to 250 mm throughout the year. As depicted in Figure 7, the monthly rainfall distribution of Sabah could allow for the implementation of rainwater harvesting and storage. In fact, the provision of policy and regulation for rainwater harvesting has been initiated by the authority after the devastating drought on 1997 – 1998 in Malaysia (Che-Ani et al., 2009; Lee et al., 2016).

The physicochemical characteristics of Malaysia harvested rainwater are tabulated in Table 8. Generally, rainwater quality is dependent on a few aspects: (i) the location and surrounding activity, (ii) harvesting system and storage material, (iii) duration of rain and (iv) period of wet and dry weather (Che-Ani et al., 2009; Despins et al., 2009; Kasmin et al., 2016; Shaheed et al., 2017; Yaziz et al., 1989). Urbanized and industrialized locations will have a more polluting element in the atmosphere which is dissolved in the rainwater. Other than atmospheric factor, water storage in plastic cisterns will influence the water to be slightly acidic whereas concrete storage makes the water to be slightly basic (Despins et al., 2009).

Rainwater Harvesting System

The main concern of rainwater quality especially in a remote area is coming from contamination at the collection system, usually, the rooftop run-off that is exposed to the environment. Settlements of atmospheric pollution, debris such as leaves and twigs and animal dropping. Based on the physicochemical characteristics of harvested rainwater across Malaysia, heavy metals content seem to exceed the limit in RWS specifically lead, zinc, cadmium, iron, aluminium and chromium (Table 8). Most heavy metals

Figure 7. Sabah's Monthly Rainfall Distribution (September 2016 – August 2017)
Source: Malaysian Meteorological Department, 2017

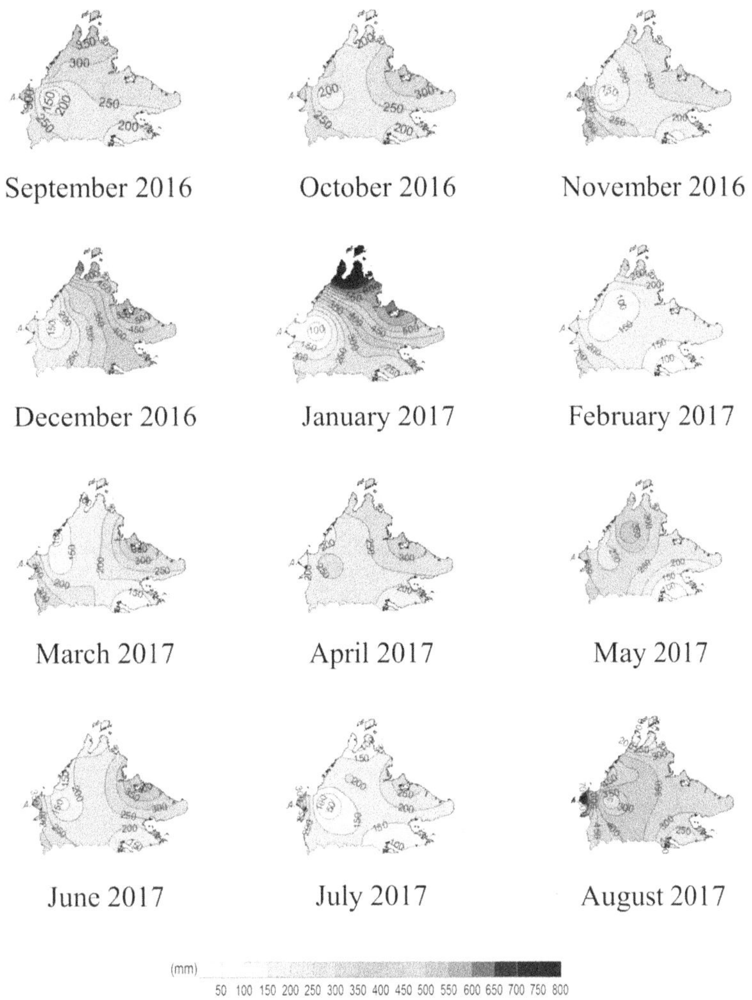

are hazardous because they tend to bio-accumulate in the human body, causing damage to the nervous system (Elango & Kannan, 2007). Atmospheric deposition on harvesting surfaces is one of the significant factors especially in urban or even semi-urban location (Sánchez et al., 2015). Combustion of coals from coal-powered plants and trace elements from car exhaust and tire wear and abrasion also contributes to heavy metal pollution to the atmosphere. Lead is the most common heavy metal to pollute harvested rainwater. On the other hand, corrosion of old rusted roofing is also associated with metal contamination in harvested rainwater. Other concern in rainwater is the pathogenic content particularly faecal bacteria makes it imperative that rainwater is not directly utilized as drinking water prior to any treatment.

Nonetheless, the harvested rainwater will be extremely important to the socio-economic of areas affected by water sources scarcity or clean water inaccessibility, which is the main problem encountered in rural regions of Sabah. The provision of potable water is one of the major issues faced by the locals even though they have abundant of (untreated) water sources due to lack of water treatment infrastructure development by the government in this field and lack of alternative water treatment knowledge that

Table 8. Physicochemical characteristics of Malaysia harvested rainwater against Malaysian Raw Water Standards (RWS)

Area	Not specified, Malaysia [a]		Urban, Selangor Malaysia [b]		Semi Urban, Selangor Malaysia [c]	Semi Urban, Johor Malaysia [d]	Semi Urban, Sabah, Malaysia [e]		RWS [f]
Roof Type/ Rainwater Type	**First Flush**	**Stored Rainwater**	**Galvanized Iron Roof**	**Concrete Tile Roof**	**Container Collection Method**	**Zinc Roof**	**Nipah-Thatched Roof**	**Galvanized Iron Roof**	
pH	6.5 - 6.9	6.54 - 7.21	6.4 - 6.6	6.8 - 6.9	n.m *	4.26 - 6.06	7.05	6.43	5.5 -9.0
Temperature	n.m	n.m	28.0 - 28.1	28.1	n.m	n.m	n.m	n.m	n.m
DO (mg/L)	7.31 - 7.92	7.05 - 7.76	n.m	n.m	n.m	6.24 - 6.84	2.37	0.8	n.m
Conductivity(μS/cm)	n.m	n.m	50.7 - 97.0	86.5 - 135.2	n.m	n.m	n.m	n.m	n.m
Turbidity (NTU)	11.2 - 44	0.4 - 3.1	10.0 - 22.0	24 - 56	n.m	57 - 68	58.4	1.53	1000
Total Solid (mg/L)	n.m	n.m	64 - 119	116 - 204	70.55 - 202.84	n.m	n.m	n.m	n.m
TDS (mg/L)	58.7 - 138	6.6 - 48.2	13 - 28	23 - 47	n.m	n.m	n.m	n.m	1500
TSS (mg/L)	15.0 - 48.0	0.7 - 2.0	52 - 91	95 - 153	n.m	14 - 43	44	42	n.m
COD (mg/L)	68.6 - 191	25.3 - 50	n.m	n.m	n.m	n.m	n.m	n.m	10
BOD (mg/L)	13.4 - 16	1.67 - 5.86	n.m	n.m	n.m	n.m	n.m	n.m	6
Faecal Coliform (/100 ml)	n.m	n.m	0 - 8	0 - 13	n.m	n.m	n.m	n.m	n.m
Total Coliform (/100 ml)	n.m	n.m	25 - 63	41 - 75	n.m	n.m	n.m	n.m	5000
Plate counts x 10^3 (/ 100 ml)	n.m	n.m	21 - 32	41 - 51	n.m	n.m	n.m	n.m	n.m
Zinc (μg/L)	n.m	n.m	294 - 497	49 - 96	32.11 - 82..65	n.m	n.m	n.m	3
Lead (μg/L)	n.m	n.m	145 - 254	102 - 271	1.03 - 7.03	0.001 - 0.004	n.m	n.m	0.05
Cadmium (μg/L)	n.m	n.m	-	-	0.22 - 1.62	n.m	n.m	n.m	0.003
Iron (μg/L)	n.m	n.m	n.m	n.m	9.80 - 44.18	n.m	n.m	n.m	1
Aliminium (μg/L)	n.m	n.m	n.m	n.m	3.91 - 27.97	n.m	n.m	n.m	-
Chromium (μg/L)	n.m	n.m	n.m	n.m	0.45 - 0.75	n.m	n.m	n.m	0.05

*n.m: not measured

Source: [a] (Kasmin et al., 2016); [b] (Yaziz et al., 1989); [c] (Alahmr et al., 2012); [d] (Rahmat et al., 2008); [e] (Ayog et al., 2016); [f] RWS is based on standard enforced by the Ministry of Health Malaysia from (MMOH, 2010)

could cater the water necessity for rural areas. In view of the abundance of precipitation across the state throughout the year, utilization of rainwater as potable water resource would be viable. This system can be easily installed and maintained for individual houses application especially isolated houses with complete inaccessibility to piped drinking water. The system can be easily operated by the locals with just simple instruction for setting it up and manoeuvring the system as it does not require thorough practical and scientific knowledge to be operated.

Rainwater harvesting is a method of gathering, storing, and using rainwater for various uses. The system works in a way that rainwater is collected from various surfaces such as house rooftop and channelled to a storage tank. The water can be directly used for applications including irrigation, washing car and so on. However, rainwater is not drinkable without considerable treatment as it may contain particulate matter and dust from the atmosphere which is unhygienic for immediate consumption. Nonetheless, this source of water can be used as potable water if properly treated to ensure its cleanliness and septicity. Figure 8 illustrates a rainwater harvesting set up for a single housing application which includes the water treatment using a membrane filtration system. The system comprised of the storage tank to receive and collect rainwater from the rooftop surface, a membrane filtration system, a pump and an overhead tank.

The rain harvesting technique requires uncomplicated unit operations and installation which makes it a practical solution for remote areas application. The rooftop acts as a collection surface for the rainwater which is carried through a pipeline network into a storage tank. Before entering the storage tank, the rainwater will flow across a simple gauze filter to screen out large particles that may clog the membrane surface and consequently reduce its efficiency. The water collected is then treated by using a membrane filtration system to eliminate particulate matters and bacteria. Microfiltration or ultrafiltration membrane can be applied for the filtration system as both membrane processes possess excellent attributes in producing clean water permeate. Ultrafiltration may produce water that is of high purity due to its smaller pores and in terms of its ability to remove suspended matters, bacteria, as well as colloid materials. Despite that, microfiltration is able to produce high-quality water for municipal uses aside from being a cheap alternative to be used in the membrane system. Pumping power is required to draw out the water permeate from the membrane filtration system and delivered to the overhead tank. The pumping work would require energy, thus an alternative would be to implement a solar pumping system by harvesting the solar energy that is abundantly available in the geographical location of Sabah. This can significantly reduce or effectively save electricity consumption for the rainwater harvesting system. The clean water stored in the overhead tank can then be channelled via gravitational mean to the piping system for household usage.

Conversely, the construction expenditure of large scale rainwater harvesting infrastructure must be taken into account. As an alternative, modification of existing domestic harvesting system could be implemented to reduce the construction cost. In fact, a qualitative study was conducted in Sandakan Township in 2009 during the early days of rainwater collection system policy implementation which has shown positive feedback. The respondents have seen a substantial reduction in dependence upon government-supplied water, a decrease in water bill and improvement in agricultural activities. Other benefits reported increased in groundwater level, reduce soil erosion, improvement of river management and potential control in climate change. However, the application of rainwater as potable water uses is highly subjected to other factors such as the support from the government in term of policy, guidelines and provision of the treatment system to ensure good management, optimum collection and storage and good quality for safe consumption.

EXAMPLE OF SMALL SCALE WATER TREATMENT DESIGN FOR RURAL APPLICATION

In this section, an example of the design capacity of a small scale water treatment plant will be discussed. The main focus is on the groundwater treatment system. By looking from the practicability sense, groundwater can be found almost in every part of Sabah such as in the level ground, hilly region, an island. Groundwater comprises of 90% of the available freshwater on earth (Mishra & Dubey, 2015). Due to its abundant nature, it is suitable to be used as the main water resource to sustain continuous operation of the water treatment. Besides, groundwater is less subjected to pollution, retains high mineral content, and is resilient to drought. However, as more water is pulled through its aquifer, then the water table level may be depleted and can affect the stability of soil and hydrology. Thus, a sustainable approach would need to be considered in the continuous utilization of groundwater.

The plant capacity design is to produce daily production that can support a minimum of 20 households on day-to-day operations. The average household size in Malaysia is 5 people per household (Department

Figure 8. Rainwater harvesting setup for individual household application

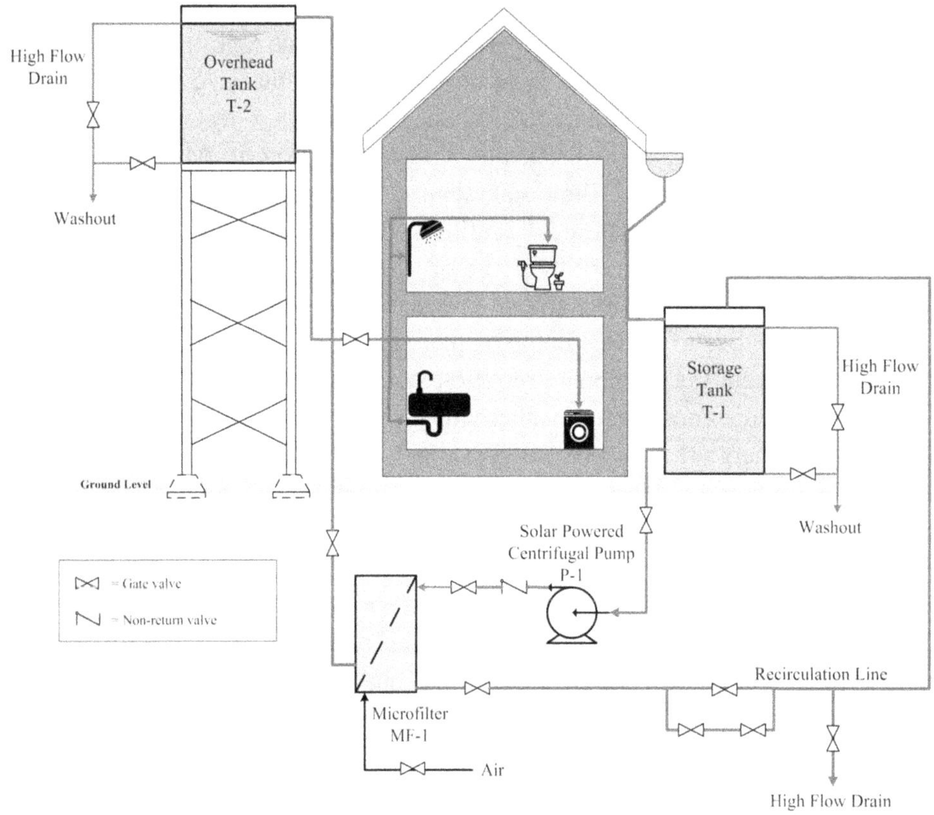

of Statistics Malaysia, 2010), whereas the average water consumption is at 209 litres per capita per day (L/c/d) (Malaysia Water Industry Guide, 2010). This gives an average water consumption per household (2010) at around 1 cubic meter per household per day. Hence, the system is designed to produce water at a minimum production of 20 cubic meters per day. Besides, it should consider a water storage availability that can sustain water demand for several days in case there are unforeseen circumstances that are faced by rural communities. These include a shortage of groundwater resources, drought, insufficient solar energy to power the plant, breakdown of equipment, or during maintenance of the treatment facilities.

For this proposed scheme, the plant capacity would be able to store water supply for the duration of 3 days. For that, the minimum capacity of the daily storage in the treated water reservoir is 60 cubic meter per day. On the other hand, the overhead tank has a volume of 10 cubic meters to hold the groundwater at one time. The day-time operation would be handled by the locals, whereby the operation should be able to produce clean water in the distribution tank at a minimum capacity of 20 cubic meters per day, and a maximum capacity of 30 cubic meters. In addition to that, the clean water storage tank can sustain additional clean water of 30 cubic meter volume which serves as the intake point for the nearby communities to collect water directly from the water treatment plant. This is especially critical to the locals living at close proximity who opts for direct clean water collection without having to invest on cost for pump and piping to their house. By referring to Figure 2, the recommended sizing dimension of the water tanks employed are tabulated in Table 9.

Table 9. Size of tanks employed in the small scale water treatment plant for groundwater (Figure 2)

Unit	Dimension
Overhead Tank, T-2	2 m × 2 m × 3 m
Treated Water Reservoir, T-3	8 m × 4 m × 2 m
Clean Water Storage Tank, T-4	5 m × 4 m × 1.5 m
Clean Water Distribution Tank, T-5	5 m × 4 m × 1.5 m

Table 10. Sabah pipe length and type

Type	Pipe Length (km)
Asbestos Cement (AC)	2,345
Mild Steel (MS)	2,622
Unplasticized polyvinyl chloride (uPVC)	1,620
High-density Polyethylene (HDPE)	796
Ductile Iron (DI) / Cast Iron (CI)	687
Others	306
Total	8,376

Source: (Salleh and Malek, 2012)

DISTRIBUTION SYSTEM

As of 2015, Sabah's total pipeline reached 8,376 km of length with various types of pipelines used as depicted in Table 10. Yet, major issues are encountered regarding the water distribution system in Sabah. Asbestos cement pipes that have been laid out underground for the last decades are starting to leak and in urgent need for replacement. These degrading pipes are the major problems in distribution since it will not handle high-pressure operation. However, the challenges that are faced with upgrading pipeline are mainly due to insufficient manpower and competent/ qualified personnel, which resulted in the replacement of old pipes and consumers meters only. On the other hand, incomplete data on the existing pipeline system and incomplete mapping system prevents further improvements of the existing system.

The water distribution system for small scale water treatment system in rural Sabah would not require a long and extensive network of pumps and piping. This is because the area covered by each water treatment plant would not be very large and only focused on certain areas. Nonetheless, we would still need to tackle the issues on some geographical conditions like the hilly terrain and mountainous region in several locations. Thus, the distribution network needs to be planned effectively to ensure efficient water delivery and management to the targeted households. The distribution system starts from the source or reservoir tank which collects and holds the clean treated water, to the main transmission pipeline and eventually to the branch pipelines which diverge to the targeted consumers. In most system, a service reservoir or overhead tank is employed after the main pipeline before water is distributed to the branch pipelines. The main purposes are to maintain the pressure inside the pipeline and also to reduce the flow resistance. The principle of the distribution system is illustrated in Figure 9.

Figure 9. Water distribution system

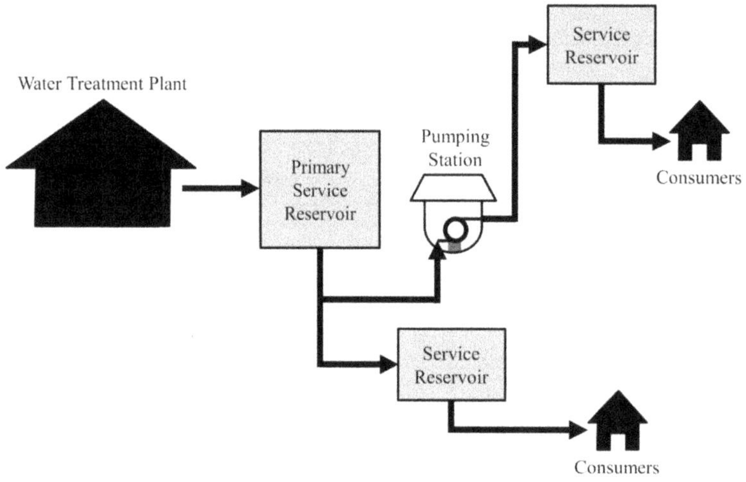

REFERENCES

Abd Razak, Y., & Abd Karim, M. H. (2009). Groundwater in the Malaysian Context, in *Groundwater Colloquium 2009, Groundwater Management in Malaysia -Status and Challenges*, 1–14.

Abdullah, M. H., & Musta, B. (1999). Phreatic Water Quality of the Turtle Islands of East Malaysia: Pulau Seling). Design of a filtration system for a small scale water treatment plant for a rural community around Maiduri and Pulau Bakkugan Kechil. *Borneo Science*, *6*, 1–9.

Alahmr, F. O. M., Othman, M., Abd Wahid, N. B., Halim, A. A., & Latif, M. T. (2012). Compositions of dust fall around semi-Urban Areas in Malaysia. *Aerosol and Air Quality Research*, *12*(4), 629–642. doi:10.4209/aaqr.2012.02.0027

Aris, A. Z., Abdullah, M. H., & Kim, K. (2007). Hydrogeochemistry of Groundwater in Manukan Island, Sabah. *The Malaysian Journal of Analytical Sciences*, *11*, 407–413.

Ayog, J., Ayog, J. L., Dullah, S., & Ramli, R. (2016). Harvested Rainwater Quality Assessment on the Effects of Roof Materials to the First Flush Runoff. *Transactions on Science and Technology*, *3*(1–2), 271–276.

Che-Ani, A. I., Shaari, N., Sairi, A., Zain, M. F. M., & Tahir, M. M. (2009). Rainwater harvesting as an alternative water supply in the future. *European Journal of Scientific Research*, *34*, 132–140.

Chu, L. H. (2004). Groundwater Utilization and Management in Malaysia. In *Proceedings of the 41st Coordinating Committee for Geoscience Programmes in East and Southeast Asia (CCOP)*, 83–94. Academic Press.

Cleophas, F. N., Isidore, F., Han, L. K., & Bidin, K. (2013). Water quality status of Liwagu River, Tambunan, Sabah, Malaysia. *Journal of Tropical Biology & Conservation*, *10*, 67–73.

Despins, C., Farahbakhsh, K., & Leidl, C. (2009). Assessment of rainwater quality from rainwater harvesting systems in Ontario, Canada. *Journal of Water Supply: Research & Technology - Aqua*, *58*(2), 117–134. doi:10.2166/aqua.2009.013

Elango, L., & Kannan, R. (2007). Rock-water interaction and its control on chemical composition of groundwater. *Concepts and Applications in Environmental Geochemistry*, *5*, 229–246. doi:10.1016/S1474-8177(07)05011-5

Faisal, M., Omang, S. A. K., & Tahir, S. (1995). Geology of Kota Kinabalu and its implications to groundwater potential, in *Annual Geological Conference. Kuala Terengganu Malaysia*, 11–20. 10.7186/bgsm38199502

Harun, S., Dambul, R., Abdullah, M. H., & Mohamed, M. (2014). Spatial and seasonal variations in surface water quality of the Lower Kinabatangan River Catchment, Sabah, Malaysia. *Journal of Tropical Biology & Conservation*, *11*, 117–131.

Harun, S., & Fikri, A. H. (2016). *Water Quality Monitoring in Sugut River and its Tributaries*. WWF Malaysia Project Report. Institute for Tropical Biology and Conservation, Universiti Malaysia Sabah.

Heng, L. Y., Chukong, L. N., Stuebing, R. B., & Omar, M. (2006). The Water Quality of Several Oxbow Lakes in Sabah, Malaysia and its Relation to Fish Fauna Distribution. *The Journal of Biological Sciences*, *2*(2), 365–369. doi:10.3923/jbs.2006.365.369

Hillel, D. (2010). *Salinity Management for Sustainable Irrigation: Integrating Science, Environment, and Economics*. Washington, DC: World Bank Publications.

Hing, T. T. (1994). Hydrochemistry of groundwater at Sahabat region, Sabah. *Newsletter of the Geological Society of Malaysia*, *20*, 229–230.

Jasin, B., & Tating, F. F. (1991). Late Eocene planktonic Foraminifera from the Crocker Formation, Pun Batu, Sabah. *Warta Geologi*, *17*(4), 187–191.

Jopony, M., & Tongkul, F. (2009). *Acid Mine Drainages at Mamut Copper Mine* (pp. 83–94). Sabah, Malaysia: Borneo Science.

Kasmin, H., Bakar, N. H., & Zubir, M. M. (2016). Monitoring on The Quality and Quantity of DIY Rainwater Harvesting System. *IOP Conference Series. Materials Science and Engineering*, *136*, 12067. doi:10.1088/1757-899X/136/1/012067

Lee, K. E., Mokhtar, M., Mohd Hanafiah, M., Abdul Halim, A., & Badusah, J. (2016). Rainwater harvesting as an alternative water resource in Malaysia: Potential, policies and development. *Journal of Cleaner Production*, *126*, 218–222. doi:10.1016/j.jclepro.2016.03.060

Lin, C. Y., Abdullah, M. H., Aris, A. Z., & Praveena, S. M. (2009). A Baseline Study on Groundwater Quality of the A Baseline Study on Groundwater Quality of the Tourist Island, Pulau. *Modern Applied Science*, *3*, 62–74. doi:10.5539/mas.v3n5p62

Manap, M. A., Sulaiman, W. N. A., Ramli, M. F., Pradhan, B., & Surip, N. (2013). A knowledge-driven GIS modelling technique for groundwater potential mapping at the Upper Langat Basin, Malaysia. *Arabian Journal of Geosciences*, *6*(5), 1621–1637. doi:10.100712517-011-0469-2

Mishra, R. K., & Dubey, S. C. (2015). Freshwater availability and its global challenge. *International Journal of Engineering Science Invention Research & Development, 2*(4), 2349–6185.

MMOH. (2010). Drinking Water Quality Surveillance Programme - Ministry of Health [WWW Document]. Retrieved from http://kmam.moh.gov.my/public-user/drinking-water-quality-standard.html

Mohammad Ali, B. N., Lin, C. Y., Cleophas, F., Abdullah, M. H., & Musta, B. (2015). Assessment of heavy metals contamination in Mamut river sediments using sediment quality guidelines and geochemical indices. *Environmental Monitoring and Assessment, 187*(1), 4190. doi:10.100710661-014-4190-y PMID:25471626

NRO. (1994). Natural Resources Office Sabah: Water Resources Master Plan, Final Report Overview.

Piratla, K. R., & Goverdhanam, S. (2015). Decentralized Water Systems for Sustainable and Reliable Supply. *Procedia Engineering, 118*, 720–726. doi:10.1016/j.proeng.2015.08.506

Praveena, S. M., Lin, C. Y., Aris, A. Z., & Abdullah, M. H. (2010). Groundwater Assessment at Manukan Island, Sabah: Multidisciplinary Approaches. *Natural Resources Research, 19*(4), 279–291. doi:10.100711053-010-9124-y

Rahmat, S. N. B., Ali, Z. M., & Musa, S. (2008). Treatment of Rainwater Quality Using Sand Filter. In *Proceedings International Conference on Environment 2008 (ICENV 2008), 2008*, 1–8. Academic Press.

Saleh, H., & Samsudin, A. R. (2013). *Geo-Electrical Resistivity Characterization of Sedimentary Rocks in Dent Peninsular* (pp. 33–43). Lahad Datu, Sabah: Borneo Science.

Sánchez, A. S., Cohim, E., & Kalid, R. A. (2015). A review on physicochemical and microbiological contamination of roof-harvested rainwater in urban areas. *Sustainability of Water Quality and Ecology, 6*, 119–137. doi:10.1016/j.swaqe.2015.04.002

Shaheed, R., Wan Mohtar, W. H. M., & El-Shafie, A. (2017). Ensuring water security by utilizing roof-harvested rainwater and lake water treated with a low-cost integrated adsorption-filtration system. *Water Science and Engineering, 10*(2), 115–124. doi:10.1016/j.wse.2017.05.002

Suratman, S. (2004). 6.6. IWRM: Managing the Groundwater Component in Malaysia. *Malaysia Water Forum, Kuala Lumpur, Malaysia*, 19–22.

Tarbuck, E. J., & Lutgens, F. K. (2015). *Earth Science* (14th ed.). England: Pearson.

Yan, X., Ward, S., Butler, D., & Daly, B. (2017). Performance assessment and life cycle analysis of potable water production from harvested rainwater by a decentralized system. *Journal of Cleaner Production, 172*, 2167–2173. doi:10.1016/j.jclepro.2017.11.198

Yaziz, M. I., Gunting, H., Sapari, N., & Ghazali, A. W. (1989). Variation in Rainwater From Roof Catchments. *Water Research, 23*(6), 761–765. doi:10.1016/0043-1354(89)90211-X

ADDITIONAL READING

Abdulkareem, H. A., Abdullahi, N., & Shehu, K. M. (2015). Design of a filtration system for a small scale water treatment plant for a rural community around Maiduri area in Borno State, Nigeria. *International Journal of Engineering Science Invention*, *4*(8), 39–43.

Afroz, R., Masud, M. M., Akhtar, R., & Duasa, J. (2014). Water pollution: Challenges and future direction for water resource management policies in Malaysia. *Environment & Urbanization Asia*, *5*(1), 63–81. doi:10.1177/0975425314521544

Ahmad, F., Siwar, C., & Begum, R. A. (2014). Water resources in Malaysia: Issues and challenges. *Journal of Food Agriculture and Environment*, *12*(2), 1100–1104.

Andrade, J. A. O. (2018). *Small-scale water supply system (SSS) for remote and rural areas in developing countries: A case study of the use of Ultra-Low Pressure Ultrafiltration (ULP-UF) as main technology on a decentralised small-scale water treatment plant for remote and rural communities in Colombia*. Germany: Kassel University Press GmbH.

Duong, H. C., Phan, N. D., Nguyen, T. V., Pham, T. M., & Nguyen, N. C. (2017). Membrane distillation for seawater desalination applications in Vietnam: Potential and challenges. *Vietnam Journal of Science and Technology*, *55*(6), 659–682. doi:10.15625/2525-2518/55/6/10715

Kisakye, V., Matte, M., & Van der Bruggen, B. (2017). Comparison of informal rainwater harvesting systems to conventional water sources in terms of microbiological water quality. *Water Resources and Rural Development*, *10*, 45–57. doi:10.1016/j.wrr.2018.10.002

Mena-Rivera, L., & Quiros-Vega, J. (2018). Assessment of drinking water suitability in low-income rural areas: A case study in Sixaola, Costa Rica. *Journal of Water and Health*, *6*(3), 403–413. PMID:29952329

Zaragoza, G., Andres-Manas, J. A., & Ruiz-Aguirre, A. (2018). Commercial scale membrane distillation for solar desalination. *npj Clean Water*, *1*, 1–6.

KEY TERMS AND DEFINITIONS

Brackish Water: Water that contains salinity between that of freshwater and seawater.

Groundwater: Water found beneath the earth's surface which is stored inside spaces and cracks in soil, rock and sand.

Hilly Area: Area with many hills or elevation of lands.

Island: A land that is entirely surrounded by water but smaller than a continent.

Physicochemical Characteristics of Water: Properties of water that covers both the physical and chemical parameters.

Rainwater: Water that is collected from fallen rain.

Small Scale Water Treatment System: System that treats and distributes water near its points of a generation where the piping system involved are comparatively shorter.

Surface Water: Freshwater that is found on the earth's surface such as rivers, ponds, and lakes.

Chapter 8
Nanofibre Membrane Distillation for Brackish Water Treatment in Offshores and Small Islands

ABSTRACT

Membrane distillation is a process driven by the vapour pressure gradient of water to allow the permeation of water vapours through the microporous hydrophobic membrane while retaining other non-volatile components present in the feed. In this chapter, the utilization of membrane distillation as promising membrane technology for the application of clean water production at offshores and small islands are discussed. One of the main challenges of membrane distillation is that the production rate is low compared to the membrane areas used as the process performance is mainly influenced by the membrane characteristics. Thus, the ideal membrane should possess distinct characteristics that are the most suitable exclusively for membrane distillation application. This chapter also highlights the nanofibre membrane as one of the excellent options as it can be fabricated to exhibit hydrophobic, thin, and open pores characteristics.

INTRODUCTION

Many areas situated in the vicinity of the sea or ocean face limited access to potable water. For instance, the offshore platforms, small islands, and isolated areas along the coastlines. The water resources in these areas have limited freshwater; thus the non-conventional water sources mainly come from saline water or brackish groundwater resources, both of which may likely contain salts and deemed unsuitable for drinking. Typically, non-conventional water sources require state of the art technologies for treatment which are often complex to operate and consume high energy.

DOI: 10.4018/978-1-7998-2645-3.ch008

Table 1. Comparison between membrane distillation and reverse osmosis

	Membrane Distillation	**Reverse Osmosis**
Phenomenon	Evaporation	Osmosis
Driving force for transport	Thermal (partial vapour pressure)	Pressure
Energy requirement [a]	1.3 kWh/m^3 for 85 L/m^{2}h at 25 °C	4 kWh/m^3 for 5 – 10 L/m^{2}h at 20 °C

Source: [a](Pangarkar, Sane, & Guddada, 2011)

Offshore platforms are ever more compelled to work in increasingly challenging and remote environments as the oil explorations progress. As a result, consistent potable water supply from the bunkering tanks carried from the mainland may be insufficient due to limited onboard normal storage capacities to support around one week of operation. Scarce water supply for drinking and operational uses may render the offshore operation to cease and faces the consequence of loss of production. Thus, water treatment technology on the offshore platform emerges as one of the most critical challenges faced by the offshore oil and gas industry. Besides, groundwater supplies which are largely relied upon at small islands and remote coastal areas exhibit brackish or saline properties and may be unsuitable for drinking without further treatment. The brackish properties are usually caused by the contamination of saltwater into the underground water reserves. Because of that, suitable treatments to produce potable water in these areas are of the interests for many researchers to come up with practical solutions. Many types of research are focusing on alternative technology approach as opposed to conventional techniques such as reverse osmosis, but are found to be less robust and less consistent.

Reverse osmosis is a membrane filtration process to separate the solvent from a solution, leaving behind a more concentrated solution. In an osmosis process with a semipermeable membrane separating a dilute solution from concentrated solution, the solvent moves through the membrane from the dilute to concentrated side in a bid to balance the concentrations. In this case, the concentration gradient drives the movement phenomenon of liquid. To prevent this flow of solvent, an opposing hydrostatic pressure is applied to the concentrated solution. The extent of pressure needed to thoroughly inhibit the solvent flow is called osmotic pressure. The reverse osmosis phenomenon is achieved when the applied hydrostatic pressure exceeds the osmotic pressure, whereby solvent flows from the concentrated solution to the dilute solution. Although reverse osmosis is capable of producing purified water, the downside is that it consumes a very large amount of energy to operate. More importantly, the high-pressure operation leads to the build-up of a concentrated layer of rejected solutes on the surface of the membrane. This phenomenon is called concentration polarization or cake formation, which is the main cause of fouling, bacteria generation, and flux decline. Moreover, high energy utilization and brine disposal issues are encountered because of limited water recovery. Also, reverse osmosis application would one day become complicated as energy demand and brine desalination will both become unsustainable in the future. Table 1 shows a general comparison between membrane distillation and reverse osmosis in terms of its principle, driving force, and energy requirement.

Promising technology with low energy consumption, high efficiency, and the prospect of utilizing alternative energy sources such as solar energy, geothermal energy, and low-grade waste heat is the main focus to address this issue. One of the interesting alternative technologies that tick the checklist is membrane distillation, which is an emerging technology for separation application that is traditionally achieved via conventional distillation or reverse osmosis. Membrane distillation and reverse osmosis are both closely associated with the desalination process with membrane application. It is attractive because it allows the retention of liquid water via surface tension while permitting the passage of water vapours across the membrane pores. Ideally, only a more volatile component passes through the membrane as permeating fluid, which allows high purity of recovered component. Currently, membrane distillation is still in the research and developmental stage. In desalination application, membrane distillation involves water vapour transport from a brackish solution across a porous hydrophobic membrane.

This process can compete with the superiority of reverse osmosis as it has high salt rejection as well as can operate with low-grade energy sources. The main drawback of this technology is low output and flux which are both originated from the lack of suitable membrane that is as good as reverse osmosis'. An excellent membrane for this application should be characterized with good permeability and hydrophobicity in terms of bubble point pressure and contact angle, low fouling, and high chemical and mechanical stability. Open-pore membranes are an excellent option for membrane distillation. Another important parameter is the membrane thickness.

Nanotechnology has an important potential for membrane distillation due to its unique layer properties. The nanofibre membranes possess open-pore characteristic which is close to the ideal membranes for membrane distillation processes. Nanofibre is defined as fibre with a diameter less than 2 μm. It is pictured as film comprises of nanofibres overlapping with each other in a completely random manner, and typically fabricated by the electrospinning method. Electrospun nanofibre is associated with high porosity and hydrophobicity, which are highly desirable properties for membrane distillation. This type of membrane can be formulated to the desired properties by incorporating various functional materials during the electrospinning process, for instance, the catalytic, biocidal, or hydrophobic attributes. Furthermore, the thin membrane is desired as the flux is inversely proportional to the membrane thickness. On the other hand, microporous membrane is also preferred despite flux being proportional to the membrane pores diameter, in order to allow only water vapours passing through it and prevent pore wetting. However, a critical challenge in producing thin nanofibre membrane lies on the fabrication skill, mainly on balancing the parameters to obtain the desired properties. The applications of nanofibres include filtration and separation, oil spills cleaning, drug delivery systems, sensors, energy storage devices, capacitors and transistors.

In this chapter, we will discuss on membrane distillation technology, some challenges surrounding it primarily on the membrane performance, and how nanofibre can improve the flux and productivity of membrane distillation. The fabrication methods of nanofibre are covered on the electrospinning and melt-blown techniques. In Universiti Malaysia Sabah, the Membrane Technology Research Group has conducted comprehensive studies on the nanofibre utilization for membrane distillation, particularly on the applications of oil-water separation, waste storage landfill, water and energy production system, groundwater treatment and brackish water desalination. Through the collaboration with Zetta Ltd., a Japanese company specializing in nanofibre membrane production, extensive research and developments have been continually carried out and successfully produced a wide range of commercialized products that are manufactured for various applications.

MEMBRANE DISTILLATION

Membrane distillation is a separation technology whereby water vapour is thermally driven across a microporous hydrophobic membrane. The driving force is the vapour partial pressure difference between the feed and permeates sides. The performance of membrane distillation is highly dependent on the membrane and module used, its configuration, and more importantly the thermal control. The operating temperature is relatively low (~70°C) which enables the use of alternative heat sources or low-grade heat energy. This technology is able to compete with conventional distillation process and reverse osmosis due to its ability to offer a better alternative in terms of energy consumption and throughput.

In the membrane distillation process, a liquid feed is directly in contact with the nanofibre membrane surface, though it cannot penetrate across the membrane pores due to the membrane having hydrophobic characteristics. It is crucial to ensure that the membrane is not wetted by the liquid as it can affect the separation efficiency. This is due to the introduction of mass transfer (diffusion) resistance in the membrane pores. Besides, this process relies on the distillation instead of size separation as seen in other conventional microporous filtration. The open membrane pores allow only the water vapour to pass through instead of liquid. The open pores concept of nanofibre membrane can increase the membrane flux as it provides passages for the water vapour to permeate across the membrane.

To prevent the phenomenon of membrane wetting, it is important to ensure that the hydrostatic pressure of the liquid feed does not go beyond the LEP_w. Apart from that, membrane fouling is also a primary cause of wetting, thus regular maintenance is recommended in this process.

Succinctly, membrane distillation should possess the following properties:

1. Using hydrophobic, thin microporous membrane;
2. The driving force is the partial pressure difference in the vapour phase of the permeating component;
3. At least one of the membrane surface must be in direct contact with the liquid feed;
4. Only vapour is transported across the membrane;
5. Membrane pores should not be wetted by liquid feed;
6. No capillary condensation occurs inside the membrane pores; and
7. The partial pressure gradient must not exceed the liquid entry pressure of the membrane.

General Applications, Advantages and Drawbacks of Membrane Distillation

The desired product from the membrane distillation application can either be the permeate or the concentrated retentate. Some other applications of membrane distillation reported are:

1. Desalination of brackish or seawater to produce pure water;
2. Extraction of ethanol from aqueous solution or fermented liquid;
3. The concentration of juice, milk, sugar, and salt; and
4. The concentration of blood.

The major advantages of membrane distillation over conventional distillation processes are the ability to use an alternative energy source (e.g. solar and geothermal energy) or low grade/waste heat source due to low operating temperature, compact unit leading to small plant footprint, and lower capital and operating costs.

Despite all that, there are some shortcomings of this method, for instances the membrane wetting and relatively high membrane costs. Membrane fouling is also a setback but is typically less critical than reverse osmosis. Another problem associated with membrane distillation is the occurrence of membrane loss of hydrophobicity, but extensive researches on the composite membrane are assumed to overcome this shortcoming.

Membrane Distillation Configuration

Membrane distillation generally consists of four configurations: direct contact membrane distillation, air gap membrane distillation, sweeping gas membrane distillation, and vacuum membrane distillation, as shown in Figure 1, Figure 2, Figure 3, and Figure 4 respectively. The major difference between these configurations rests on the different ways vapour pressure difference is generated and permeate recovered.

Direct Contact Membrane Distillation

In the direct contact membrane distillation, evaporated vapours in the hot feed move across the liquid-membrane interface, through the membrane layer, into the permeate side driven by the partial pressure difference. Cool water is used in the permeate side as condensing fluid, upon which water vapours are condensed back into liquid when in contact with the cooling water. Both feed and permeate streams are liquids in direct contact with the membrane surfaces.

Direct contact membrane distillation is widely used in membrane distillation applications as it is simple to set-up, operate, and can give high flux when operated at optimum operating conditions. However, the main drawback of this configuration is that it experiences heat loss by conduction due to the poor conductivity of the polymeric membrane as well as having low thermal efficiency, which is the portion of heat energy utilized for evaporation. This is because, the evaporator (hot feed) and condenser (cooling water) surfaces are very close separated by only a membrane layer, causing low effective driving force. This ultimately leads to low thermal energy efficiency.

Direct contact membrane distillation has been used in many applications primarily in the desalination of brackish water or seawater to produce water at high purity. It is also used in groundwater treatment for drinking water production, wastewater treatment, salt crystallization, fruit juice concentrations, and dye removals in textile industries (Ashoor et al., 2016).

Figure 1. Schematic of a direct contact membrane distillation

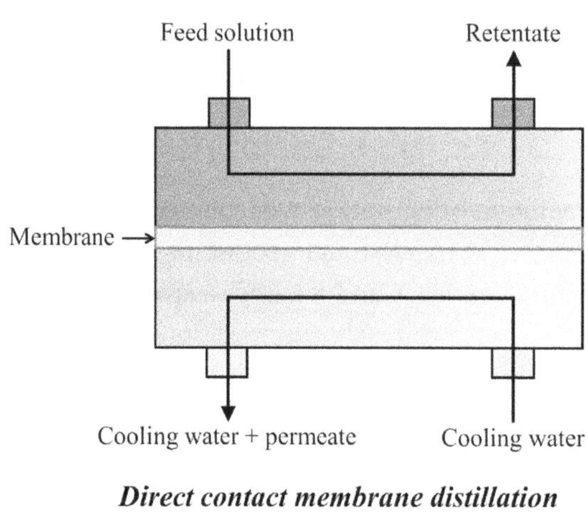

Air Gap Membrane Distillation

The air gap is employed in the permeate side of the membrane distillation, separating the membrane from the cooling plate. Thus, only one membrane surface is in direct contact with the liquid in the feed solution. The liquid boundary layer on the feed-membrane interface allows the water vapours moving across the membrane pores to the other side. The permeating water vapours cross the air gap and condensed on the cold surface of the cooling plate.

The presence of stagnant air gap significantly reduces the heat loss via conduction by the membrane material, since air has poor thermal conductivity. However, it creates an additional mass transfer resistance which is considered a disadvantage. Additional resistance results in lower permeate flux.

This type of membrane configuration is commonly used in desalination of seawater to produce high-quality water, treatment of oil-produced water, dye removal from textile wastewater, and heavy metal removal (Attia et al., 2017).

Sweeping Gas Membrane Distillation

In sweeping gas membrane distillation, water vapours generated at the hot feed-membrane interface diffuse through the membrane and then swept alongside the sweeping gas flowing at the permeate stream to an external condenser. Cold inert gas is used in the permeate side to sweep and transfer the vapours to the outside of the membrane distillation module where condensation occurs. Due to the heat transfer from hot feed to the membrane, the temperature of the sweeping gas increases as it moves along the membrane module length. Typically, dry air or nitrogen gas is used as the sweeping gas.

The advantages of using sweeping gas are the low conductive heat loss as a result of a gas barrier on the permeate side like in air gap configuration, but also reduced mass transfer resistance leading to higher permeate flux as opposed to the air gap configuration. Nonetheless, the small volume of permeates diffuses into a large volume of sweep gas which requires large condenser. Furthermore, sweeping gas configuration incurred higher capital and operational cost as an external condensation unit is required, making it the least popular choice (Onsekizoglu, 2012).

Figure 2. Schematic of an air gap membrane distillation

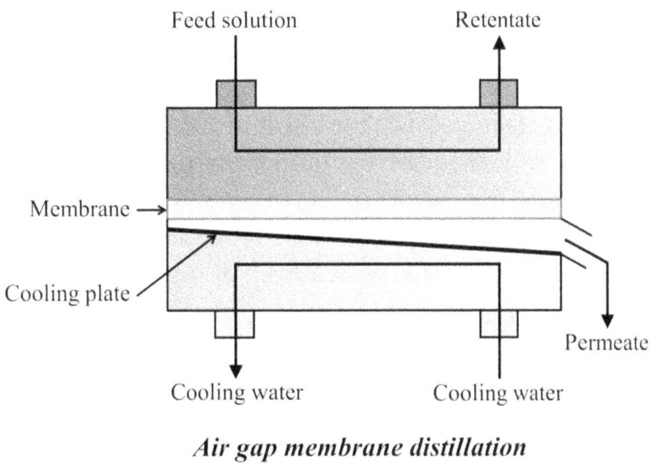

Figure 3. Schematic of a sweeping gas membrane distillation

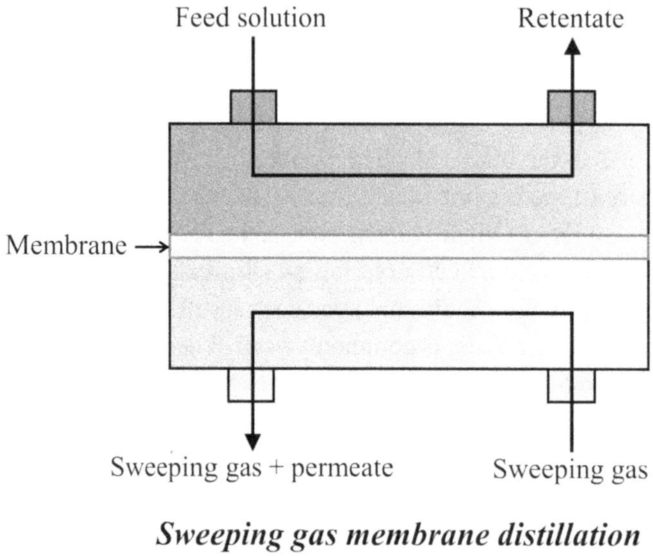

The application of sweeping gas membrane distillation is commonly to remove volatile components from aqueous solution (Alkhudhiri, Darwish, and Hilal, 2012) and concentrating non-volatile compounds (Shirazi & Kargari, 2015).

Vacuum Membrane Distillation

Partial pressure gradient can alternatively be generated in the membrane distillation by applying a vacuum on the permeate side by means of a vacuum pump. The operating vacuum pressure should be lower than the saturation pressure of water. Furthermore, the vacuum pressure should be maintained to prevent the pressure differences from exceeding the liquid entry pressure of water, LEP_w which leads to the undesired membrane wetting. Similar to the sweeping gas configuration, the condensation of water vapours also take place outside the membrane module.

In vacuum membrane distillation, the permeate flux is considered high and heat loss through conduction is very low and negligible due to the very low pressure. However, there is a high probability of wetting the membrane pores since the vacuum created a high absolute pressure difference across the membrane. The occurrence of membrane pore wetting will contaminate the purity of permeate and leads to membrane separation process failure. Since the effect of membrane pore wetting is quite severe, a membrane with smaller pore size than other types of membrane distillation should be used in vacuum configuration.

Vacuum membrane distillation can be applied in the removal of volatile compounds from aqueous solution, the concentration of juices or extracts, treatment of wastewater, and desalination of brackish or seawater (Abu-Zeid et al., 2015).

Figure 4. Schematic of a vacuum membrane distillation

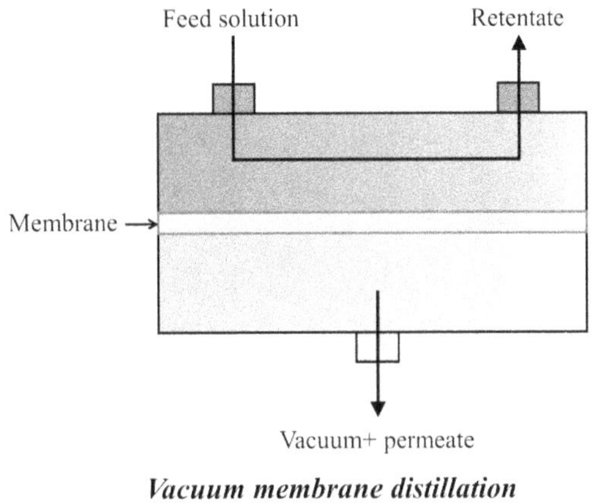

Table 2. Comparison of the major configurations of membrane distillation

Configuration	Schematic of Operation	Principle of Operation	Advantages	Disadvantages	Applications
Direct contact membrane distillation	See Figure 5.	Cooling water is used in the permeate side as a condensing agent to condense water vapours back into a liquid.	Simple to set-up. Simple to operate. High permeate flux.	Heat loss by conduction. Low thermal efficiency.	Desalination. Groundwater treatment Wastewater treatment. Juice concentrations. Dye removal.
Air gap membrane distillation	See Figure 6.	A stagnant air gap exists between the membrane sheet and the condensing (cooling) plate on the permeate side.	Low heat loss as a result of high thermal efficiency.	Additional air gap resistance results in lower permeate flux.	Desalination. Treatment of oil-produced water. Dye removal. Heavy metal removal.
Sweeping gas membrane distillation	See Figure 7.	Sweeping gas (e.g. dry air, inert gases) is used in the permeate side to sweep and carry the vapours to the outside of the membrane system where condensation occurs.	Low conductive heat loss. High permeate flux.	Requires large condenser. Higher capital and operational cost for the external condensation unit.	Remove volatile components from aqueous solution. Concentrating non-volatile compounds.
Vacuum membrane distillation	See Figure 8.	Vacuum pressure is introduced at the permeate side, but a condensation of vapours takes place outside the membrane system.	Low conductive heat loss. High permeate flux.	High probability of wetting the membrane pores	Removal of volatile compounds from aqueous solution. The concentration of juice or extract. Treatment of wastewater. Desalination

Figure 5. Direct contact membrane distillation operation

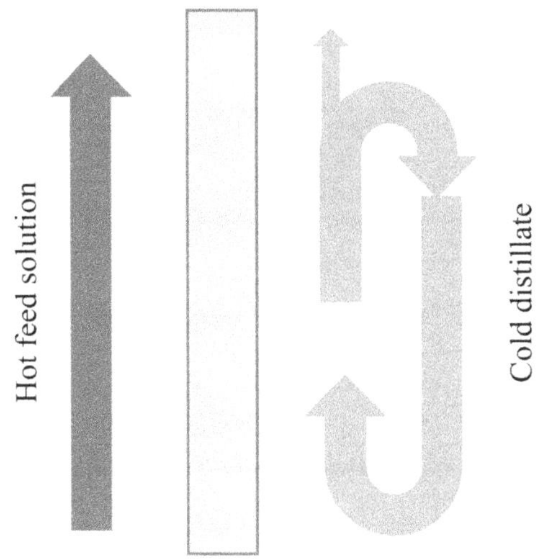

Comparison of the Different Configurations of Membrane Distillation

In summary, the major difference between the four main configurations of membrane distillation lies on the method used in the permeate side (i.e. cooling water, air gap, sweeping gas, or vacuum) to create the vapour pressure gradient between the two sides of the membrane to drive the permeation flux. The schematics of operation for each of the methods are illustrated in Figure 5, Figure 6, Figure 7, and Figure 8. The method used should control the water vapours concentration on the permeate stream as vapour concentration affects the partial pressure of water vapours. A general comparison between these configurations is described in Table 2.

Membrane System Design and Arrangement

Membrane System Design

A membrane system may be comprised of several units of modules or pressure vessels. There are a series of steps taken to design the membrane system and to find the optimum arrangement of the system. The steps to design the arrangement of the membrane system are (The Dow Chemical Company, 2019):

Step 1: Determine the separation objective, feed source, flow rate and quality, and the permeate flow and quality.

Step 2: Determine the type of flow configuration to be used (i.e. plug flow, concentrate recirculation. In plug flow, the feed only passes through the system once, whereas concentrate recirculation requires a recycle stream to recirculate the retentate back into the system to achieve sufficient permeate recovery.

Step 3: Determine the membrane module type and size depending on the feed quality, feed fouling tendency, the rejection requirement, and energy requirement. The module configuration and size consideration is very important in membrane system design to ensure that the system can cope with

Figure 6. Air gap membrane distillation operation

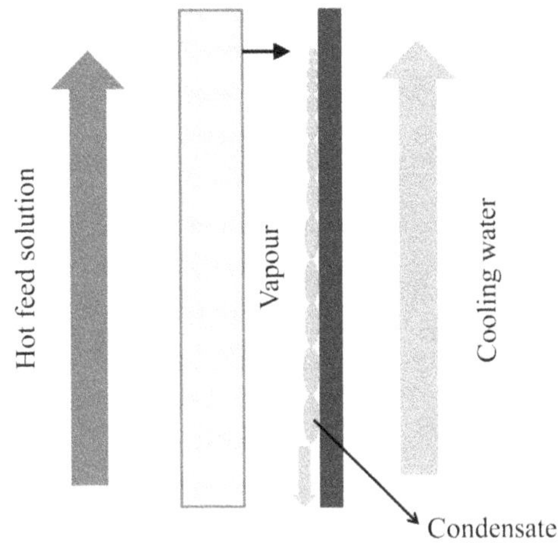

the operating conditions. For example, if the membrane module is too large but the feed flow rate is too low or the pressure is not strong enough to fill the module with liquid feed, less permeate will be obtained. Besides, it leads to inefficient separation performance.

Step 4: Determine the average membrane flux. The membrane flux value can be determined from experimental research or directly from the membrane supplier.

Step 5: Determine the number of membrane element needed. The number of elements required in the depends on the flow rate of the feed, membrane flux, and the membrane surface area required. Spiral wound membrane element is commonly used, where a large membrane sheet is wound into a compact structure element and finally fitted into a housing component to form a membrane module. Figure 9 illustrates the schematic of a typical spiral wound membrane element as well as the cross-section of a pressure vessel with three membrane element. The number of membrane element can be calculated using the formula:

Figure 7. Sweeping gas membrane distillation operation

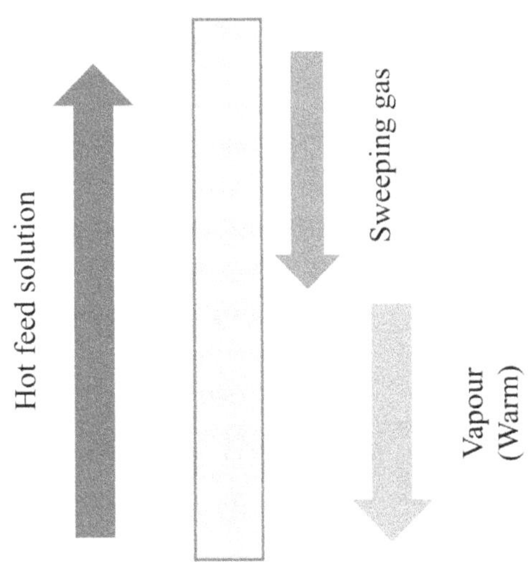

Figure 8. Vacuum membrane distillation operation

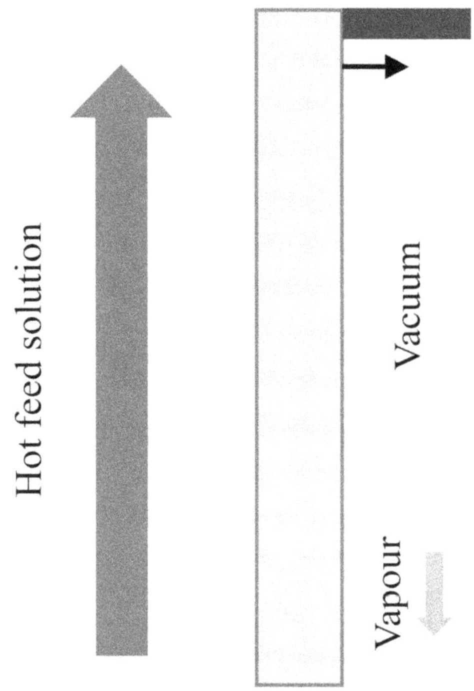

$$N_E = \frac{Q_F}{J.A_m} \quad (1)$$

where N_E is the number of membrane element, Q_F is the feed flow rate, J is the membrane flux, and A_m is the membrane surface area.

Step 6: Determine the number of pressure vessel needed. The pressure vessel is the membrane modules/ elements house with one element fitted into for small or compact system, and even up to 8 elements for a large system. The number of pressure vessel required can be calculated using the formula:

$$N_V = \frac{N_E}{N_{E/V}} \quad (2)$$

where N_V is the number of pressure vessel, N_E is the total number of membrane element, and $N_{E/V}$ is the number of elements per pressure vessel.

Step 7: Determine the number of stages required. The membrane system can be arranged either in a single stage of multiple stages. One stage of a membrane system shows how many elements are arranged in series, and each stage comprised of a number of membrane vessels. The number of

Figure 9. Schematic of (a) a spiral wound membrane element and (b) a pressure vessel with three membrane element

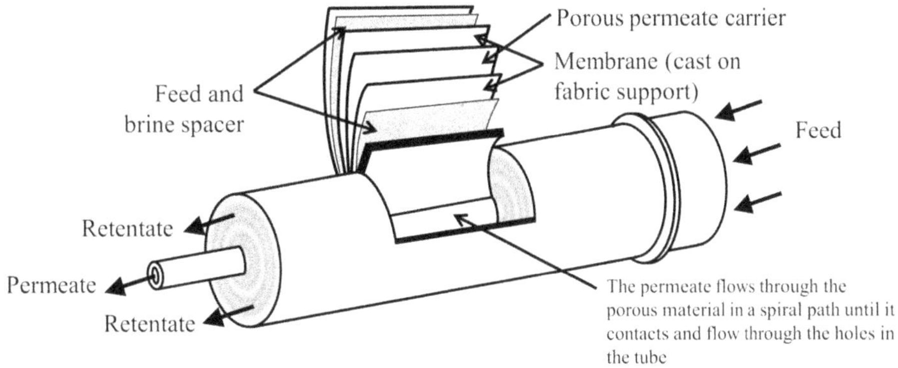

(a) Cutaway view of spiral wound membrane element

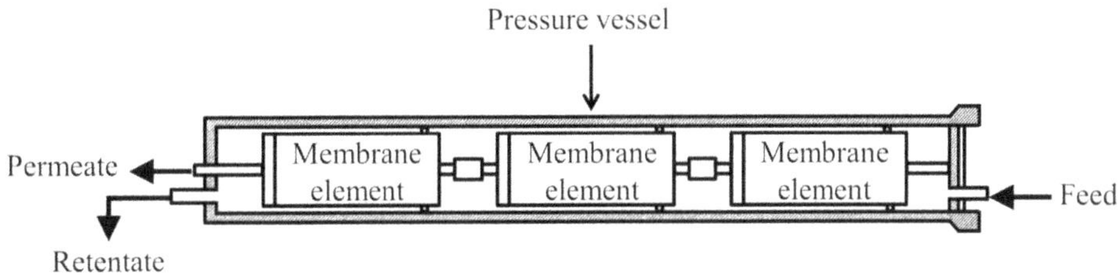

(b) Cross section view of pressure vessel with 3 membrane element

stages depends on the system recovery needed, a number of elements fitted in a module, and the quality of feed. A high system recovery will require more stages in a certain arrangement, i.e. the number of pressure vessels placed in each stage. This can be seen from for an example in Table 3 for the number of stages required in brackish water desalination with respect to the desired system recovery for a fixed 6-element per module system.

Step 8: Determine the staging ratio. Staging ratio describes the relationship between the number of pressure vessels in the succeeding stages. For example, a two-stage membrane system with six vessels in the first stage and three vessels in the subsequent stage has a staging ratio of 2:1. This also applies to a system with four membrane vessels in the first stage and two vessels in the second stage, and so on. Another example is for a three-stage membrane system with four, three, and two membrane vessels in the first, second, and third stages respectively, the staging ratio of the system is 4:3:2.

The staging ratio between two adjacent stages can be calculated using:

$$R_s = \frac{N_V(i)}{N_V(i+1)} \tag{3}$$

Table 3. Number of stages for a brackish water desalination system

System Recovery (%)	Number of Elements in Series	Number of Stages (6-Element Module)
40 – 60	6	1
70 – 80	12	2
85 – 90	18	3

where R_s is the staging ratio, $N_V(i)$ is the number of pressure vessel in the *i*th stage, and $N_V(i+1)$ is the number of pressure vessel in the (*i*+1)th stage.

For an ideal stage membrane system, each stage should operate at a similar fraction of system recovery only if the modules hold the same number of elements. The staging ratio, R_s for a membrane system with n stages and a fraction of system recovery, Y can be calculated as:

$$R_s = \left(\frac{1}{1-Y}\right)^{\frac{1}{n}} \tag{4}$$

Consequently, the number of pressure vessels in stage one can be calculated using the staging ratio and the total number of pressure vessels. Equation 5, 6 and 7 shows the calculation for two-stage, three-stage and n-stage membrane system.

$$N_V(1) = \frac{N_V}{1+R_s^{-1}}, for\, n = 2 \tag{5}$$

$$N_V(1) = \frac{N_V}{1+R_s^{-1}+R_s^{-2}}, for\, n = 3 \tag{6}$$

$$N_V(1) = \frac{N_V}{1+R_s^{-1}+R_s^{-2}+\ldots+R_s^{-(n-2)}+R_s^{-(n-1)}}, for\, n\text{-}stage \tag{7}$$

where $N_V(1)$ is the number of pressure vessels in stage 1, N_V is the total number of pressure vessels required, R_s is the staging ratio, and n is the number of stages.

The number of pressure vessels in stage 2 and so on can then be calculated using Equation 3.

Membrane System Arrangement

The membrane system arrangement during installation is a critical process as it affects the majority of the process performance. The optimization of membrane system design to produce effective operation

Figure 10. 1-stage membrane system

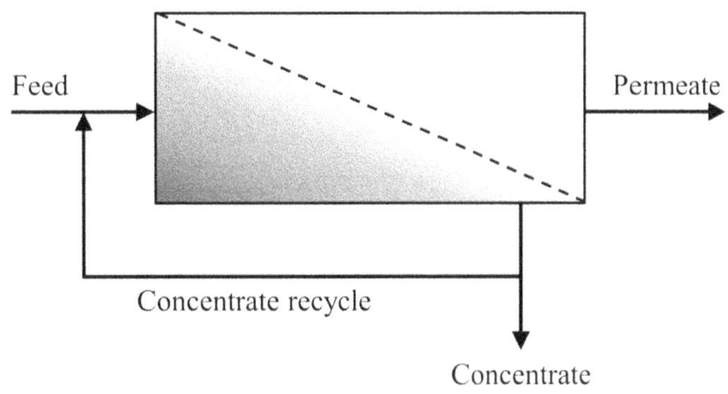

should consider two important aspects, which are the technical and economic qualities. There are several established ways to design the membrane system arrangement, which take into account the fraction of recovery, feed and permeate qualities, space limitation, costing, etc. In general, the arrangement of the membrane system consists of the stage and pass considerations. In a system with more than one stage arrangement, the retentate streams are passed through to the next stages to further recover the permeate volume, whereas membrane system with more than 1 pass will pass the permeate streams to the next passes to further increase the permeate quality.

A membrane system can be designed to consist of 1-stage (single-stage) or 2-stage or more (multi-stage) separation as shown in Figure 10 and Figure 11 respectively. This type of arrangement is usually employed when the permeate quality is not of the highest interest. In the 1-stage membrane separation, the feed passes through the pressure vessel in a single flow, where the permeate is recovered, while the retentate is discharged or alternatively recycled back to the feed stream to recover more permeate. Typically, single-stage membrane distillation is used for laboratory testing or mini-pilot plant.

In the other multistage membrane separation (two or more stages), the permeate recovery is increased through the means of using two or more pressure vessel, whereby the feed flows through a single pass

Figure 11. 2-stage membrane system

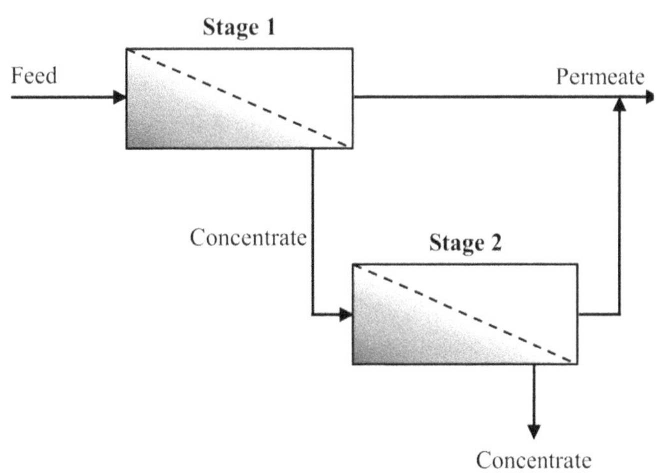

Figure 12. 2-stage (2:1 array) membrane system

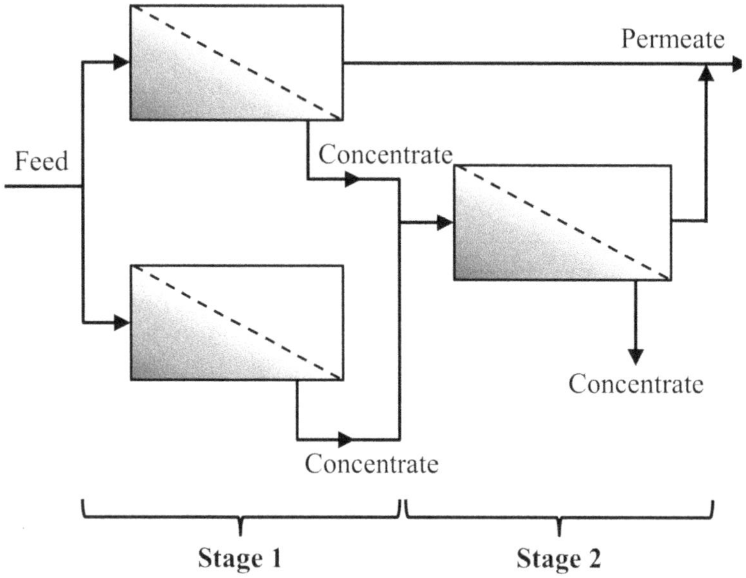

across the vessels arranged in series. The retentate from one vessel becomes the feed stream for the subsequent vessel, and finally, the permeate recovered by each vessel is accumulated together. Increasing the number of stages will normally increase the recovery rate. In multistage membrane system, the vessels can be arranged in numerous ways depending on the optimum operating conditions that fit the operation.

The multistage membrane system can be arranged in arrays at a certain staging configuration to further increase the recovery rate. Typical staging ratio used in array configuration is 2:1, where the retentate streams from the two pressure vessels in the first stage are combined and become the feed stream for the successive one pressure vessel in the second stage. This type of arrangement further improves the recovery to up to 90%. This type of arrangement is illustrated in Figure 12.

Membrane separation processes which require high permeate qualities commonly employs the membrane pass consideration. In a single-pass membrane system arrangement, the process is similar to the 1-stage arrangement as the feed stream is filtered only once through a single membrane vessel. Thus, the permeate quality is the same as the 1-stage operation. Both single-pass and 1-stage membrane system may comprise of several vessels arranged in parallel. This is because, the essence is that all vessels receive the same feed stream and technically produce the same permeate quality and recovery, given that similar membrane vessels are used. Figure 13 shows an example of a single-pass membrane system with two membrane vessels arranged in parallel.

In operations where high permeate quality is preferred, a membrane system with two or more passes is employed. Normally, the double-pass configuration is sufficient for most processes, by which permeate recovered from the first vessel becomes the feed for the second vessel. This results in higher permeate quality, as the feed basically passes through "two" membrane system although both are configured as a single system. Similar to membrane staging configuration, the multi-pass membrane system can also be arranged in arrays as shown in Figure 14.

In conclusion, the most optimum membrane system arrangement relies on many considerations which affect the overall performance. Once the best system has been designed, in practice the system may not

Figure 13. Single-pass membrane system

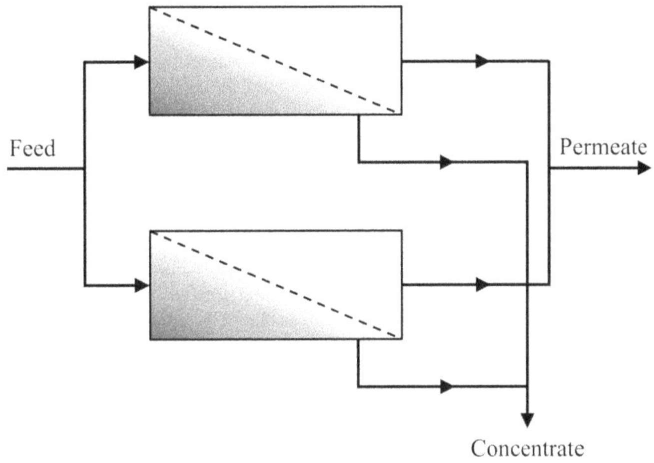

work very well due to external constraints that are not anticipated during the design stage. In membrane distillation especially, the system is sensitive to changes in partial pressure difference which are mainly controlled by the operation employed in the permeate side and some by the feed pump. For example, multiple vessels in the array membrane distillation system are commonly arranged horizontally stacked upwards, whereby each membrane vessel experiences different pressure drop due to greater hydrostatic pressure experienced at the bottom vessels. Higher pressure may be required to pump the feed to reach the top vessel, but this might come at the price of membrane wetting at the bottom vessels due to higher pressure experienced. On the other hand, the system may also be arranged vertically or slanted. Therefore, extensive research is required when developing any membrane system, because though a system may work theoretically, it may fail in real practice. The membrane distillation system design requires experiments, tests, and optimization until the best system can be obtained.

Figure 14. Double-pass (2:1 array) membrane system

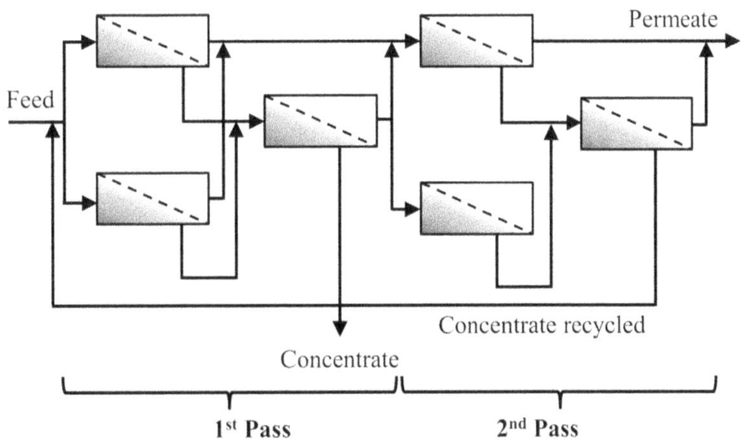

Table 4. Critical characteristics of membrane for membrane distillation application

Membrane Characteristics	Description
Thickness	[a] Thin membrane has lower mass transfer resistance; however, it possesses poor heat efficiency as a result of temperature polarization and loss of conduction through the membrane. The opposite is also true in this context. Some researchers have found the optimum membrane thickness to lie between 30 – 60 µm.
Pore size	Porous membrane typically does not have a single pore size, instead, it has a pore size distribution in some range. Pores with larger size can produce higher flux, yet highly likely to encounter pore wetting. Besides, it reduces the membrane performance by allowing the brackish liquid feed to penetrate into the distillate. The typical pore size for nanofibre is between 10s – 100s nanometers.
Porosity	The membrane porosity affects the flux and thermal efficiency of the membrane. This is because the heat conductivity of air inside the pores is lower than the polymeric membrane.
Hydrophobicity	[b] Hydrophobic polymeric membranes are typically made up from polypropylene, polyethene, polytetrafluoroethylene, and polyvinylidene fluoride. Polytetrafluoroethylene gives the highest hydrophobicity but is challenging to manage, thus polyvinylidene fluoride is mostly used to make a hydrophobic membrane.

Source: [a](Lagana, Barbieri, & Drioli, 2000); [b](Drioli, Ali, & Macedonio, 2015)

NANOFIBRE MEMBRANE FOR MEMBRANE DISTILLATION

Characteristics of Nanofibre

Membrane distillation performance is strongly subjected to the membrane structure of the membrane used, for example, its thickness, porosity, and pore size distribution. It is crucial for the membrane to provide an interface for the feed and distillate while creating a resistance barrier that prevents them from mixing. The 'perfect' membrane should possess hydrophobic (nonwetted) characteristic with high flux (thin membrane, large pores), low heat loss (thick membrane, small pores), and ability to produce pure distillate (high liquid entry pressure of water (LEP_w), high contact angle). Yet, it would be nearly impossible to fulfil all these conflicting necessities, thus an optimum balance would need to be obtained. The critical characteristics of membrane for membrane distillation application are further elaborated in Table 4.

Nanotechnology-based nanofibre is emerging nanotechnology for an extensive scale, from research and development to commercialization applications. This stems from its distinctive physicochemical attributes and properties. Nanofibre normally has a cross-sectional diameter in the range of tens to hundreds of nanometers. Due to this, it holds a very high specific surface area as well as the surface area to volume ratio. Furthermore, they can create networks of extremely porous fibre mesh that has incredible interconnectivity between the pores. Aside from this, nanofibre can be fabrication from an extensive variety of materials, such as natural and synthetic polymers, a semiconducting material, and composite material. Nanofibre has the characteristics of high porosity, interconnected open pore structure, and highly permeable to vapour which are rarely found in conventional fibres.

Fabrication of Nanofibre Membrane

Nanofibre membranes are commonly produced as 'nonwoven' webs through the conventional techniques of solvent electrospinning or melt-blown. Nonwoven web implies a sheet or film of fibres linked together by entanglement between each fibre without the need for stitching or knitting. Nonwoven nanofibres

Table 5. Comparison between electrospinning and meltblowing

Criteria		Electrospinning	Meltblowing
Filament size		0.05 → 0.8 μm	0.7 → 2.5 μm
Fibre length		Continuous	Continuous
Size deviation		Low, 30%	High, ≥ 50%
Final fibre		Dry	Dry
Production throughput	Costs	0.5 – 2.5 kg/hr	10 – 20 kg/hr
	Energy consumption	Medium	High
Other criteria	Markets	Filtration, textile, medical	Filtration, medical
	Polymers	Wide range	Limited to melt polymers
	Other	Mainly used as a surface coating technology	Commercial mass production

Source: (Gangwal & Wright, 2013)

are notable for their characteristics of having a high specific surface area, large ratio of surface-area-to-volume, exceptionally small pore size, and superb mechanical features

Table 5 shows a broad comparison between electrospinning and melt-blown methods. Generally, the electrospinning process can produce fibres with a smaller diameter with a low deviation. However, the production rate of the melt-blown process is 8 – 20 times higher than electrospinning. In terms of costs of the machine, both processes require high capital costs, but lower energy is consumed in electrospinning. As for the polymer used, electrospinning can use both polymer solution or polymer melt, while meltblowing can only use polymer melt in its process.

Electrospinning Technique

Electrospinning technique is commonly used to fabricate the nanofibre membrane for the application of membrane distillation. Very fine nanofibre can be fabricated using this method in the range of 50 – 800 nm. Generally, the setup of electrospinning comprised of a syringe with a nozzle containing the polymer solution or melt, high voltage power supply, a grounded collector, and a pump. It can be configured either through downward, upward, or horizontal setup. The upward electrospinning setup is shown in Figure 15. Electrospinning utilizes the electrostatic repulsion forces in a large electrical field to synthesis the nanofibre. This process can produce nanofibre either from polymer solution or melt which is housed in the syringe.

Commonly, a polymer solution is used instead of the polymer melt in the electrospinning process, thus a solvent is employed in the process. As the polymer solution is ejected, it is being electrospun by the big electrical field generated between the syringe nozzle and grounded collector. The polymer solution droplet at the nozzle tip forms a cone-shaped distortion due to the potential difference between the nozzle and grounded collector. Whilst the cone-shaped polymer solution moves toward the grounded collector, evaporation of the solvent occurs, which subsequently forms the solid continuous nanofibre on the grounded collector.

By operating the electrospinning using polymer solution, typically <30 wt% of polymer pellets are dissolved in solvent(s) to form a homogenous polymer solution. It is important to select a solvent that

Figure 15. Schematic view of the upward electrospinning setup

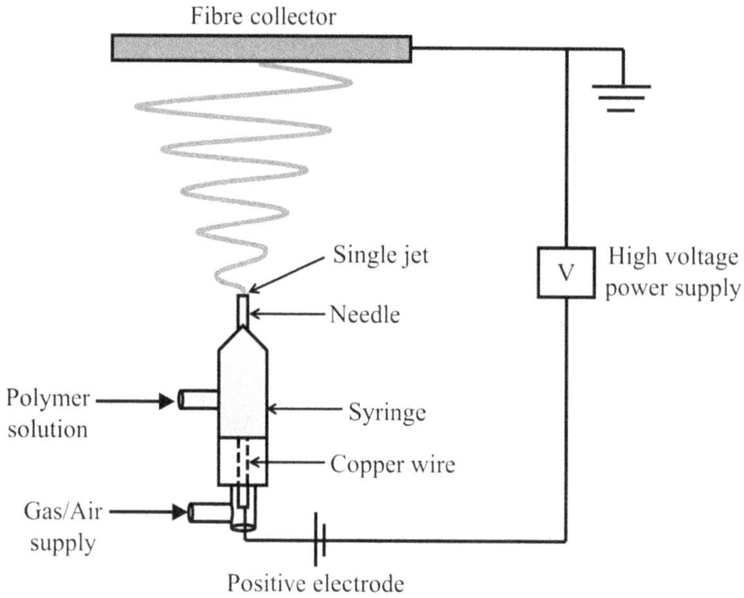

has a high degree of conductivity to increase the quality of the fibres. Otherwise, an organic or inorganic salt can be employed to spike the polymer solution. Yet, this presents impurity on the final product. The most commonly used solvents for the electrospinning application are dimethylformamide, dimethyl sulfoxide, chloroform, tetrahydrofuran, dichloromethane, and hexafluoroisopropanol. As the polymer is electrospun, around 70 wt% of the solvent is evaporated and discharged to the surrounding. The solvents are generally expensive and hazardous, hence they must be captured and recovered.

The physical properties of nanofibre produced through electrospinning process are reliant on an array of affecting parameters. These include solution properties (i.e. conductivity, viscosity, surface tension, and solvent volatility), environmental influences (i.e. operating temperature, and humidity), and technical factors (i.e. nozzle-collector distance, electrical potential, and polymer solution flow rate).

Meltblowing Technique

Meltblowing is a simple one-step process for the production of self-bonded nonwoven nanofibre membrane directly from polymer resin. The nanofibre size produced by melt-blown technique is comparatively bigger than electrospinning, generally in the range of 700 nm – 2.5 μm. The setup of a melt-blown machine normally consists of four major elements: (1) extruder, (2) metering pump, (3) melt-blown die assembly, and (4) drum collector, as illustrated in Figure 16. The equipment can be configured either in a horizontal or vertical setup.

In meltblowing process, the polymer melt is solely used as the feed to make the nanofibre. The polymer in the shape of pellet, granule, or powder is gravity-fed from the resin chamber to the extruder at a controlled flow rate. The extruder is a heated compartment which functions to heat and subsequently melts the polymer to attain the desired viscosity. The polymer melt is then carried towards the die assembly by the action of the rotating screw(s). The metering pump ensures that the polymer melt is distributed

Figure 16. Schematic diagram of a meltblowing equipment

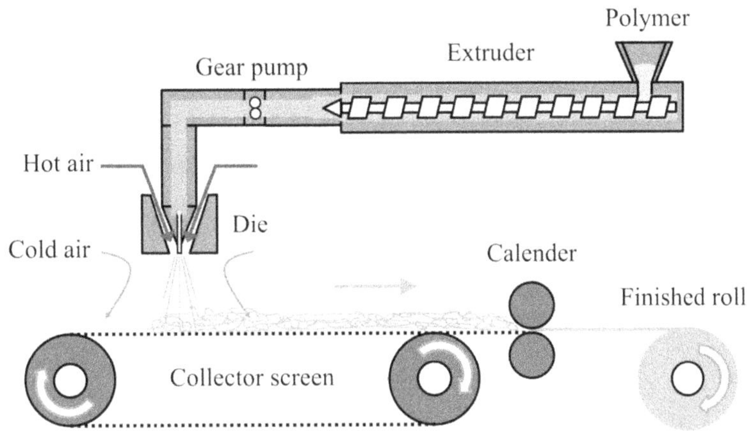

evenly and consistently to the die assembly. Here, the nanofibre is ejected while hot air is blown to prevent it from solidifying instantly. The resulting nanofibres are then collected as web in the drum collector.

The nanofibre diameter formed is affected by several parameters associated with the material properties and process. For the material properties, these include the type of polymer, molecular weight or melt flow index, polymer form used (e.g. pellet, granule, or powder), and usage of additive. On the other hand, the process parameters affecting the nanofibre produced include the distance from tip to the collector, flow rate and temperature of the polymer, flow rate and temperature of the air, and die profile.

Nanofibre Production Through UMS-Zetta Partnership

Mass Production of Nanofibre as Industrial Raw Material

Mass production of nanofibre by Zetta Ltd. has accomplished through their novel Zs methods on the basis of electrospinning and melt-blown. These methods are (1) Zs (Solvent) and (2) Zs (Melt) methods. Through their novel methods, Zetta is able to mass-produce nanofibre at low cost with a diameter less than 100 nm and production capacity of 100 kg/hr under development.

Zetta's methods produce distinguished features in their nanofibre, such as high electrical charge, high volume generation, small diameter fibres, and safer operation (i.e. explosion is prevented by air blowing charged nanofibre in between nozzle and substrate, and by diluting flammable solvent using air to less or equal to 2000 ppm).

Face Masks and Filters

There are wide applications suitable for the nanofibre produced by Zetta. Face masks and filters are produced from Zs (Solvent) method, with an outstanding capability of filtering fine particulate matter (PM2.5). Moreover, since this method can generate a 3D structure, it is also effective for microparticulate matter (PM0.5). The filters have high collection efficiency and also low-pressure loss. Furthermore, the filters produced are effective against Ebola haemorrhagic fever since it can be sized to trap Ebola viruses. By utilizing these features, this product can be expanded into other application such as protective clothing.

Oil-Water Separation

The nanofibre is fabricated from polypropylene or polyethene terephthalate which are characterized as lipophilic (oil-loving) and exceptionally water-repelling. Because of that, the nanofibre has large oil absorption capability (up to 100 times of its own dead weight), has absorption-type retention and storage type absorption rate, low influence by viscosity, and fast absorption rate. Various applications for oil-water separation are applicable using this nanofibre, including (1) oil recovery from accidental leakage of oil from marine or off-shore oil platform, (2) palm oil recovery from palm oil mill effluent (POME), (3) oil-water separation for oil emulsion created from shale oil, and (4) oil recovery from oily smoke or grease trap.

Desert Greening

By utilizing the water-repellent characteristic of nanofibre made from polypropylene or polyethene terephthalate, water-storing can be performed by using the capillary phenomenon of sand in the desert. Nanofibre sheet placed on desert sand can reflect sunlight which consequently preserves thermal retention of sand. Besides, it helps to prevent water from freezing at night and keeping the water from evaporating during the day. Since nanofibre is very light, only a small amount is needed for this application. For instance, 1 kg of the nanofibre sheet can cover the desert area of 30 m^2. The installation can be done manually without the usage of heavy machinery. By retaining water in the sand, water-lodged sand will not be blown away, while sand flown from other areas can be used to retain water. This can give measures of the sandstorm in the area affected.

REFERENCES

Abu-Zeid, M. A. E., Zhang, Y., Dong, H., Zhang, L., Chen, H. L., & Hou, L. (2015). A comprehensive review of vacuum membrane distillation technique. *Desalination, 356*, 1–4. doi:10.1016/j.desal.2014.10.033

Alkhudhiri, A., Darwish, N., & Hilal, N. (2012). Membrane distillation: A comprehensive review. *Desalination, 287*, 2–218. doi:10.1016/j.desal.2011.08.027

Ashoor, B. B., Mansour, S., Giwa, A., Dufour, V., & Hasan, S. W. (2016). Principles and applications of direct contact membrane distillation (DCMD): A comprehensive review. *Desalination, 398*, 222–246. doi:10.1016/j.desal.2016.07.043

Attia, H., Osman, M. S., Johnson, D. J., Wright, C., & Hilal, N. (2017). Modelling of air gap membrane distillation and its application in heavy metals removal. *Desalination, 424*, 27–36. doi:10.1016/j.desal.2017.09.027

Drioli, E., Ali, A., & Macedonio, F. (2015). Membrane distillation: Recent developments and perspectives. *Desalination, 356*, 56–84. doi:10.1016/j.desal.2014.10.028

Gangwal, S., & Wright, M. (2013). Nanofibres: New scalable technology platform for producing polymeric nanofibres. *Filtration & Separation, 50*(Mar/Apr), 30–33. doi:10.1016/S0015-1882(13)70077-6

Lagana, F., Barbieri, G., & Drioli, E. (2000). Direct contact membrane distillation: Modelling and concentration experiments. *Journal of Membrane Science*, *166*(1), 1–11. doi:10.1016/S0376-7388(99)00234-3

Onsekizoglu, P. (2012). Membrane Distillation: Principle, Advances, Limitations, and Future Prospects in Food Industry. In S. Zereshki (Ed.), *Distillation – Advances from Modelling to Applications* (pp. 233–266). Rijeka, Croatia: InTech Open. doi:10.5772/37625

Pangarkar, B. L., Sane, M. G., & Guddad, M. (2011). Reverse osmosis and membrane distillation for desalination of groundwater: A review. *ISRN Materials Science*, *2011*, 1–9. doi:10.5402/2011/523124

Shirazi, M. M. A., & Kargari, A. (2015). A review on applications of membrane distillation (MD) process for wastewater treatment. *Journal of Membrane Science and Research*, *1*, 101–112.

The Dow Chemical Company. (2019, March 29). *FILMTEC Membranes – Steps to Design a Reverse Osmosis System*. Retrieved from http://msdssearch.dow.com/PublishedLiteratureDOWCOM/dh_0043/0901b803800435ce.pdf

ADDITIONAL READING

Alkhudhiri, A., Darwish, N., & Hilal, N. (2012). Membrane distillation: A comprehensive review. *Desalination*, *287*, 2–218. doi:10.1016/j.desal.2011.08.027

Feng, C., Khulbe, K. C., Matsuura, T., Gopal, R., Kaur, S., Ramakrishna, S., & Khayet, M. (2008). Production of drinking water from saline water by air-gap membrane distillation using polyvinylidene fluoride nanofibre membrane. *Journal of Membrane Science*, *331*(1-2), 1–6. doi:10.1016/j.memsci.2007.12.026

Jonsson, A. S., Wimmerstedt, R., & Harrysson, A. C. (1985). Membrane distillation: A theoretical study of evaporation through micro-porous membranes. *Desalination*, *56*, 237–249. doi:10.1016/0011-9164(85)85028-1

Khayet, M., & Matsuura, T. (2011). *Membrane Distillation: Principles and Applications*. Oxford, UK: Elsevier. doi:10.1016/B978-0-444-53126-1.10012-0

Martinez, L. (2004). Comparison of membrane distillation performance using different feeds. *Desalination*, *168*, 359–365. doi:10.1016/j.desal.2004.07.022

Schofield, R. W. A., Fane, A. G., & Fell, C. J. D. (1987). Heat and mass transfer in membrane distillation. *Journal of Membrane Science*, *33*(3), 299–313. doi:10.1016/S0376-7388(00)80287-2

Tomaszewska, M. (2000). Membrane distillation – examples of applications in technology and environmental protection. *Polish Journal of Environmental Studies*, *9*(1), 27–36.

Wirth, D., & Gabassud, G. (2002). Water desalination using membrane distillation: Comparison between inside/out and outside/in permeation. *Desalination*, *147*(1-3), 139–145. doi:10.1016/S0011-9164(02)00601-X

KEY TERMS AND DEFINITIONS

Air Gap Membrane Distillation: One of the four types of configurations of membrane distillation, which is characterized by the stagnant air gap between the membrane and cooling plate at the permeate side to reduce the heat loss through conduction.

Direct Contact Membrane Distillation: One of the four types of configurations of membrane distillation which uses cooling water in the permeate side as the condensing fluid.

Electrospinning: A technique to produce fibres and is widely employed to fabricate nanofibres, which uses electric force to draw the polymer solution or melt from the nozzle in a completely random manner up to fibre diameter of tens to hundreds nanometer.

Meltblowing: A one-step fabrication process to produce self-bonded nonwoven nanofibre from polymer melt with a diameter in the range of hundreds to several thousands of nanometer.

Nanofibre: Fibre with the diameter in the nanometer range.

Sweeping Gas Membrane Distillation: One of the four types of configurations of membrane distillation that uses sweeping gas in the permeate side to sweep and carry the permeate vapours to the outside of the membrane system where condensation occurs.

Universiti Malaysia Sabah: A public university situated at Kota Kinabalu, Sabah, East Malaysia, which was established in the year 1994.

Vacuum Membrane Distillation: One of the four types of configurations of membrane distillation that employs vacuum pressure at the permeate side with the condensation of vapours taking place outside the membrane system.

Zetta: A Japanese company specializing the production of nanofibre membrane and other nanofibre membrane-based products such as industrial membranes and face masks.

Section 3

Chapter 9
Conventional Wastewater Treatments

ABSTRACT

Conventional wastewater treatment consists of chemical, biological, physicochemical, and mechanical processes to remove organic loading, solids, and nutrient contents from wastewater. Biological processes are more commonly used in wastewater treatment as secondary or tertiary treatments, as it is more effective and more economical than chemical and mechanical processes. In this chapter, several types of wastewaters generated from municipal or industrial activities are discussed. Wastewater has different pollutant contents depending on the point of generation which consequently requires different ways of treatment. Some commonly used conventional wastewater treatment technologies are introduced. A particular focus is given to both aerobic and anaerobic treatments.

INTRODUCTION

Wastewater is any water which has been adversely affected in quality by anthropogenic influences. The wastewaters are originated from human wastes, cesspit leakage, septic tank discharge, sewage treatment plant leakage, washing water, surface water contaminated by sewage, groundwater infiltrated into sewage and industrial wastes. In general, the wastewaters can be categorised into domestic or sanitary wastewater, wastewaters from institutions, industrial wastewater, infiltration into sewers, stormwater, leachate and septic tank wastewater. The composition of wastewater changes widely. The wastewater may contain about 90% of water, microorganisms, organic materials, nutrients, metals, odour and taste and radioactivity. The characteristics of wastewaters may vary according to their point sources.

Wastewater needs to be treated as they pose harmful hazards to human and the environment. Several major sources of wastewater in Sabah will be discussed, which includes sewage and palm oil mill industry wastewater which are discussed in Chapter 10 and Chapter 11 respectively, and oil and gas industry wastewater, urban wastewater, nonpoint source, wastewater from floating residential areas, and other

DOI: 10.4018/978-1-7998-2645-3.ch009

industrial wastewaters which are discussed in the subsequent sections. A particular focus is given to the wastewater characteristics and status of its treatment and management.

The most common wastewater treatment methods are using conventional wastewater treatments. Conventional wastewater treatments plants generally combine the chemical processes, physicochemical processes, biological processes, and mechanical processes. In this chapter, the discussion is centralized to the biological treatment processes as the chemical and physicochemical processes are already discussed in Chapter 6. Some of these processes can be used interchangeably between water and wastewater treatment.

Biological wastewater treatment is a critical part of any wastewater treatment plants for treating both municipal and industrial wastewaters. Biological processes possess economic value in both capital and operating expenses as compared to some other chemical processes (Mittal, 2011). Biological treatments can be operated under anaerobic or anaerobic condition. Aerobic refers to a condition where the microbial reactions occur in the presence of free oxygen, while anaerobic denote a biological reaction that takes place in the absence of free oxygen. These reactions treat the wastes present and produce by-products such as biogas, water, heat, and compost. These conditions are determined depending on the microorganisms involved, i.e. aerobes or anaerobes. Table 1 shows some differences between aerobic and aerobic treatment.

The biological wastewater treatment is discussed in terms of the technologies commonly used to treat various types of wastewater. This includes open ponding, activated sludge, sequencing batch reactor, oxidation ditch, extended aerator, anaerobic digester, trickling filter, rotating biological contactor, septic tank, and Imhoff tank.

SOURCES OF WASTEWATER

Oil and Gas Industry Wastewater

The increase of population, industrial development and the depletion of natural resources have invited a scientist to explore other sources for energy demand especially resources of fossil fuel. Being a naturally diversified state, Sabah has always been cautious about its natural resources and has put a full effort into protecting its natural environment. For the same reason, Sabah has become a popular target for exploration of natural resources and industrial development for opportunity seekers.

After the discovery of an abundant source of fossil fuel on offshore of Sabah, the state has gained an increasing amount of attention from some oil and gas company around the world. Consequently, in order to fully utilize the natural occurring resource, more infrastructures for exploration, extraction, transfer, storage and processing is expected to be constructed. It is therefore anticipated that Sabah's oil and gas industry to boom in the near future.

The oil and gas industry is generally divided into three main operations, namely upstream, midstream and downstream. Exploration of hydrocarbon resources, well drilling and hydrocarbon recovery including crude oil and liquefied natural gas (LNG) are conducted in the upstream operation. Midstream operations connect the upstream and downstream operation. The midstream operations mostly include transportation and storage of recovered hydrocarbon. Facilities in midstream may include pipelines and gathering systems. The downstream operation of oil and gas is where the recovered hydrocarbon is processed, refined and distributed as marketable products such as petrol, kerosene, diesel, lubricants, waxes, asphalt, natural gas, and liquefied petroleum gas (LPG) as well as numbers of petrochemicals.

Table 1. Differences between aerobic and anaerobic treatments

Parameter	Aerobic Treatment	Anaerobic Treatment
Principle of operation	Microbial reactions take place in the presence of free oxygen	Microbial reactions take place in the absence of free oxygen
Reaction rates	Relatively rapid and fast	Relatively time consuming and slow
Suitability	Wastewater with low organic concentration	Wastewater with medium and high organic concentrations
By-products	Carbon dioxide, water, excess biomass	Carbon dioxide, methane, excess biomass
Biomass yield	Relatively high	Relatively low
Specific substrate utilization rate	Relatively low	Relatively high
Post-treatment	Typically followed by filtration, disinfection, or direct discharge	Typically followed by aerobic treatment
Footprint	Relatively large	Relatively small
Capital and operating expenses	Relatively high	Relatively low
Example of technologies	Activated sludge process, sequencing batch reactor, oxidation ditch, aerobic pond	Anaerobic digester, up-flow anaerobic sludge blanket reactor, anaerobic pond

Characteristics of Oil and Gas Wastewater

In Sabah, the development of the oil and gas industry has begun more a decade ago. Malaysia's very own oil and gas company, Petroleum Nasional Berhad (PETRONAS) is the key to the development of Sabah oil and gas industry. PETRONAS has established several projects in Sabah that has significantly changed the state's human capital and economic development. As of 2015, newly operated projects in Sabah include Sabah Oil and Gas Terminal (SOGT) in Kimanis, Sabah-Sarawak Gas Pipeline (SSGP), Kimanis Power Plant (KPP) and Kinabalu Non-Associated Gas (NAG). These projects are in addition to the existing Sabah Gas Terminal (SBGAST) located in Tuaran, Sabah. Another Liquefied Natural Gas (LNG) plant is expected to be built in Lahad Datu.

Apart from these projects, PETRONAS is also opening up a petrochemical plant, which is the Sabah Ammonia-Urea Plant Project (SAMUR). This plant is to be expected to begin its operation in 2017. This plant is expected to produce 1.2 million tons of products on an annual basis. With the estimation of 0.3 tons of water formed in every ton of produced urea (Rahimpour et al., 2010), 0.36 million tons of wastewater will potentially be released to Sabah sea each year.

Furthermore, the establishment of Sipitang Oil & Gas Industrial Park (SOGIP) in 2011 which is located south-west of Sabah has already attracted local and international investors. First mega project in SOGIP, initiated by SAMUR will definitely invite other industries into the area to cater to the supply and services demand.

With the opening of these infrastructures, the production of wastewaters is inevitable. Especially with a large capacity of product handling, the wastewater generated will also be in big volume. Even though in the current planning, there is no refinery plant to be built in Sabah, some amount of environmental polluting discharge will still be produced. The potential generated wastewaters will be occurring mostly from the cleaning process of the storage areas and transferring pipelines. This wastewater will contain mostly cleaning detergent and oil debris. Leakage of any units will also contribute to the generation of liquid waste and may contain a certain concentration of heavy metals such as cadmium, chromium,

Table 2. General characteristics of oil and gas industry wastewater

Parameter	Values	Heavy Metal	Values (mg/L)
Density (kg/m3)	1014–1140	Calcium	13–25800
Surface Tension (dynes/cm)	43–78	Sodium	132–97000
TOC (mg/L)	0–1500	Potassium	24–4300
COD (mg/L)	1220	Magnesium	8–6000
TSS (mg/L)	1.2–1000	Iron	<0.1–100
pH	4.3–10	Aluminium	310–410
Total oil (IR; mg/L)	2–565	Boron	5–95
Volatile (BTX; mg/L)	0.39–35	Barium	1.3–650
Base/neutrals (mg/L)	<140	Cadmium[a]	<0.005–0.2
Total non-volatile oil and grease (g/L)	275	Chromium	0.02–1.1
Chloride (mg/L)	80–200,000	Copper	<0.002–1.5
Bicarbonate (mg/L)	77–3990	Lithium	3–50
Sulfate (mg/L)	<2–1650	Manganese	<0.004–175
Ammoniacal nitrogen (mg/L)	10–300	Lead[a]	0.002–8.8
Sulfite (mg/L)	10	Strontium	0.02–1000
Total polar (mg/L)	9.7–600	Titanium	<0.01–0.7
Higher acids (mg/L)	<1–63	Zinc[a]	0.01–35
Phenols (mg/L)	0.009–23	Arsenic[a]	<0.005–0.3
VFA's (volatile fatty acids) (mg/L)	2–4900	Mercury	<0.001–0.002
		Silver[a,b]	<0.001–0.15
		Beryllium	<0.001–0.004

[a]Analysed by atomic absorption.
[b]Value should be regarded as a minimum due to poor solubility.
Source: (Tibbetts et al., 1992)

copper, lead, mercury nickel, silver and zinc. Table 2 summarizes the general produced wastewater characteristic in the oil and gas industry. Apart from the wastewater from the infrastructure main unit, municipal wastewater will also be produced in the workplaces and staff quarters.

Treatment and Management Status

Every industrial facility onshore in Malaysia that produces effluent discharges must conform to the existing effluent limits specified in the fifth and seven schedules of Environmental Quality (Industrial Effluent) Regulation 2009. In the other hand, the offshore discharge is more stringent as to ensure the marine water quality meet the Interim Marine Water Quality Standard for Malaysia (IMWQSM).

Therefore, in order to meet the requirement, treatment of the produced wastewater is compulsory. Treatment applied in oil and gas sectors are typically by physical process or chemical precipitation and biological treatment by using microorganism for nutrients uptakes. In an offshore extraction infrastructure, the physical and chemical process is much preferred because of the space constraint and the compact size of the treatment system.

Physical treatment would include the process of adsorption, sand filtration, cyclone system, evaporation and a few others. The alternative chemical treatment involves precipitation or an oxidation process. However the capital cost of physical methods and cost of chemicals for chemical treatment of hazardous sludge is high, the application of these methods is limited. On the other hand, biological treatment is utilized in removing dissolved and suspended compounds from oil and gas wastewater in onshore extraction facilities.

Urban Wastewater

Characteristics of Urban Wastewater

Urban wastewater refers to the sewerage produced both by residential areas and wastewaters generated from public facilities such as schools, local business, offices and shops lots. In every town and districts especially the ones with a large number of population, a proper wastewater treatment system is crucial to prevent environmental and health threats that can be leaden by the raw discharge of sewage such as water pollution, spreading of waterborne diseases, and a nuisance to the aesthetic value as well as bad odour.

Due to Sabah's districts distribution, the population trend in Sabah is sparsely distributed especially in rural areas. The wastewater treatment system in Sabah can be considered as one of the undeveloped among all the states in Malaysia. This could be due to the unawareness of the public as well as the lack of regulations and enforcement. However, in recent years, the public awareness upon the management of wastewater, sewerage, in particular, has increased and the enforcement of regulations has become stricter.

Treatment and Management Status

The common wastewater treatment applied in Sabah is divided into two categories, the first category which includes collection and treatment of the wastewater and applied in major towns such as Kota Kinabalu, Sandakan, Lahad Datu, Keningau and part of Tawau. This treatment is comprised of a common Wastewater Treatment Plant (WWTP), containing a simple oxidation pond. Figure 1 depicts the Kinabutan oxidation pond located in Tawau.

Some WWTP are applying a Sequencing Batch Reactor (SBR). An SBR system has replaced an oxidation pond for sewage treatment in Lyon, Sabah since 2002. The design capacity of the SBR is 60,000 population equivalent (pe) which has been increased from 45,000 pe of the oxidation pond (Source: Environmental Conservation Department Sabah, Wastewater treatment in Sabah, Survey Report, 2002). More recent WWTP in Sabah applies a more intensified version of SBR which is Intermittent Decanted Extended Aeration (IDEA) technology.

The second category of wastewater treatment only involves the collection of wastewater and channelling it into rivers and seas and currently being applied by most districts especially in the rural areas of Sabah. Some residential areas including terrace houses, apartments and condominium, which is not connected to any main sewer lines, will construct its individual WWTP that is collecting and treating its own produced wastewater. This WWTP is commonly applying Extended Aeration process and managed by its own Management Corporation and monitored by the Local Authority.

Figure 1. Oxidation pond located at Kinabutan, Tawau
Source: Environmental Conservation Department Sabah, Wastewater treatment in Sabah, Survey Report, 2002

In rural residential areas, individual septic tanks are more common in Sabah. Due to the sparse distribution of the private housing trend, construction of the main sewer system is very difficult in rural areas. In this treatment, solid in the wastewater is simply settled to form sludge at the bottom of the tank and fluid wastewater will overflow into drains. However, proper and regular maintenance is crucial for the system to be working properly. In terms of sludge management, even though almost all WWTP in Sabah produces sludge as a by-product from its treatment, the absence of sludge management in Sabah is causing it to be left untreated and unmanaged.

Over the years, wastewater treatment and discharge issues in Sabah have gained more attention both from the public and the local authority. The public becomes more sensitive upon the importance of wastewater management and the consequences of discharging raw sewerage. Complaints were forged on the mismanagement and faulty of the operating WWTP and the discharge of raw sewerage into water bodies.

Due to this matter as well as the apparent deteriorating environmental quality caused by the low-quality discharge of wastewater, a survey was done in 2002 to investigate the status of the wastewater treatment system in Sabah, which was conducted by an independent company. Based on the survey, even though the treatment plant has been constructed in every district in Sabah at the time, most of this WWTP is non-functional mainly caused by irregular maintenance. The finding of the survey, we can conclude that more efficient treatment of wastewater with the intensified process in terms of cost of construction, maintenance, footprint and treatment period should be highly considered to improve the status of wastewater management in Sabah.

Nonpoint Source Pollution

Sabah being a state with high annual precipitation should also consider the nonpoint source pollution aside from sewage and urban wastewater source. The nonpoint source pollution differs from sewage and urban wastewater such that it is originated from various and diffused sources. Rainfalls accumulate these pollutants and flows as surface or groundwater which will finally be discharged into water bodies.

Nonpoint Source Pollution Contributors

The nonpoint source pollution is caused by numbers of human activities such as land development that require land clearing and will lead to soil erosion. This is especially significant in sites that are improperly managed during its development phases that may include vegetation clearing, land levelling and building construction. Sediments from the soil erosion will be accumulated in rivers, lakes or sea and negatively affect the aquatic life by increasing the turbidity, reducing visibility and reducing the water depth.

In urban areas, buildings and businesses such as workshops, restaurants, factories and power plants are prone to release the oil, grease, toxic chemicals and fugitive air particles. Even though regulations for such pollutant has been enforced by the Malaysian Department of Environment by drafting the Environmental Quality Act 1974, the regulation and monitoring are only mainly focused on factories and plants with big production capacities. Small businesses and workshops are easily excluded from this regulation, making them a greater threat of nonpoint source pollution.

The transportation industry is also one of the major contributors to nonpoint source pollution. Increase in population, life affordability and increase in goods transportation demand have considerably escalated the number of cars and lorry. This has increased the release of exhaust gas which consists of hydrocarbons, carbon monoxide, NO_x, carbon dioxide and particulate matters. Research has shown that emissions from diesel-powered engines are even higher in pollution with increased concentration of these pollutants with additional toxic pollutants such as polycyclic aromatic hydrocarbons, volatile organic compounds, dioxins and dioxin-like compounds (Mwangi et al., 2015). For Sabah especially, the use of diesel-powered engine is higher than a petrol-powered engine due to the necessity for a higher-powered vehicle such as 4x4 trucks and lorry. This is necessary for transportation in hilly areas as well as transportation of goods via land to other districts.

However, the nonpoint source pollution is not exclusively caused by industrialization but also by agriculture activities such as farming, animal rearing and fish farming which are abundant in Sabah. The treat to nonpoint source pollution is especially high from those with uncontrolled used by fertilizers and waste management. Polluting elements such as chemicals, rich nutrients and bacteria in the fertilizer, pesticides, livestock and animal waste will be accumulated by rainfall and carried into water bodies. These threats are intensified because regulation or monitoring of the releases from these activities is not strictly enforced.

The common household is also one of the contributors of nonpoint source pollution when the individual septic tanks are not designed or maintained properly and fault are found causing groundwater to carry the pollutants. Moreover, solid waste generated from residential and commercial sources that are disposed of improperly will cause in the generation of leachate when contact with water. Leachate generally consists of a high concentration of COD, pH, ammonia nitrogen and heavy metals as well as strong colour and bad odour.

Due to less awareness of the nonpoint source pollution and loose regulation and monitoring, it may be the most significant source of pollution and may cause high negative impacts in our water bodies that directly affects our aquatic life, water supply and ecology.

Management of Nonpoint Source Pollution

Due to its various sources, the management of nonpoint source pollution varies greatly from that of sewage and urban wastewater. Management of such pollutant should be addressed by their unique sources and prevention of any toxic release is the best method.

Taking into consideration the soil erosion in construction sites, a ponding system to catch any runoff water before any discharge should be constructed. This small pond will allow sedimentation of particles carried by the runoff water and prevent it to be discharged into water bodies. This ponding system is also suitable to be utilized in urban areas where drainage systems are often directly discharging water to water bodies. Ponds or lakes should be constructed to allow pollutant to be trapped before it is released into the sea. However, maintenance of these ponds should be conducted regularly so that its performance is not compromised.

Rules and regulations are important in order to reduce the nonpoint source pollutant. In line with the existing regulations, more stringent limits should be applied to not only the high capacities industries but also small businesses to ensure even with high numbers of business and industries arising, the environmental quality of Sabah is not compromised. In addition, clear and stringent emission limits for car exhaust gas must be enforced to control the emission of toxic gases especially with the increasing number of cars. It is also important that Sabah's public transportation quality is improved to encourage the public to use public transportation and reduce emission by personal cars.

Other than that, general awareness of activities that leads to nonpoint pollution should be raised among the public. Proper management of solid waste and responsible agriculture management should be promoted to ensure that pollution can be reduced especially in rural areas. Improvement of services such as a centralized sewage treatment system and regular solid waste collection should take place to avoid consequences of an improper septic system in individual homes and illegal solid waste dumpsites or open burning.

Wastewater from Floating Residential Areas

Current Scenario of the Wastewater from Floating Residential Areas in Sabah

Unique to the island of Borneo, some of the indigenous tribes of Sabah, Sarawak and Brunei build their homes above the seawater, often on the shorelines. These homes are known as floating houses and are commonly populated by people of Bajau, Suluk and Brunei descendant who tend to reside on coastlines. The livelihoods of the population living in these homes are mostly fisherman and their lives typically revolve on water. However, in recent years, more and more of the younger generation are seen to migrate into the city to find better education and jobs.

The existence of this unique culture has become one of tourist attraction in Sabah, along with the clear seawater and the beauty of Sabah's undisturbed coral reefs. Especially in the eastern part of Sabah, the floating houses have always been a tourist destination due to the undisturbed culture and nature by the modernization. These homes are constructed interconnected and very close to each other which reflects the warm and welcoming culture of these tribes. Populations living on these floating houses are known to be among of the few tribes in Sabah that still preserves their ancestral rich tradition and culture.

Despite the unique culture of these houses, many challenges are faced to maintain their existence, especially in this modern era. Because of these community are untouched by modernization, the domestic management in these houses are also excluded from improvement to modern practice. Most of these homes especially in Sabah do not practise good sanitation and disposal system. These homes are still implementing traditional waste disposal method by dumping their waste directly into the sea, including solid waste, sewage and wastewater coming from local businesses. Therefore, the coexistence of these homes in urban places such as in Kota Kinabalu, Sandakan and Tawau has been a great challenge. Despite being a tourist attraction, it is also becoming one of the biggest pollution contributors in the area.

The disposal of solid waste and sewage into seawater are highly polluting and could cause severe negative impact into the aquatic life from a high concentration of BOD, COD, pH and ammonia from the wastes disposed of. On the other hand, the health of the local community is threatened directly by the high concentration of bacteria and viruses from human waste such as Escherichia coli, salmonella and rotavirus. Other than that, the high concentration of nutrients, chemicals, oil, grease and heavy metals from kitchen and workshops waste can equally cause deterioration of human and animal health.

Without proper management of waste disposal and the awareness of these community, it is highly potential that Sabah will have a great loss in terms of money. A significant amount of money will be required for solid waste clean-up and seawater treatment due to improper solid waste and sanitary system. The tourism sector of Sabah also has the possibility to face a great income loss due to the deteriorating natural attraction that is caused by irresponsible and uncontrolled waste disposal. The aquaculture industry that depends highly on sea cultivation method will face problem due to the low quality of seawater. Increase medical expenses and loss in human capital is also very likely due to the decline in health condition caused by exposure to an unhealthy environment.

Proposed Solutions to Tackle the Problem

The solution for the challenges with wastewater from floating residential areas should be taken into serious consideration and should be implemented as soon as possible. Nevertheless, the destruction of these floating houses and migrating the community into modern areas is not the right solution as it will also diminish the cultural traits of this community. Therefore, it is important that the measures taken will also ensure their cultural practices are not threatened and is preserved.

Most importantly, the responsible authorities should implement that every household in this area will be required to build a proper sewage and wastewater system. This system should ensure that all wastewater coming from these houses will not be discharged into the sea without prior treatment. Regular inspection and maintenance should also be conducted to ensure no leakage occurs. However, implementation of these plans is always challenging especially in the beginning. Therefore, supports should be given to the community not only in the form of financial but also consultancies in design, construction and operational procedures.

With the sanitary system in place, general awareness of the importance of the good sanitary system, wastewater treatment as well as solid waste management is also important. The community should have knowledge of the consequences that they may face if their poor cleanliness practices are continued. Simple education on danger of these pollutants should be given by the responsible authorities to help them gain the understanding and awareness of environmental cleanliness in general.

Sarawak and Brunei can be used as a good example as they have taken the lead to face the challenges of floating residential areas. The floating residential areas in Sarawak and Brunei have implemented a proper sewage system. The cleanliness of the neighbourhood is also maintained by the help of a committee formed by the community. The enthusiasm of the community is boosted by organizing annual

Table 3. Untreated effluent discharged from the SFI pulp and paper mill

Parameter	Value
Temperature (°C)	38.9 (±3%) [1]
pH	8.4 (±7%) [1]
BOD_5 (20°C) (mg/L)	160 (±13%) [1]
COD (mg/L)	550 (±110%) [1]
Conductivity (µS/cm)	1 650 (±7%) [1]
Colour (Pt-Co units)	1 130 (±23%) [1]
Chlorine (mg/L)	580 (±29%) [1]
Sodium (mg/L)	380 (±26%) [1]
Suspended solid (SS) (mg/L)	230 (±43%) [1]
Total dissolved solids (mg/L)	1 280 (±7%) [1]
1,1-dichlorodimethyl sulfone (µg/L)	286 – 580 [2]
2,4,6-trichlorophenol (µg/L)	0.95 – 4.46 [2]
4,5-dichloroguaiacol (µg/L)	1.10 – 2.45 [2]
2,3,4,6-tetrachlorophenol (µg/L)	0.80 – 2.22 [2]
3,4,5-trichloroguaiacol (µg/L)	7.13 – 18.70 [2]
4,5,6-trichloroguaiacol (µg/L)	0.34 – 0.67 [2]
3,4,5-trichlorocatechol (µg/L)	2.65 – 17.50 [2]
Tetrachlorouaiacol (µg/L)	2.55 – 14.19 [2]
Trichlorosyringol (µg/L)	12.04 – 21.90 [2]
Tetrachlorocatechol (µg/L)	4.65 – 16.70 [2]
Dichloroacetic acid (µg/L)	1 102 – 1 747 [2]
Trichloroacetic acid (µg/L)	853 – 1 775 [2]

Sources: [1](Mohamed & Landner, 1993); [2](Mohamed et al., 1989)

cleanliness and house attractiveness competition. Sabah should take these practices as an example to improve the overall quality of the environment as well as the quality of their lives.

However, education and knowledge alone will not be enough to ensure pollution is prevented. Strict enforcement should always be conducted together with a penalty system. This will ensure that not only those who are aware of the environmental quality importance is taking part to keep their areas clean but also the entire community.

Other Wastewater Sources in Sabah

Pulp and Paper Mill Effluent

Sabah Forest Industries (SFI) Pulp and Paper Mill is the only mill in Malaysia produces printing/writing paper, with an annual capacity of 165 000 metric tons (Source: Food and Agriculture Organisation). The SFI Mill is located 7 km south of the town of Sipitang and about 1.5 km from the Brunei Bay coastline.

Table 4. Treated effluents discharged in the Brunei Bay and the standard values

Parameter	Value	Standard
Temperature (°C)	38.9 (±3%) [1]	40 [2]
pH	7.3 (±5%) [1]	5.5 – 9.0 [2]
BOD_5 (20°C) (mg/L)	16 (±28%) [1]	50 [2]
COD (mg/L)	270 (±15%) [1]	-
COD (kg/tp)	40 [2]	65 [2]
Conductivity (μS/cm)	1 750 (±5%) [1]	-
Colour (Pt-Co units)	1 050 (±29%) [1]	-
Chlorine (mg/L)	590 (±27%) [1]	-
Sodium (mg/L)	380 (±24%) [1]	-
Suspended solid (SS) (mg/L)	10 – 21 [2]	100 [2]
	95 (±22%) [1]	-
Total organic carbon (TOC) (mg/L)	3 – 14 [2]	-
Total dissolved solids (mg/L)	1 250 (±6%) [1]	-
Total organochlorine (TOCl) (kg/tp)	1.4 [2]	1.5 [2]
1,1-dichlorodimethyl sulfone (μg/L)	86 – 232 [3]	-
	27 000 – 33 000 [2]	-
2,4,6-trichlorophenol (μg/L)	0.45 – 3.71[3]	-
4,5-dichloroguaiacol (μg/L)	0.21 – 2.2 [3]	-
2,3,4,6-tetrachlorophenol (μg/L)	0.10 – 0.95 [3]	-
3,4,5-trichloroguaiacol (μg/L)	2.86 – 11.20 [3]	-
4,5,6-trichloroguaiacol (μg/L)	0.14 – 0.44 [3]	-
3,4,5-trichlorocatechol (μg/L)	2.12 – 10.50 [3]	-
Tetrachlorouaiacol (μg/L)	1.79 – 12.30 [3]	-
Trichlorosyringol (μg/L)	5.49 – 18.35 [3]	-
Tetrachlorocatechol (μg/L)	0.27 – 10.70 [3]	-
Dichloroacetic acid (μg/L)	14 – 18 [3]	-
Trichloroacetic acid (μg/L)	8.38 – 994 [3]	-

Sources: [1](Mohamed & Landner, 1993); [2](Mohamed et al., 1990). Distances of samplings collected from effluent discharge point are between 0 and 2.7 km and depths of 16 – 22 m; [3](Mohamed et al., 1989)

Each process involves in the pulp and paper production uses large amounts of water for washing and screening generates a significant quantity of effluent. The wastewater of the SFI Mill is originated mainly from wood preparation, digester, pulp washing, pulp bleaching and papermaking. In the stage of wood preparation, the wastewater produced from wood cleaning contains suspended solids, BOD, dirt, grit, fibres, etc. The effluent released from the digester system is called 'black liquor'. This effluent comprises resins, fatty acids, colour, BOD, COD, AOX and VOCs. The wastewater from the pulp washing contains high pH, BOD, COD and suspended solids and dark brown in colour. The wastewater generated from pulp bleaching contains dissolved lignin, carbohydrate, colour, COD, AOX and inorganic chlorine compounds. The wastewater produced from papermaking essentially consists of particulate waste, organic compounds, inorganic dyes, COD and acetone.

Table 3 shows the major chlorinated organics in the untreated mill effluent. The effluents are treated in a biological treatment system and the Brunei Bay is the treated effluent receiver. Table 4 presents the compositions contained in the treated effluent discharged into the Brunei Bay and some standard values.

Rubber Industry Effluent

Malaysia is the third-largest rubber producer in the world. Sabah Rubber Industry Board (SRIB) totally owns four rubber factories; two factories operated in Turan and Tenom produce Standard Malaysia Rubber (SMR) with capacities of 60,000 and 12,000 tonnes per year, respectively, one factory processes Latex Concentrate High Ammonia (LCHA) which is located in Tuaran with annual capacity of 1,800 tonnes, and one factory with a capacity of 2,500 tonnes/year produces Specialty Rubber (SR) which is situated in Beaufort.

The wastewater produced from the rubber industry in Malaysia contains skim, latex serum, uncoagulated latex and washings from various stages of rubber processing. Typically, 20 tonnes of rubber can generate 410 thousand litres of effluent per day (Mohammadi et al., 2010). Table 5 lists the characteristics of the rubber effluent and the corresponding standard value.

Aquaculture Wastewater

Aquaculture practices in Malaysia include freshwater pond aquaculture, brackish water aquaculture and marine aquaculture. Brackish water aquaculture is the predominant practice, with a total production of 144,189 tonnes covering an area of 17,357 ha. Sabah possesses the biggest area of land-based earthen ponds. The total area covered is 7,879 and the ponds are mainly used for culturing black tiger shrimp and marine fish (Source: Food and Agriculture Organisation of the United Nations). A few years ago, recirculating aquaculture system (RAS) has been introduced in Sabah. KO-NELAYAN has built the RAS system in Tuaran. An area of about 150 m^2 is able to produce five metric tonnes of fish in a year.

The water treatment in the RAS is essential because the excess organic materials and nutrients may be harmful to the fish. The organic materials and nutrients comprise of feed wastage, faecal solids and detached bacterial flocs. The nutrients content consist of nitrogen and phosphorus. About 85% of the phosphorus and 52 – 95% of the nitrogen in the feed may be lost into the environment through excretion, feed wastage and faecal production. The faecal production in RAS is about 26 – 46% of the ingested feed, whereas the uneaten feed and waste excretion are about 11 – 38% of the applied feed. The dissolved nutrients can cause alga blooms which can reduce the dissolved oxygen in the water. Mineralisation and leaching of uneaten feed and waste excretion particulates produce the dissolved ammonia-nitrogen which is toxic to the fish (Chiam & Sarbatly, 2012).

Alkaline-surfactant-polymer (ASP) Produced Water

Alkaline-surfactant-polymer (ASP) flooding technology is well known as a cost-effective chemical EOR process which yielding high recovery rates at approximately 20% (Chang et al., 2006). In Malaysia, PETRONAS and Royal Dutch Shell have signed two 30-year production sharing contracts for ASP EOR projects offshore Sarawak and the North Sabah development areas offshore Sabah, i.e. the Baram Delta

Table 5. Characteristics of effluent from rubber processing and the standard values

Parameter	Typical Range	Standard, 1984 and Later
pH	3.7 – 5.5	-
BOD (mg/L)	1,500 – 7,000	100
COD (mg/L)	3,500 – 14,000	400
Suspended solid (SS) (mg/L)	200 – 700	150
Total nitrogen (mg/L)	200 – 1,800	-
Ammonia nitrogen (mg/L)	-	300
Sulphate (mg/L)	500 – 2,000	-

Source: (Mohammadi et al., 2010)

oil fields. The projected increase in the average recovery factor by the ASP EOR in the Baram Delta operations and North Sabah fields will be 36 – 50%.

Nevertheless, the handling of produced water from the ASP EOR process is challenging. The produced water is oily wastewater which is generated by pumping directly from the oil wells. For example, the amount of produced water is up to about 3×10^8 tons per year in the Daqing oil field, China. Because the use of alkali, surfactant and polymer in the injected aqueous solution in ASP flooding technology, the produced water from the ASP EOR process contains large numbers of residual chemicals. As an example, the properties of the produced water from ASP flooding in the Daqing oil field are: < 2,000 mg/L of oil with a median diameter of oil droplets 3 – 5 μm; 312 – 630 mg/L of partially hydrolysed polyacrylamide; < 630 mg/L of surfactant; and < 1,500 mg/L NaOH at temperature 42 – 45°C (Deng et al., 2002).

The surfactant is responsible to stabilize the oil droplets by reducing the oil-water interfacial tension and zeta potential on the surfaces of the oil droplets. The skin barrier surrounding the tiny droplets in the oil-water emulsion prevents the water droplets from uniting. Hence, the produced water from ASP flooding is more complex and stable emulsion system which is more difficult to treat than that the produced water from water flooding (Foyeke & Diane, 1998).

The management of the produced water would become more challenging for offshore ASP EOR projects, especially considering the environmental issues and project economics. Construction of disposal wells for the wastewater is unfavourable because it is expensive. Occasionally, the produced waters are shipped to onshore facilities (Goh et al., 2010). Commonly, overboard disposal is practised especially for remote offshore locations. However, de-oiling is difficult to work because of the stringent regulatory standards for discharging produced water. The permitted oil and grease limits for treated produced water discharge offshore in China, Australia, US and North-East Atlantic ranges between 10 and 50 mg/L (Fakhru'l-Razi et al., 2009).

Leachate Wastewater

Leachate refers to the fluid that percolates or leaches from municipal solid wastes that are mostly found in a landfill. Leachate generally contains a high amount of dissolved organic and inorganic compounds, and suspended solids which are characterized by high values of COD, BOD, pH, heavy metals, and ammonia nitrogen. Leachate is also associated with strong colour and unpleasant odour. The general

Table 6. Characteristics of landfill leachate

Parameter	Typical Range (mg/L, Unless Otherwise Stated)	Upper Limit (mg/L, Unless Otherwise Stated)
Total alkalinity (as $CaCO_3$)	730 – 15,050	20,850
Calcium	240 – 2,330	4,080
Chloride	47 – 2,400	11,375
Magnesium	4 – 780	1,400
Sodium	85 – 3,800	7,700
Sulfate	20 – 730	1,826
Specific conductance	2,000 – 8,000 μmhos/cm	9,000 μmhos/cm
Total dissoved solid (TDS)	1,000 – 20,000	55,000
Chemical oxygen demand (COD)	100 – 51,000	99,000
Biological oxygen demand (BOD)	1,000 – 30,300	195,000
Iron	0.1 – 1,700	5,500
Total nitrogen	2.6 – 945	1,416
Potassium	28 – 1,700	3,770
Chromium	0.5 – 1.0	5.6
Manganese	Below detection level – 400	1,400
Copper	0.1 – 9.0	9.9
Lead	Below detection level – 1.0	14.2
Nickel	0.1 – 1.0	7.5

Source: (Mukherjee, Mukhopadhyay, Hashim, & Gupta, 2014)

characteristics of leachate are shown in Table 6. The characteristics of leachate produced may vary in different locations of the landfill as it is affected by multiple factors such as the type and composition of solid waste, the solubility of solid waste components, hydrological influences, and the landfill design and operations.

Improper leachate management can cause contamination of natural resources. Leachate pollution is a serious issue which affects mainly the groundwater and surface water supplies. Consequently, this leads to negative effects on human health due to the high number of heavy metals as well as decomposed waste materials. Thus, proper leachate wastewater management is a crucial practice that should consider its unique properties that are highly varied from site to site. The typical approach towards landfill leachate treatment include biological treatment (e.g. membrane bioreactors (MBR), aerobic lagoons sequencing batch reactors (SBR), activated sludge process (ASP), and constructed wetlands), physical/chemical treatment (e.g. coagulation/flocculation, filtration, oxidation, and activated carbon), and developing technology such as membrane technology (e.g. reverse osmosis (RO), and pervaporation).

However, some rural regions do not have a municipal landfill site to properly dispose of their solid wastes. Instead, some rural communities resort to open burning their garbages that includes domestic wastes and plastic wastes. This practice is commonly seen in isolated villages in Sabah, in which the population is highly scattered and limited access to municipal facilities. This act would not only lead

Table 7. Characteristics of wet market wastewater at Public Market Parit Raja, Batu Pahat and Pasar Ampangan, Seremban

Parameter	Public Market Parit Raja	Pasar Ampangan [a]
pH	6.0 ± 0.1	5.6 – 5.8
Turbidity (NTU)	66.0 ± 8.9	–
Biological oxygen demand (BOD)	89 ± 3.61	71 – 122
Chemical oxygen demand (COD)	456 ± 8.19	381 – 560
Total suspended solids (TSS)	132.3 ± 21.7	60 – 122
Sulfate, SO_4^{2-} (mg/L)	32.3 ± 0.78	–
Total chlorine, Cl^- (mg/L)	32.0 ± 0.69	–
Total nitrogen, TN (mg/L)	36.9 ± 0.5	30.3 – 37.3
Total phosphorus, TP (mg/L)	1.61 ± 0.13	ND – 22.2
Total organic carbon, TOC (mg/L)	118.67 ± 2.89	–
Oil and grease	5.22 ± 0.07	13 – 43
Zinc, Zn (mg/L)	0.312 ± 0.0021	–
Iron, Fe (mg/L)	1.071 ± 0.001	–

ND – Not detected
Source: (Jais et al., 2015; Zulkifli, Roshadah & Tunku Khalkausar, 2011)

to leachate contamination to the water bodies, but it also releases dangerous chemicals to the air which ultimately contributes to air pollution.

Wet Market Wastewater

Malaysia has abundant wet markets which house Malaysia's fresh and local produce of meats, vegetables, and kinds of seafood. Many activities conducted at the wet market produce a copious amount of wastewater. These activities mainly comprise of preparation and sales of meat and seafood, preparation of fruit and vegetables, food stalls operation, and public restrooms.

Wet market wastewater mainly contains high values of organic material, fats, oil and grease, nutrients, and suspended solids, which are typically discharged directly to drainage without prior treatment. Consequently, bad odour and high strength wastewater presence within the surrounding shopping area become a conducive environment for waterborne diseases to propagate which also attracts disease-spreading insects such as fly. These problems are worsened by the lack of stringent regulations to control the wastewater discharge.

Studies conducted by Jais et al. (2015) and Zulkifli, Roshadah and Tunku Khalkausar (2011) reveal the characteristics of the wet market at two locations situated at Batu Pahat and Seremban (Table 7, both of which are in Malaysia. Generally, the wastewater has low pH, moderate turbidity, BOD, COD, and TSS levels. Nonetheless, the nutrients and TOC levels in the wastewater are rather high.

Recently, the impact of wet market wastewater and others as serious river pollutant in Malaysia has been gaining more awareness from the government. A number of wastewater treatment plants had been constructed in the vicinity of wet markets through a project known as River of Life to rehabilitate and restore the Klang River. Previously, the wet market wastewater was not treated and directly discharged

Table 8. General characteristics of slaughterhouse wastewater

Parameter	Range	Average
pH	4.90 – 8.10	6.95
Chemical oxygen demand (COD)	1,250 – 15,900	4,221
Biological oxygen demand (BOD)	610 – 4,635	1,209
Calcium	32 – 316	67
Potassium	0.01 – 100	90
Sodium	62 – 833	621
Plumbum	0.21 – 34	4
Total nitrogen	50 – 841	427
Total organic carbon (TOC)	100 – 1,200	546
Total phosphorus	25 – 200	50
Total suspended solids (TSS)	300 – 2,800	1,164

Source:(Bustillo-Lecompte, Mehrvar, & Quiñones-Bolaños, 2016)

to the surrounding water body. The wastewater was classified under Class V which is not suitable for any purposes. As a result, harmful pollutants like the BODs, suspended solids, oil and grease, and *E. coli* bacteria have contaminated the river. The wastewater treatment will produce treated water under the Class II(b) which can be used for recreational purposes surrounding the wet markets, such as to clean the market.

Slaughterhouse Wastewater

The increasing demand for meat products (e.g. beef, lamb and pork) by consumers has led to an increasing number of slaughterhouse facilities around the globe. The meat processing industry consumes a high volume of freshwater to be used mainly for slaughtering of animals and cleaning the slaughterhouse facilities. With a projected steady growth of this industry, it is expected that there will be a growing volume of slaughterhouse wastewater to be treated. Improper disposal of untreated slaughterhouse wastewater can directly contribute to public health hazards due to many pollutants that are deemed unsafe for us and the environment. Groundwater pollution and river deoxygenation are some consequences caused by an improper discharge of slaughterhouse wastewater.

The slaughterhouse wastewater comes from the multiple activities conducted in the slaughterhouse and meat processing plant. Slaughterhouse wastewater needs to undergo thorough treatment processes before it can be sustainably released to the environment. Generally, slaughterhouse wastewater contains a complex mixture of fats, proteins, fibres, pathogens, high organic contents, and pharmaceuticals used in veterinary purposes. In some cases, it may contain toxic, non-biodegradable matters, and bio-resistant. The characteristics of slaughterhouse wastewater are tabulated in Table 8.

The composition of slaughterhouse wastewater typically varies depending on the type and amount of animals being slaughtered in the meat processing plant. Nonetheless, the effluent generated usually presents a high concentration of organic matter with moderate soluble residues, a high number of pathogens, and cleaning agents which can cause severe polluting effects.

Biological treatment processes are usually employed to treat slaughterhouse wastewater due to its effectiveness in treating high strength slaughterhouse wastewater and simple operation. Anaerobic and aerobic treatments are the two biological processes used to treat the wastewater. Anaerobically treated slaughterhouse effluent is generally effective to treat the slaughterhouse wastewater, but it requires further post-treatment to adhere to the discharge regulation. This is because, treatment with the anaerobic process only could not entirely stabilize the organic matter, which is required to comply with the regulated discharge limit.

Hence, slaughterhouse wastewater treatment using aerobic operation is more preferred as it can produce higher operation rates than anaerobic operation. Furthermore, aerobic processes can also be integrated as a post-treatment of the anaerobic process, considering the fact that higher strength wastewater is directly proportional to the oxygen uptake and treatment duration for the aerobic process alone. Using this treatment scheme would potentially reduce some operating cost and treatment duration while also increasing the treatment efficiency.

WASTEWATER EFFLUENT STANDARDS

In the view of maintaining the quality of water in Malaysian water bodies, the Malaysian Authority has enforced that all industries discharging wastewater should comply with the discharge standard before any effluent can be released. The water quality standards employed in Malaysia are listed in Table 9.

The wastewaters must be treated before discharging to the environment. The Environment Quality (Sewerage and Industrial Effluents) Regulations 1979 set out two standards of effluent quality, Standard A and B. Effluent that is discharged upstream of a water supply intake should meet Standard A, while an effluent that is discharged downstream has to meet Standard B. Table 10 presents the standard values of the parameters.

CONVENTIONAL WASTEWATER TREATMENT TECHNOLOGIES

This section discusses several types of Conventional wastewater treatments namely the open ponding system, activated sludge system, sequencing batch reactor, oxidation ditch, extended aeration, anaerobic digester, up-flow anaerobic sludge blanket reactor, expanded granular sludge bed reactor, trickling filter, rotating biological contactor, septic tank, Imhoff tank, roughing filter, and sand filter. Figure 2 – Figure 14 illustrated the schematic of each of these technologies.

Open Ponding

Oxidation ponds are classified into aerobic ponds, anaerobic ponds, facultative ponds, maturation ponds and aerated lagoons.

Aerobic ponds are shallow which typically measure less than 1 m in depth. The wastewater treatment mechanism starts as sunlight penetrates into the aerobic ponds and photosynthesis takes place where oxygen is produced. Aerobic microbes consume the oxygen to decompose the organic matter contained in the wastewater.

Table 9. Malaysian water quality standards

<table>
<tr><th rowspan="2">Parameters</th><th rowspan="2">Unit</th><th colspan="6">Classes</th></tr>
<tr><th>I</th><th>IIA</th><th>IIB</th><th>III</th><th>IV</th><th>V</th></tr>
<tr><td>Ammoniacal nitrogen</td><td>mg/l</td><td>0.1</td><td>0.3</td><td>0.3</td><td>0.9</td><td>2.7</td><td>> 2.7</td></tr>
<tr><td>BOD</td><td>mg/l</td><td>1.0</td><td>3.0</td><td>3.0</td><td>6.0</td><td>12.0</td><td>> 12.0</td></tr>
<tr><td>COD</td><td>mg/l</td><td>10.0</td><td>25.0</td><td>25.0</td><td>50.0</td><td>100.0</td><td>> 100.0</td></tr>
<tr><td>DO</td><td>mg/l</td><td>7.0</td><td>5.0 – 7.0</td><td>5.0 – 7.0</td><td>3.0 – 5.0</td><td>< 3.0</td><td>< 1.0</td></tr>
<tr><td>pH</td><td>-</td><td>6.5 – 8.5</td><td>6.0 – 9.0</td><td>6.0 – 9.0</td><td>5.0 – 9.0</td><td>5.0 – 9.0</td><td>-</td></tr>
<tr><td>Colour</td><td>TCU</td><td>15.0</td><td>150.0</td><td>150.0</td><td>-</td><td>-</td><td>-</td></tr>
<tr><td>Electrical conductivity*</td><td>Umhos/cm</td><td>1000.0</td><td>1000.0</td><td>-</td><td>-</td><td>6000.0</td><td>-</td></tr>
<tr><td>Floatables</td><td>-</td><td>n</td><td>n</td><td>n</td><td>-</td><td>-</td><td>-</td></tr>
<tr><td>Odour</td><td>-</td><td>n</td><td>n</td><td>n</td><td>-</td><td>-</td><td>-</td></tr>
<tr><td>Salinity</td><td>%</td><td>0.5</td><td>1.0</td><td>-</td><td>-</td><td>2.0</td><td>-</td></tr>
<tr><td>Taste</td><td>-</td><td>n</td><td>n</td><td>n</td><td>-</td><td>-</td><td>-</td></tr>
<tr><td>TDS</td><td>mg/l</td><td>500.0</td><td>1000.0</td><td>-</td><td>-</td><td>4000.0</td><td>-</td></tr>
<tr><td>TSS</td><td>mg/l</td><td>25.0</td><td>50.0</td><td>50.0</td><td>150.0</td><td>300.0</td><td>300.0</td></tr>
<tr><td>Temperature</td><td>°C</td><td>-</td><td>Normal + 2°C</td><td>-</td><td>Normal + 2°C</td><td>-</td><td>-</td></tr>
<tr><td>Turbidity</td><td>NTU</td><td>5.0</td><td>50.0</td><td>50.0</td><td>-</td><td>-</td><td>-</td></tr>
<tr><td>Faecal coliform **</td><td>counts/100 ml</td><td>10.0</td><td>100.0</td><td>400.0</td><td>5000.0 (20000.0)[a]</td><td>5000.0 (20000.0)[a]</td><td>-</td></tr>
<tr><td>Total coliform</td><td>counts/100 ml</td><td>100.0</td><td>5000.0</td><td>5000.0</td><td>50,000.0</td><td>50,000.0</td><td>>50,000.0</td></tr>
<tr><td>Iron</td><td>mg/l</td><td rowspan="5">Natural level or absent</td><td>1.0</td><td>1.0</td><td>1.0</td><td>1.0 (leaf)
0.5 (others)</td><td rowspan="5">Levels above IV</td></tr>
<tr><td>Manganese</td><td>mg/l</td><td>0.1</td><td>0.1</td><td>0.1</td><td>0.2</td></tr>
<tr><td>Nitrate</td><td>mg/l</td><td>7.0</td><td>7.0</td><td>-</td><td>5.0</td></tr>
<tr><td>Phosphorus</td><td>mg/l</td><td>0.2</td><td>0.2</td><td>0.1</td><td>-</td></tr>
<tr><td>Oil & grease</td><td>mg/l</td><td>0.04; N</td><td>0.04; N</td><td>N</td><td>-</td></tr>
</table>

Notes:
n : No visible floatable materials or debris or no objectionable odour, or no objectionable taste.
* : Related parameters, only one recommended for use.
** : Geometric mean.
[a]: Maximum not to be exceeded.
N : Free from visible sheen, discolouration and deposits.
Class Uses
Class I : Conservation of natural environment.
Water Supply I – practically no treatment necessary.
Fishery I – very sensitive aquatic species.
Class IIA : Water Supply II – conventional treatment required.
Fishery II – sensitive aquatic species.
Class IIB : Recreational use with body contact.
Class III : Water Supply III – extensive treatment required.
Fishery III – common, of economic value and tolerant species; livestock drinking.
Class IV : Irrigation.
Class V : None of the above.

Table 10. Parameter limits of the effluent of Standard A and B

Parameter	Unit	Standard A	Standard B
Temperature	°C	40	40
pH	-	6.0 – 9.0	5.5 – 9.0
BOD at 20°C	mg/l	20	50
COD	mg/l	50	100
Suspended solids	mg/l	50	100
Mercury	mg/l	0.005	0.05
Cadmium	mg/l	0.01	0.02
Chromium, hexavalent	mg/l	0.05	0.05
Arsenic	mg/l	0.05	0.10
Cyanide	mg/l	0.05	0.10
Lead	mg/l	0.10	0.50
Chromium, trivalent	mg/l	0.20	1.0
Copper	mg/l	0.20	1.0
Manganese	mg/l	0.20	1.0
Nickel	mg/l	0.20	1.0
Tin	mg/l	0.20	1.0
Zinc	mg/l	1.0	1.0
Boron	mg/l	1.0	4.0
Iron (Fe)	mg/l	1.0	5.0

Anaerobic ponds are between 2 and 4 m in depth. The surface area of the anaerobic ponds is small resulting in low oxygen diffusion. The anaerobic ponds are used to treat high strength industrial wastewater. Anaerobic microbes break down biodegradable matter in the absence of oxygen. The microbes hydrolyse the insoluble organic polymers such as carbohydrates, fats and proteins into sugars, fatty acids and amino acids. Then, acidogenic microbes convert the soluble sugars, fatty acids and amino acids into carbon dioxide, hydrogen, ammonia, acetic acid and volatile fatty acids. These microbes further digest the fatty acids into acetic acid, carbon dioxide, ammonia and hydrogen. Finally, methanogens transform these products into methane and carbon dioxide.

Facultative ponds consist of an aerobic upper zone, an anaerobic bottom zone and a facultative middle zone. The ponds are between 1 and 2.5 m in depth. Biodegradable mechanisms occur differently in the facultative ponds because the concentration of dissolved oxygen decreases with the pond depth.

Maturation ponds are also known as polishing ponds. These ponds receive the effluents from aerobic ponds, anaerobic ponds, facultative ponds or other secondary biological treatment systems. The maturation ponds are between 1.5 and 3 m in depth. These ponds further reduce BOD, solids and nutrients.

Aerated lagoons are oxygenated through artificial aeration. The methods for aerating the lagoons include motor-driven submerged or floating jet aerators, motor-driven floating surface aerators, motor-driven fixed-in-place surface aerators and injection of compressed air through submerged diffusers.

Oxidation ponds are widely used in Malaysia because this treatment system is simple, low maintenance requirement and cheaper. Figure 2(a) shows the schematic diagram of the ponding system which is typi-

Figure 2. Conventional biological wastewater treatment processes

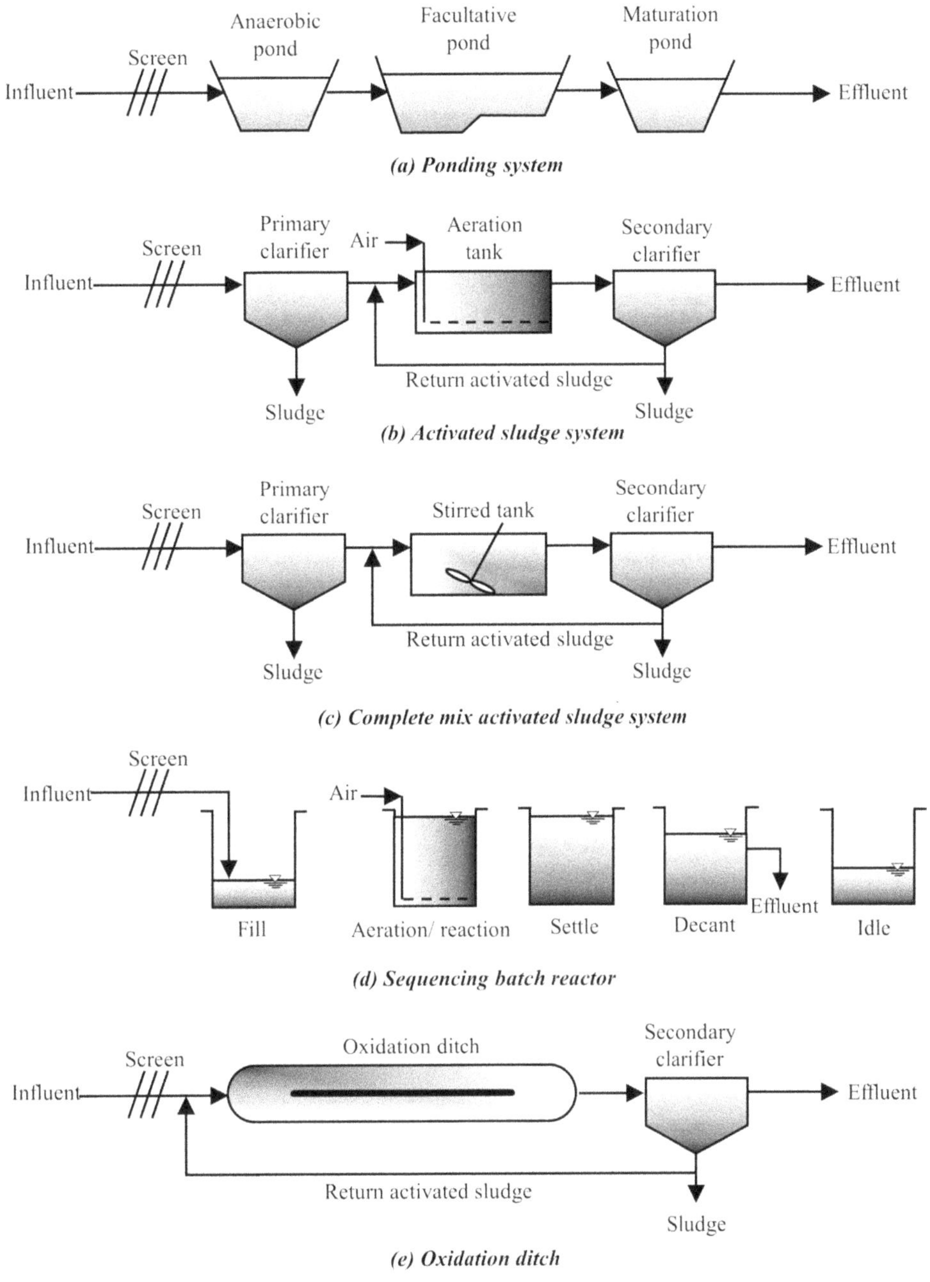

cally installed in Malaysia for sewage treatment. For example, in Sabah, up to 2002, there are about six units of oxidation ponds built in Kota Kinabalu, Penampang and Tawau for sewage treatment (Source: Environmental Conservation Department Sabah, Wastewater treatment in Sabah, Survey Report, 2002). The current application of sewage and urban wastewater treatment in Sabah is discussed in Section 9-2.2.

Malaysian Palm Oil Board (MPOB) reported more than 85% of the palm oil mills in Malaysia have employed anaerobic ponds to treat the POME (Tong and Jaafar, 2004). Generally, the raw POME flows

Figure 3. Conventional biological wastewater treatment processes (Continued)

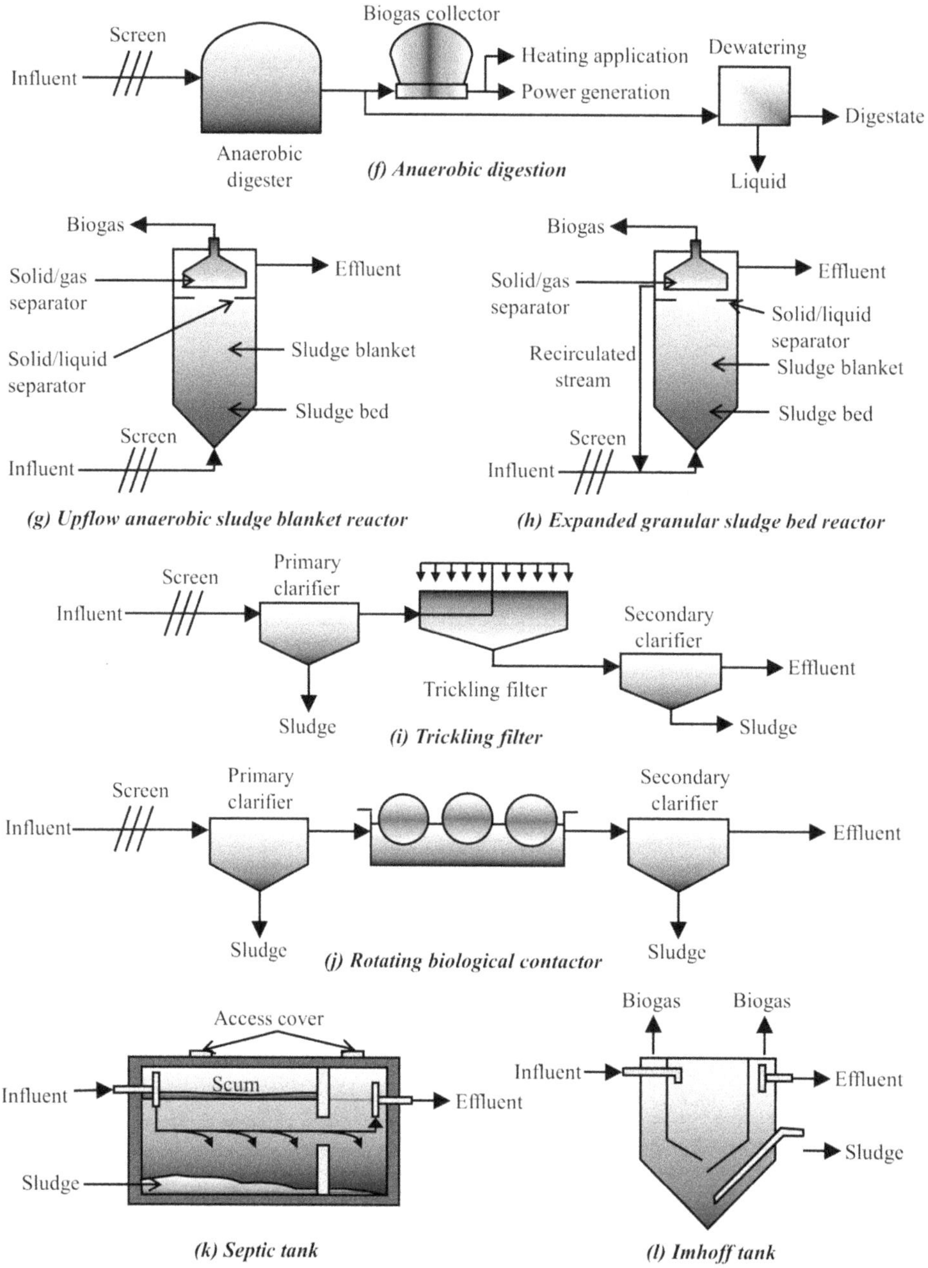

through a tank or pond where the POME is cooled and de-oiled with 1 day hydraulic retention time (HRT). The POME is transferred to an acidification pond with 2–4 days HRT and then the POME is fed into the 5–7 m depth of anaerobic ponds with 30–45 days HRT. The methane produced is released to the atmosphere. The treated effluent is further subject to the anaerobic or facultative pond, which is 1.5 m in depth for achieving the required discharge standards.

Oxidation ponds are the most popular treatment method for rubber wastewater in Malaysia (Wong, 1980; Mohammadi et al., 2010). The oxidation ponds are installed as the primary treatment system. The ponding technology can remove over 95% BOD from rubber wastewater. Rubber wastewater treatment is discussed in Section 9-2.5.

Activated Sludge System

An activated sludge process uses aerobic microbes to feed on organic contaminants in wastewater, to produce a high-quality effluent. Air or oxygen is continuously injected into the primary treated wastewater in an aeration tank as shown in Figure 2(b) or the wastewater can be completely stirred such in Figure 2(c). Consequently, the microbes grow and form particles which clump together. These particles, also known as flocs are allowed to settle to the bottom of a secondary clarifier, leaving a clear liquid which is free of suspended solids. Most of the sludge is returned to the aeration tank.

When the incoming primary treated wastewater is mixed with an amount of the return sludge, the mixture is called mixed liquor. The aeration and stirring of the mixed liquor keep the particles in suspension. The mixed liquor flows into the secondary clarifier where the particles to settle. The settled solids are called the activated sludge. The activated sludge is returned to the aeration tank to start the process again. The most common variations of the activated sludge process are sequencing batch reactor, oxidation ditch and extended aeration.

The activated sludge process employed by the SFI Pulp and Paper Mill can remove 20 – 70% of chlorinated phenols, 60% of 1,1-dichloro dimethyl sulphone, and 30 – 99% of chlorinated acetic acids (Mohamed et al., 1989).

Sequencing Batch Reactor

A sequencing batch reactor (SBR) is considered a fill-and-draw activated sludge system. The SBR operates on a batch basis. Installation of SBR is chosen when space is limited and insufficient for establishing the oxidation ponds. The treatment process involves five stages as illustrated in Figure 2(d): (1) anoxic fill, the influent wastewater is pumped into the reactor through a distribution manifold to provide a good contact between the microbes and the substrate; the mixed liquor is mixed without aeration, (2) aeration and reaction, the mixed liquor is aerated by using fixed or floating mechanical pumps, or transferring air through fine bubble diffusers fixed to the floor of the tank; the microbes reduce the BOD and nitrogen, (3) settle, no mixing or aeration is provided; no liquid entering or leaving the tank; settling of suspended solids starts, (4) decent, the treated effluent or the supernatant is withdrawn, (5) idle, waste the sludge and backwash the jet aerator. All stages of the process are taken place in the same tank.

Oxidation Ditch

An oxidation ditch is an extremely effective variation of the activated sludge process which comprises a ring or oval-shaped channel, as shown in Figure 2(e). The oxidation ditches are completed with brush rotors or disc aerators which maintain the solids in suspension when the mixed liquor circulates around the ditch. The HRT of the oxidation ditches ranges from 24 to 48 hours and the sludge ages are between 12 and 20 days. Figure 4 shows the oxidation ditch installed in Kingfisher Park, Kota Kinabalu for sewage treatment with a plant design capacity of 1,000 – 1,200 pe.

Figure 4. Oxidation ditch installed at King Fisher Park, Kota Kinabalu
Source: Environmental Conservation Department Sabah, Wastewater treatment in Sabah, Survey Report, 2002

Extended Aeration

Extended aeration is a type of activated sludge process without primary treatment. It is commonly used for sewage treatment. An extended aeration plant comprises two main components; they are an aeration chamber and a settling tank. After initial screening, the raw sewage flows directly into the aeration chamber where all the solids are aerobic digested. The sewage is aerated for a minimum retention time of 24 hours. The treated sewage flows to the settling tank for clarification. The supernatant may be subject to filtration or disinfection in order to meet the discharged requirement.

Up to 2002, approximately 65 extended aeration plants have been installed for sewage treatment in Sabah which covers the districts of Kota Kinabalu, Penampang and Tawau (Source: Environmental Conservation Department Sabah, Wastewater treatment in Sabah, Survey Report, 2002). Figure 5 displays several real extended aerations activated sludge plants built for sewage treatment.

Anaerobic Digester Tank

Anaerobic digester tanks are made of concrete or steel. The tanks can be designed as cylindrical or egg-shaped. The typical ratio of diameter to height of the tanks is 2:1. The waste is retained in the digester tank for 20 and 40 days, and the digester is fed continuously. The anaerobic digestion process is illustrated in Figure 3(f). The treated effluent of the anaerobic digestion contains solid, liquid and gas. The solid-liquid mixture is separated through the de-watering process. The digestate (solids) is reused as fertiliser and the liquid is further treated prior to discharge. The biogas is captured and reused as renewable energy.

About 5 to 10% of palm oil mills in Malaysia have built open-top digester tanks rather than oxidation ponds because of land limitation (Tong & Jaafar, 2004). However, the biogas (methane) produced from the anaerobic digestion is released directly to the atmosphere.

Keck Seng (Malaysia) Berhad at Masai, Johor is the first palm oil mill constructs a complete-mixed, closed tank anaerobic system in 1984. The anaerobic digester is a continuous flow stirred tank reactor (CSTR). The biogas is captured and reused as the boiler fuel as well as to generate power. In the early

Figure 5. Extended aeration systems; (a), (b) and (c) located at Kepayan Ridge (Kota Kinabalu), Leeka Industrial Estate (Tawau) and Taman Muhibbah (Tawau), respectively
Source: Environmental Conservation Department Sabah, Wastewater treatment in Sabah, Survey Report, 2002

years of the 2000s, the CSTR-type anaerobic digesters have replaced the anaerobic open ponds in several palm oil mills located at Kunak and Lahad Datu, owned by TSH Resources Berhad. The effluent from the anaerobic digesters is then further treated in SBR and holding tanks to permit the treated effluent to achieve BOD < 20 mg/L and COD < 500 mg/L before discharging to the environment.

Up-flow Anaerobic Sludge Blanket Reactor

An up-flow anaerobic sludge blanket (UASB) reactor is a methanogenic digester. Figure 3(g) shows a typical UASB reactor. The UASB reactor is basically a tank comprises a sludge bed where organic matter is degraded. The sludge bed contains granular sludge which is formed from diverse microbes. Wastewater enters at the bottom of the reactor. Methane gas and treated effluent leave from the top of the reactor. First, the organic matter passes through the expanded sludge bed which contains a high concentration of biomass. Next, the remaining organic matter which passes through the less dense biomass and suspended above the sludge bed is called the sludge blanket. The UASB is a three-phase separator as it can separate gas, water and sludge.

Rinwood Palm Oil Mill located at Mukah, Sarawak has installed a UASB to replace the existing anaerobic open pond in order to reduce methane emission. The biogas is captured for internal heat use and the excess biogas is flared.

Expanded Granular Sludge Bed Reactor

An expanded granular sludge bed (EGSB) reactor is a modification of a conventional UASB reactor. The EGSB reactor has an effluent recirculation, as depicted in Figure 3(h), which aims to improve the mixing at the bottom of the reactor. The up-flow velocity in the EGSB is higher than in the UASB, and the velocity is usually greater than 4 m/h. The high effluent flow rates cause the sludge to expand and this has eliminated the dead zone. The biomass appears as compact granules with 2 – 4 mm in diameter, which is formed through the rolling mechanism under continuous up-flow conditions.

A palm oil mill owned by Teck Guan Development Sdn. Bhd. at Pulau Sebatik located at Tawau, Sabah has used an EGSB to treat the POME. A pilot-scale EGSB has been tested with POME collected from Sungai Burung Palm Oil Mill, Tawau, Sabah. The pilot EGSB can convert about 41 – 43% of the organic matter to biogas and reduce about 91 – 93% of COD (Zhang et al., 2008 a,b).

Trickling Filter

A trickling filter is filled with a high specific surface area material such as rock, gravel, plastic media or slag. These media are filled from 1 to 3 m deep, but plastic media, which are lighter can be filled up to 12 m in depth. Figure 3(i) illustrates a general trickling filter system. Primary treatment of the wastewater is required to avoid the media from clogging and to ensure efficient treatment. The wastewater is then trickled over the surface of the media. Microbes grow on the surface of the media and form a thin biofilm over the surface. Oxygen is depleted within the biomass and the inner layers may become anoxic or anaerobic. The incoming wastewater is sprayed over the media by using a rotating sprinkler. When the biomass grows with time and the biofilm becomes thicker, the attached layer will be deprived of oxygen and finally, the biomass will slough off.

About three trickling filters have been installed in cities of Kota Kinabalu and Penampang in Sabah to treat the sewage wastewater. The trickling filters are designed to serve from 1,000 to 16,000 population. Figure 6 shows an example of the trickling filter installed in Taman Indah Permai, Kota Kinabalu.

Figure 6. Trickling filter system installed at Taman Indah Permai, Kota Kinabalu
Source: Environmental Conservation Department Sabah, Wastewater treatment in Sabah, Survey Report, 2002

Rotating Biological Contactor

A rotating biological contactor (RBC) is a type of mechanical secondary treatment system. Figure 3(j) depicts the wastewater treatment using the RBC system. The wastewater passes through a screening process followed by a period of sedimentation to remove grit and course materials. The wastewater then passes through one or more RBC units. One RBC unit consists of a series of closely spaced and parallel discs mounted on a horizontal rotating shaft. The discs are typically made from a plastic material. The discs are circular with diameter ranges from 2 to 4 m and about 10 mm in thickness. The discs are partially submerged in the wastewater and they are slowly rotated through the wastewater. The microbes present in the wastewater attach on the rotating discs. Biological degradation of the organic matter takes place when the microbes grow on the surface of the discs. The microbes consume oxygen when the rotating discs are partially exposed to the atmosphere. The microbes grow and form a thin biofilm on the discs. The biomass slough off when the biofilm becomes thicker and the inner layer is lacking oxygen. The effluent from the RBC enters a clarifier where the biomass is settled as sludge. The sludge is withdrawn periodically.

There have been about seven RBC plants installed in Kota Kinabalu and Penampang in the state of Sabah until 2002 for sewage treatment purpose with plant capacity, ranging from 180 to 2,300 pe (Source: Environmental Conservation Department Sabah, Wastewater treatment in Sabah, Survey Report, 2002). In 2013, there are approximately 40 RBC plants installed in Malaysia. Figure 7 shows the RBC system installed in Taman Khidmat, Kota Kinabalu.

Septic Tank

An individual septic tank (IST) is a small-scale sewage treatment system and it is the simplest method to treat the sewage wastes, as presented in Figure 3(k). The IST comprises two chambers connected in a series. The chambers are separated by a dividing wall which has openings located about midway between the floor and the roof of the chamber. In the first chamber, the solids from the incoming sewage settle as a sludge layer while the greases and oils float as a scum layer. The liquid between the scum and sludge layers passes through the openings and flows into the second chamber where further sedimentation

Figure 7. Rotating biological contactor installed at Taman Khidmat, Kota Kinabalu
Source: Environmental Conservation Department Sabah, Wastewater treatment in Sabah, Survey Report, 2002

takes place. The effluent from the second chamber is discharged into a drain or is eliminated through percolation into the soil.

Anaerobic microbes grow in the sludge layers settled in the chambers. The microbes decompose and mineralise the sewage waste while producing more stabilised organic compounds and gases such as methane, carbon dioxide and hydrogen sulphide. The sludge is removed for up to two years to maintain the ISTs operate efficiently.

The volume of an IST is between 4,000 and 7,500 litres. There are now more than 1.2 million septic tanks in Malaysia in 2010. The ISTs are suitable for a single building with a population equivalent up to 150.

Imhoff Tank

The Imhoff tank is named for its inventor, Dr Karl Imhoff, a German engineer. The Imhoff tanks are simple forms of sewage treatment methods which require less operation skill. An Imhoff tank consists of an upper sedimentation chamber and a lower digestion chamber, as displayed in Figure 3(l). The two chambers are unconnected and they are separated by a sloping partition which contains narrow slots. In the upper sedimentation chamber, solids settle and pass through the slots into the lower digestion chamber. Scum floats in the upper chamber. The solids settled and form sludge in the lower chamber. Anaerobic microbes digest the sludge while biogas is released into the air.

The Imhoff tanks are normally used to serve for a population equivalent up to 1,000. About 24% of sewage treatment plants in Malaysia are operating the Imhoff tanks. Up to 2002, there have been five Imhoff tanks installed in Kota Kinabalu and Tawau with a capacity of approximately 5,000 pe (Environmental Conservation Department Sabah, Wastewater treatment in Sabah, Survey Report, 2002).

Roughing Filter

Roughing filter is usually used as an efficient pre-treatment process prior to the sand filtration. It removes suspended solids from the water and thus helps to reduce the clogging problem in the sand filter. There are several materials that can be used as the filter media, such as gravel, pebbles, broken clay bricks, burnt charcoal, and coconut fibre. Several types of roughing filters include down-flow roughing filter, up-flow roughing filter and horizontal roughing filter.

Figure 8. Pre-treatment stages in roughing filter

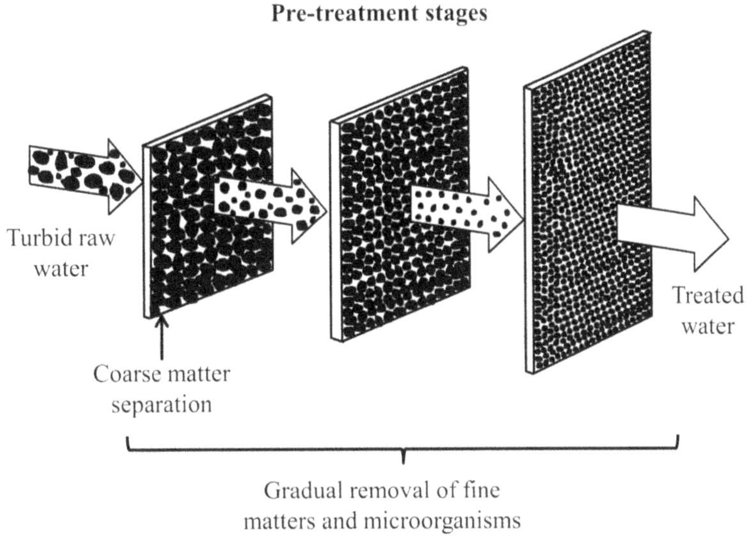

Generally, the filtration rate plays an important role in treatment removal. Lower filtration rate can achieve better removal efficiency in the roughing filter. This is because the slow flow rate is important to retain particles that are gravitationally deposited to the surface of the media.

Figure 8 shows the stages involved in a roughing filter. There are three step-by-step processes used in this pre-treatment process in order to remove different size impurities. The three steps work by separation of the coarse solids, finer particles removal, and finally the removal or destruction of small solids and microorganisms respectively. These pre-treatment steps are also used to reduce pathogenic microorganisms.

Sand Filters

Sand filtration is one of the simplest and the best method for water treatment. There are two types of sand filter that are commonly used which are slow sand filter and rapid sand filter. Filtration speed is not the only difference between these two types of filters since the fundamental theory of treatment processes are also distinct. The slow sand filter is basically a biological process while a rapid sand filter is a physical treatment process. Other types of sand filters are also available such as the pressure filter and bio-sand filter

Since sand filter doesn't require electricity, have simple function and precision, it is well adapted in rural areas. The filters require the use of flocculant chemicals to work efficiently except for slow sand filters. Slow sand filters can produce high quality of water by removing pathogens by 90 – 99% depending on the strains, as well as taste and odour without the need for chemical aids.

Slow Sand Filter

The slow sand filter is considered one of the most simple and best methods of water treatment because of its comprehensibility, low cost, simple function and effectiveness. Besides that, the sand filter is also not driven by electricity (if constructed as a gravity-flow filter) and no chemical usage. Figure 9 shows

Figure 9. Slow sand filter

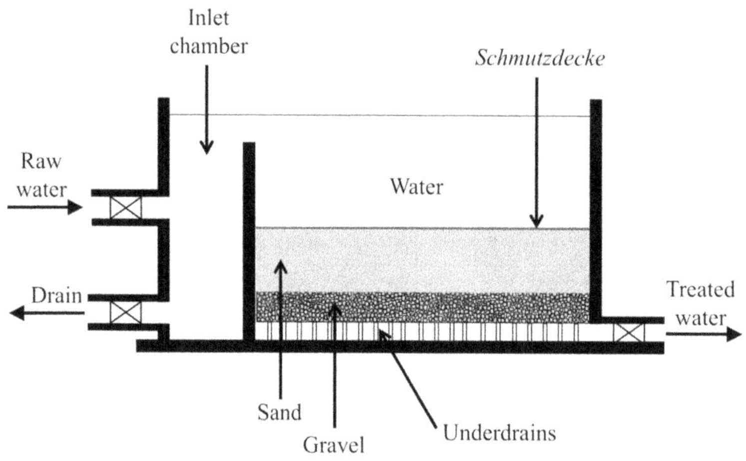

the typical schematic drawing of a slow sand filter. Basically, a sand filter consists of a tank which contains a supernatant layer of raw water, a bed of fine sand, a system of underdrains, an inlet and outlet structure, and valves.

Slow sand filtration technology has been used in Europe since the early 1800s. The slow sand filter is usually designed as the tertiary treatment stage of the water purification in the conventional water and wastewater treatment sector. Turbidity and pathogens in wastewater can be effectively removed by slow sand filter over numerous biological, physical and chemical processes in a single treatment step if the slow sand filtration is well-designed and has proper maintenance.

Slow filtration speed of around 0.1 – 0.3 m/h with the fine sand size of 0.1 – 0.3 mm resulted in the accomplishment of the high efficiency of water treatment by slow sand filtration. Slow sand filtration can be classified into the down-flow method and up-flow method sand filter which refers to the direction of water flow through the filter media.

The fundamental principle of a slow sand filter is that contaminated water will pass through a layer of sand to be physically filtered and biologically treated, which lead to the removal of both sediments and pathogens. The formation of a microbial community called *schmutzdecke* on the uppermost layer of the sand made it becomes biologically active. These microbes are typically originated from the source water and founding a colony within a few days with the aid of slow filtration rate and fine sand size.

The advantages of slow sand filter are no pre-treatment required, good bacterial removal efficiency, no chemicals used, less corrosive effluent, less expert supervision and long periods of acceptable operation between the cleanings. On the contrary, the disadvantages are it needs a vast land cover area, poor turbidity removal, poor colour removal efficiency, and a large workforce requirement for construction and cleaning.

Rapid Sand Filter

During the 1920s, the rapid sand filter was mostly used as a potable water filtration technique as it essentially requires less land area in contrast with slow sand filters. Figure 10 illustrates the typical schematic of a rapid sand filter. The rapid sand filter consists essentially of an open-topped box that is usually made

Figure 10. Rapid sand filter

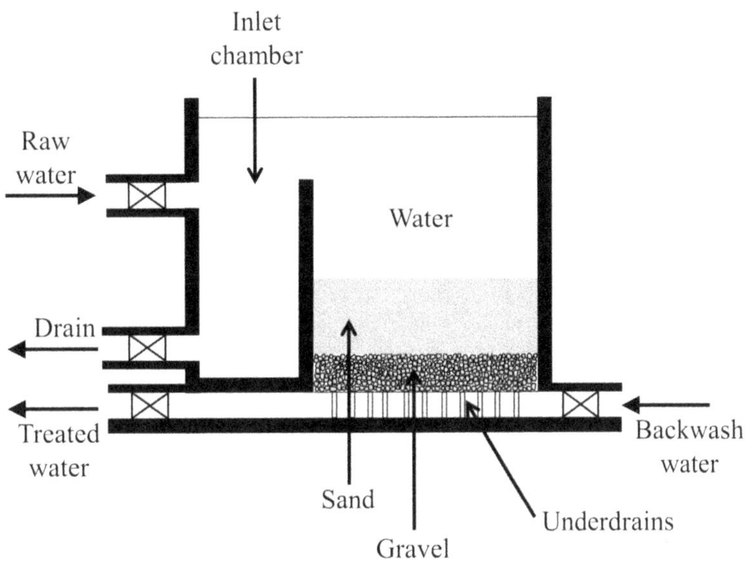

of concrete, partly filled with clean sand and drained at the bottom. The function of a rapid sand filter is usually to purify the water and normally used in municipal drinking water facilities. Rapid filtration is not very effective in removing microorganisms since it is primarily a physical treatment process.

Water is provided to the top of the sand bed which was supported by layers of gravel. Water with a depth of about 1.5 to 2 meter is maintained beyond the sand bed in order to allow the water flow through the sand. The purifying activity of the sand is completely mechanical and suspended matter is collected in the spaces between sand grains until the output of water drops too low. At this point, the filter is drained and cleaned by backwashing. A well-operated rapid sand filter will decrease turbidity to less than 1 NTU and often less than 0.1 NTU.

The rapid sand filter requires small land areas compared to slow sand filters. However, this type of filters is not appropriate for small communities and developing lands because these systems are relatively more complex and more expensive to be operated and maintained. The purifying process takes place during the downward route and the treated water for gravity filter is discharged through underdrains. The rapid sand filter also requires frequent cleaning by backwashing, which involves reversing the direction of the water and adding compressed air.

Pressure Filter

Figure 11 illustrates the schematic of a typical pressure filter which contains the bed of sand or the combination of media in a closed vessel through which water is forced under pressure. It is used extensively in iron and manganese removal plants. The first step in pressure filter operation is that the raw water undergoes an aerated process to oxidize the iron or manganese. Then, the water is pumped to the filter to remove the suspended material. Air binding will not occur in this filter because the water is under pressure. Nonetheless, the pressure filters have a major disadvantage in which is the backwash process

Figure 11. Pressure filter

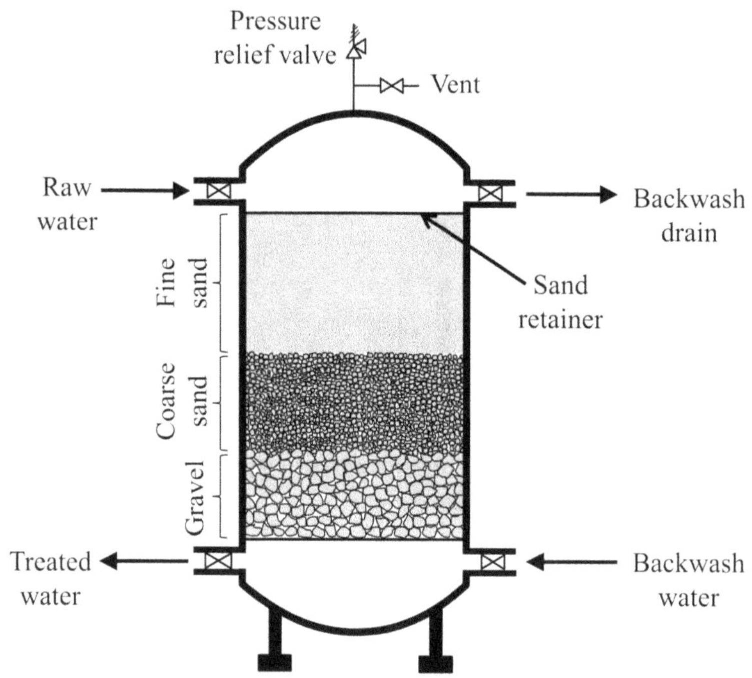

cannot be observed. Because of that, the filter bed can crack easily which can allow the iron and manganese particles to go straight through the filter.

Bio-sand Filter

Bio-sand filter is a water treatment system that is adapted from the slow sand filter. It is suitable to be used in household applications. In general, the bio-sand filter is a filter container that contains a few layers of selected sand and gravel. It is firstly developed by Dr David Manz in the 1990s at the University of Calgary, Canada. However, there is little data available from the field studies that describe the bio-sand filter's performance and sustainable use over time.

The feed water for the bio-sand filter can range in several kinds such as well water, borehole water, pond or river water, tap-stand water, or rainwater. However, the feed water should not contain chlorine as it will kill the bio-layer. Not only that, the feed water should not contain dangerous chemicals since the bio-sand filter is unable to remove most of the chemicals from water.

The working principle of bio-sand filter is the same as the large scale slow sand filtration systems. The filter purifies the feed water by a combination of several mechanisms. The processes that occurred in the filtration include mechanical trapping, predation, adsorption and natural death as shown in Figure 12. The biological active layer in the filter is beneficial to the filter as it develops spontaneously in the micro-environment that exists near to the interface of the sand and water. The organic nutrients that can be obtained through the wastewater are often trapped in the upper 5 – 10 cm of sand. Aerobic respiration occurs as the oxygen diffuses through the standing water from the air to the bio-layer. The community of organisms that exist in the filter improves the filter's ability to remove the bacteria and parasites.

Figure 12. Bio-sand filter

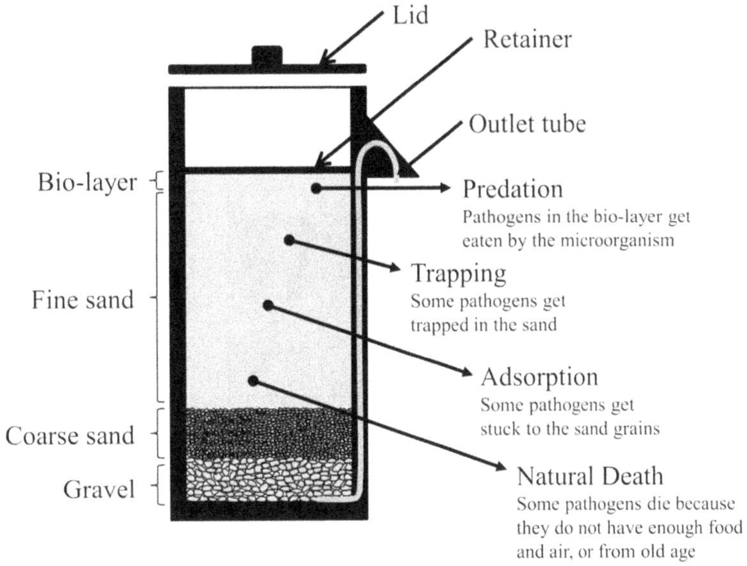

The operation of bio-sand filtration is divided into two periods which are running period and pause period. During the running period, the high water level will push the water through the diffuser and go through the filter media. The water level in the reservoir will slowly go down and flows evenly through the sand media. After that, the flow rate will decreases over time as less pressure is available to force the water through the filter. Many of the suspended particles and pathogens are trapped on the top of the sand media due to their large size. Besides, the inlet water consists of dissolved oxygen, nutrients and contaminants. It provides oxygen to the microorganisms that grow in the bio-layer.

During the pause period, the microorganisms in the bio-layer are allowed to consume the pathogens and nutrients in the water. This period restores the flow rate through sand media as the microorganisms consumed the pathogens and nutrients that are trapped in the sand media. However, the pause period cannot be too long or the bio-layer will consume all of the pathogens and nutrients and finally die-off. The reduction of bio-layer will decrease the efficiency of the bio-sand filter. Therefore, the pause period should be set to be a minimum of 1 hour to a maximum of 48 hours.

There are many advantages of the bio-sand filter such as it can be built locally, easy to use and maintain, does not require electricity to work, and also economically viable to be constructed.

Sand Filter Flow Regime

Down-flow Sand Filter

Most conventional sand filters are operated using the downward flow mechanism. This flow regime is governed by the gravitational force, where raw water flows through the filter media from the top to the bottom of the filter as depicted in Figure 13. Down-flow filtration requires shorter sand depth than up-flow filtration. Based on the established practice, the effective size of sand media for down-flow filtration varies between 0.4 - 0.8 mm, which are fairly smaller compared to the effective size of sand media for up-flow filtration. Because of that, turbulent flow or high-velocity flow rate of the feed causes gradual

Figure 13. Down-flow sand filtration

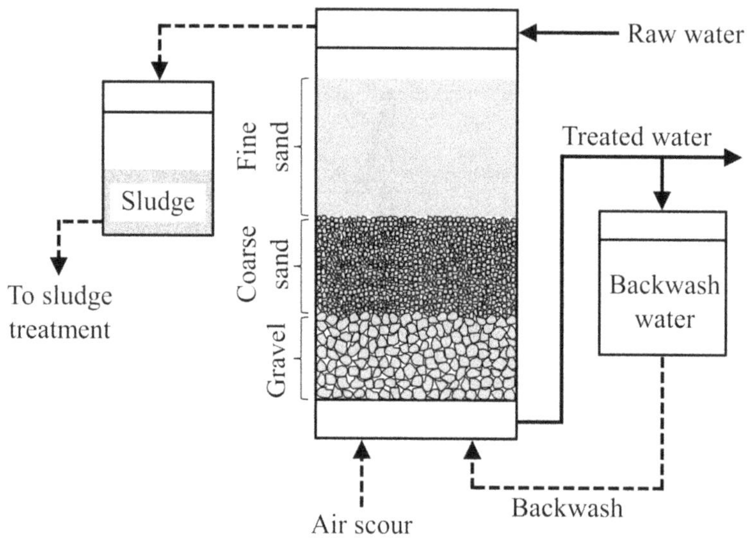

compression of the sand media. Thus, the suspended solids removal efficiency for down-flow filtration is higher than up-flow filtration, producing a higher quality of effluent.

However, the smaller size of sand media used produces higher resistance to flow and are more prone to clogging. Over time, increased resistance to flow can reduce the flux across the sand media. Besides, down-flow filtration will cause a higher rate of heat loss compared to up-flow filtration. Since the down-flow sand filter is more prone to clogging, it requires frequent backwashing. It is also more difficult to monitor, operate and maintain compared to up-flow filter. Installing a pre-treatment system prior to the down-flow sand filter unit can reduce the needs for frequent backwashing. This is because pre-treatment such as screening or roughing filter can remove larger size suspended solids which are the main source of clogging.

Up-flow Sand Filter

Figure 13Figure 14 shows an up-flow sand filter which consists of a packed bed of sand through which raw water flows upward from the filter bottom and emerges upward. During filtration operation, the filter remains packed, capturing solids within or on the filter surface, returning clean effluent water. The effectiveness of the filtration process depends on the influent suspended solid variable such as the type, amount of the solids and the physical filtration variables such as filtration rate, head loss, size, depth and type of filter media. The effective size of sand media for up-flow sand filter varies between 0.8 – 2 mm.

When the fluid passes through the bed of sand particle, it is actually encountering resistance to flow and a resultant pressure drop. As the water velocity increased, a point is reached at which the pressure drop is sufficient to bear the weight of the solid particles. As the flow rate is further increased, the bed will expand and the effective pressure drop remains unchanged. At this stage, the particles of the bed are in equilibrium even though they are moving and it is called fluidization of bed particles. One critical factor in operating of the up-flow sand filter is obtaining the minimum velocity above which fluidization of sand media occur. Maintaining the filtration operation at or slightly below this minimum velocity is crucial to obtain optimum operation, it can prevent the sand bed expansion.

Figure 14. Up-flow sand filtration

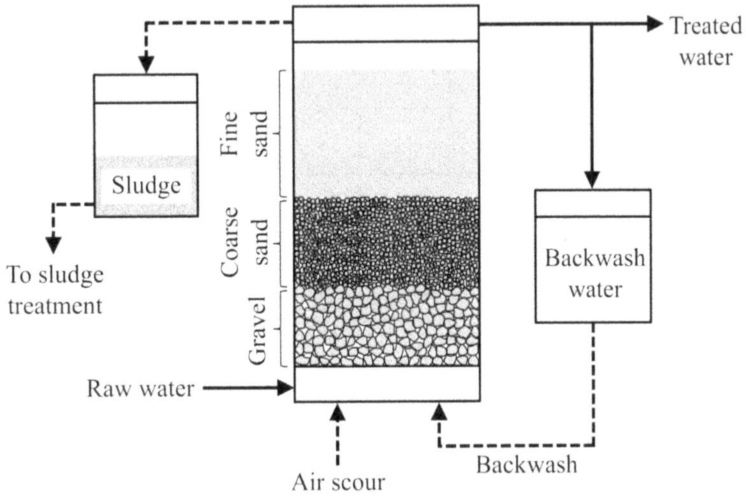

For an up-flow sand filter, the cleaning process of the sand bed occurs when the sand expands or fluidization happens because as the flow rate increases, trapped solids within the filter bed are removed by fluid shearing and particle abrasion. During the backwashing process, the water flow tends to expand the bed that causes the sand media to shake and rub against each other. This will result in unwanted solid particles to left the sand media. For up-flow configuration, the cleaning process is done without reversing the flow of the water but in a higher flow rate of clean water to expand and fluidize the sand bed.

The up-flow sand filter is good at treating high turbidity wastewater and also for the apparent colour and coliforms removal. Furthermore, it is also considered to be very efficient and economical. It can operate at high flow rates and is less susceptible to clogging. Therefore, it is suitable to be used to treat wastewater. One of the disadvantages of using up-flow filtration is that the total bed depth used in up-flow filtration is large and this will increase the structural cost. Besides, the suspended solids removal efficiency for up-flow filtration is much lower compared to down-flow filtration. Other than that, the backwash period is longer, which is typically around 15 – 60 minutes.

INVESTMENT COSTS OF WASTEWATER TREATMENT PLANT

The cost of building a wastewater treatment plants is affected by several factors such as the plant capacity, quality of the plant's influent and effluent, treatment level, the design flow, and technologies used. Prior to constructing the wastewater treatment plant, an estimate of the costs needed are calculated using the cost equations. According to Rodríguez-Miranda, García-Ubaque & Penagos-Londoño (2015), the cost equation used to estimate the cost of building municipal wastewater treatment plant is mainly related to the flow of the wastewater through an exponential equation of,

$$CI = aQ_{inf}^{\ b} \tag{1}$$

Table 11. Cost functions of different wastewater treatment technologies

Treatment Technology	Cost Function (CI)	No. of Data	Reliability Limit (L/s)	Correlation Coefficient	Source
Secondary treatment	$8{,}988Q^{0.71}$	37	16.20 – 1,388.9	0.908	Friedler & Pisanty, 2006
Secondary advanced and nitrification	$2{,}790Q^{0.84}$	11	34.7 – 173.6	0.938	
Activated sludge	$0.0031Q^{0.881}$	6	115.7 – 289.3	0.979	Singhirunnusorn &Stenstrom, 2010
Oxidation pits	$0.0017Q^{0.910}$	8	11.6 – 902.8	0.604	
Aerated ponds	$0.0143Q^{0.681}$	11	11.6 – 902.8	0.822	
Oxidation ponds	$0.0004Q^{1.060}$	23	11.6 – 902.8	0.790	
Conventional secondary treatment	$0.116Q^{0.854}$	9	NA	0.935	Tsagarakis, Mara & Angelakis, 2003
Extended mechanical aeration	$0.206Q^{0.775}$	35	NA	0.829	
Extended aeration dissolved air	$0.153Q^{0.71}$	32	NA	0.808	
Primary treatment	$15.75Q^{0.684}$	NA	1 – 4,000	1.000	Engin & Demir, 2006
Secondary treatment	$23.46Q^{0.763}$	NA	1 – 5,000	1.000	

NA – Not available
Source: (Retrieved from Rodríguez-Miranda, García-Ubaque & Penagos-Londoño, 2015)

where CI is the cost of investment or construction, Q_{inf} is the design flow of influent, and a and b are the calculated coefficients.

Many types of research have been conducted to predict the investment costs of the wastewater treatment plant through the cost equations. In Table 11, the cost equations with the calculated coefficients for different treatment technologies are tabulated. The cost equations may be different in different locations depending on the analyses of exponential regression of various factors being considered for the calculation. In the research performed by Rodríguez-Miranda, García-Ubaque & Penagos-Londoño (2015), the researchers forecasted new cost functions based on the previous data of investment costs collected (Table 12), by taking into account the variables of design flow or capacity, BOD, nitrogen content, TSS, and phosphorus content. In short, the cost equation can be generated depending on how many variables are taken into consideration.

Table 12. Investment costs of some municipal wastewater treatments in Cundinamarca using different technologies

Design Population (Inhab)	Flow (L/s)	Investment Cost (US$)	Investment Cost per Capita (US$/Inhab)	Treatment Technology
Municipal Wastewater Treatment Plants with Anaerobic Technology				
108	0.4	108,151	1,001	(a), (b)
999	6.2	1,067,206	1,068	(a), (c)
2,343	4.9	431,103	184	(a), (d)
10,000	38.0	520,000	52	(a)
42,796	181.2	6,969,656	163	(a)
Municipal Wastewater Treatment Plants with Activated Sludge Technology				
1,984	38.9	2,778,560	1,400	(e), (f)
8,787	95.3	1,664,186	189	(e), (g)
16,000	32.0	2,162,500	135	(e)
106,099	188.6	451,274	4	(e)
108,000	187.0	6,975,000	65	(e)
Municipal Wastewater Treatment Plants with Oxidation Pond Technology				
5,000	20.0	516,000	103	(h)
11,936	20.0	552,000	46	(h)
32,234	117.0	3,955,000	123	(h)
43,000	70.0	1,914,000	45	(h)
141,349	958.7	2,227,683	16	(h)
Municipal Wastewater Treatment Plants with Aerated Pond Technology				
4,100	17.0	480,000	117	(i)
12,000	20.0	1,183,000	99	(i)
22,899	123.5	969,199	42	(i)
56,458	355.6	459,500	8	(i)
122,814	862.6	1,299,548	11	(i)
Municipal Wastewater Treatment Plants with Advanced Dissolved Air Flotation Technology				
1,746	6.5	766,224	439	(j)
1,924	7.4	890,240	463	(j)
3,324	12.3	1,031,838	310	(j)
17,086	25.0	1,647,634	96	(j)

(a) Anaerobic reactor; (b) Anaerobic filter; (c) Upflow anaerobic sludge blanket (UASB); (d) Anaerobic baffled reactor (RAP); (e) Activated sludge; (f) Extended aeration; (g) Oxidation pit, (h) Oxidation pond; (i) Aerated pond; (j) Advanced dissolved air flotation (DAF)

Source: (Rodríguez-Miranda, García-Ubaque & Penagos-Londoño, 2015)

REFERENCES

Bustillo-Lecompte, C., Mehrvar, M., & Quiñones-Bolaños, E. (2016). Slaughterhouse Wastewater Characterization and Treatment: An Economic and Public Health Necessity of the Meat Processing Industry in Ontario, Canada. *Journal of Geoscience and Environment Protection*, *4*, 175–186.

Chang, L., Zhang, Z. Q., Wang, Q. M., Xu, Z. S., Guo, Z. D., Sun, G. Q., ... Qiao, Q. (2006). Advances in polymer flooding and alkaline/surfactant/polymer processes as developed and applied in the People's Republic of China. *Journal of Petroleum Technology*, *58*, 84–89.

Chiam, C. K., & Sarbatly, R. (2011). Purification of aquacultural water: Conventional and new membrane-based techniques. *Separation and Purification Reviews*, *40*, 126–160.

Deng, S., Bai, R., Chen, J. P., Yu, G., Jiang, Z., & Zhou, F. (2002). Effects of alkaline/surfactant/polymer on stability of oil droplets in produced water from ASP flooding. *Colloids and Surfaces. A, Physicochemical and Engineering Aspects*, *211*, 275–284.

Engin, G. O., & Demir, I. (2006). Cost analysis of alternative methods for wastewater handling in small communities. *Journal of Environmental Management*, *79*(4), 357–363. PMID:16307841

Environmental Conservation Department Sabah. (2002). *Wastewater Treatment in Sabah, Survey Report*.

Fakhru'l-Razi, A., Pendashteh, A., Abdullah, L. C., Biak, D. R. A., Madaeni, S. S., & Abidin, Z. Z. (2009). Review of technologies for oil and gas produced water treatment. *Journal of Hazardous Materials*, *170*, 530–551. PMID:19505758

Foyeke, O. O., & Diane, J. B. (1998). Influence of interfacial properties of lipophilic surfactants on water-in-oil emulsion stability. *Journal of Colloid and Interface Science*, *197*, 142–150. PMID:9466854

Friedler, E., & Pisanty, E. (2006). Effects of design flow and treatment level on construction and operation costs of municipal wastewater treatment plants and their implications on policymaking. *Water Research*, *40*(20), 3751–3758. PMID:17028062

Goh, K. C., Narayasamy, D., Briers, J., de Boer, F., Ibrahim, K., & Jaberi, M. Y. (2010). *Asset-wide reconciled production monitoring – a key enabler to successful real-time field management*. SPE paper No. 128654.

Jais, N. M., Radin Mohamed, R. M. S., Wan Mohamad Apandi, W. A., & Peralta, H. M. M. (2015). Removal of Nutrients and Selected Heavy Metals in Wet Market Wastewater by Using Microalgae *Scenedesmus* sp. *Applied Mechanics and Materials*, *773–774*, 1210–1214.

Mittal, A. (2011, August). Biological wastewater treatment. *Water Today*, 32–44.

Mohamed, M., Chin, L. L., & Lim, T. S. (1990). The Brunei Bay as an effluent receiving waterbody: Observations during the start-up period of a kraft pulp and paper mill. *Pertanika*, *13*, 123–128.

Mohamed, M., & Landner, L. (1993). Environmental impact assessment of a mixed tropical hardwood integrated pulp and paper mill – A case study. *Environmental Impact Assessment Review*, *13*, 353–374.

Mohamed, M., Matayun, M., & Lim, T. S. (1989). Chlorinated organics in tropical hardwood kraft pulp and paper mill effluents and their elimination in an activated sludge treatment system. *Pertanika*, *12*, 387–394.

Mohammadi, M., Man, H. C., Hassan, M. A., & Yee, P. L. (2010). Treatment of wastewater from rubber industry in Malaysia. *African Journal of Biotechnology*, *9*, 6233–6243.

Mukherjee, S., Mukhopadhyay, S., Hashim, M. A., & Gupta, B. S. (2014). Contemporary environmental issues of landfill leachate: Assessment & remedies. *Critical Reviews in Environmental Science and Technology*, *45*(5), 472–590.

Rodríguez-Miranda, J. P., García-Ubaque, C. A., & Penagos-Londoño, J. C. (2015). Analysis of the investment costs in municipal wastewater treatment plants in Cundinamarca. *Dyna (Bilbao)*, *82*(192), 230–238.

Singhirunnusorn, W., & Stenstrom, M. K. (2010). A critical analysis of economic factors for diverse wastewater treatment process: Case studies in Thailand. *Sustainable Environmental Research*, *20*(4), 263–268.

Tibbetts, P. J. C., Buchanan, I. T., Gawel, L. J., & Large, R. (1992). A comprehensive determination of produced water composition. In J. P. Ray, & F. R. Engelhardt (Eds.), *Produced Water: Technological/Environmental Issues and Solutions* (pp. 97–113). New York: Plenum Publishing.

Tong, S. L., & Jaafar, A. B. (2004). Waste to energy: Methane recovery from anaerobic digestion of palm oil mill effluent. *Energy Smart*, Issue 14, February 2004.

Tsagarakis, K. P., Mara, D. D., & Angelakis, A. N. (2003). Application of cost criteria for selection of municipal wastewater treatment systems. *Water, Air, and Soil Pollution*, *142*(1–4), 187–210.

Wong, K. K. (1980). Application of ponding systems in the treatment of palm oil mill and rubber mill effluents. *Pertanika*, *3*, 133–141.

Zhang, Y., Yan, L., Chi, L., Long, X., Mei, Z., & Zhang, Z. (2008a). Startup and operation of anaerobic EGSB reactor treating palm oil mill effluent. *Journal of Environmental Sciences (China)*, *20*, 658–663. PMID:18763558

Zhang, Y., Yan, L., Qiao, X., Chi, L., Niu, X., Mei, Z., & Zhang, Z. (2008b). Integration of biological method and membrane technology in treating palm oil mill effluent. *Journal of Environmental Sciences (China)*, *20*, 558–564. PMID:18575108

Zulkifli, A. R., Roshadah, H., & Tunku Khalkausar, T. F. (2011). Department of Environment, Malaysia.

ADDITIONAL READING

Al-Rekabi, W. S., Qiang, H., & Qiang, W. W. (2007). Improvements in wastewater treatment technology. *Pakistan Journal of Nutrition*, *6*(2), 104–110.

Aljuboury, D. A. D. A., Palaniandy, P., Abdul Aziz, H. B., & Feroz, S. (2017). Treatment of petroleum wastewater by conventional and new technologies – a review. *Global NEST Journal*, *19*(3), 439–452.

Ashrafi, O., Yerushalmi, L., & Haghighat, F. (2015). Wastewater treatment in the pulp-and-paper industry: A review of treatment processes and the associated greenhouse gas emission. *Journal of Environmental Management*, *158*, 146–157. PMID:25982876

Awasthi, A., Dalal, P., & Srivastava, J. K. (2017). Dye industry wastewater treatment by coagulation process: Review paper. *Imperial Journal of Interdisciplinary Research*, *3*(8), 260–267.

Campos, J. L., Valenzuela-Heredia, D., Pedrouso, A., Val del Rio, A., Belmonte, M., & Mosquera-Corral, A. (2016). Greenhouse gases emissions from wastewater treatment plants: Minimization, treatment, and prevention. *Journal of Chemistry, 2016*, 1–12.

Dhote, J., Ingole, S., & Chavhan, A. (2012). Review on wastewater treatment technologies. *International Journal of Engineering Research & Technology*, *1*(5), 1–10.

Dracea, D., Tronac, A., & Mustata, S. (2018). Current trends in biological wastewater treatment. *"Agriculture for Life, Life for Agriculture" Conference Proceedings*, *1*(1), 373–376.

Liu, F., Zhang, S., Luo, P., Zhuang, X., Chen, X., & Wu, J. (2018). Purification and reuse of non-point source wastewater via Myriophyllum-based integrative biotechnology: A review. *Bioresource Technology*, *248*, 3–11. PMID:28803062

Ostad-Ali-Askari, K., Gholami, H., Eslamian, S., Crusberg, T. C., Singh, V. P., Dalezios, N. R., ... Cheng, A. (2018). Review urban wastewater contamination in agronomical soils. *International Journal of Research Studies in Science. Engineering and Technology*, *5*(8), 27–33.

Rajasulochana, P., & Preethy, V. (2016). Comparison on efficiency of various techniques in treatment of waste and sewage water – a comprehensive review. *Resource-Efficient Technologies*, *2*(4), 175–184.

Seow, T. W., Lim, C. K., Md Nor, M. H., Muhammad Mubarak, M. F., Lam, C. Y., Yahya, A., & Ibrahim, Z. (2016). Review on wastewater treatment technologies. *International Journal of Applied Environmental Sciences*, *11*(1), 111–126.

Shah, M. P. (2016). Industrial wastewater treatment: A challenging task in the industrial waste management. *Advances in Recycling & Waste Management*, *2*(1), 1–11.

KEY TERMS AND DEFINITIONS

Activated Sludge Process: Biological process which utilizes aerobic microbes to feed on organic contaminants in wastewater to produce a high-quality effluent.

Biological Wastewater Treatment: Process that uses natural processes to treat wastewater which involves the microbial activity of decomposing organic and other substances.

Microbes: Microscopic-size organisms found in water, air, and soil which are responsible for many natural processes on earth such as decomposing organic matter and nutrients which is particularly significant in wastewater treatment.

Nonpoint Source Pollution: Pollution resulting from a number of human activities such as land development that require land clearing, and will lead to soil erosion carrying the human-made pollutants into water bodies.

Oil and Gas Wastewater: Wastewater that is originated from the oil and gas sector, including flow back and produced water containing toxic chemicals, and other contaminants like heavy metals, hydrocarbons, and salts.

Palm Oil: A vegetable oil extracted from the mesocarp of the oil palm fruits.

Ponding System: Oxidative treatment of wastewater which is classified into aerobic ponds, anaerobic ponds, facultative ponds, maturation ponds and aerated lagoons.

Urban Wastewater: Consists of sewerage produced by residential areas and wastewaters generated from public facilities such as schools, local business, offices and shops lots.

Wastewater: Any water which has been adversely affected in quality by anthropogenic influences.

Chapter 10
Sewage Management and Treatment

ABSTRACT

Accumulation of sewage is a global issue that occurs primarily due to the rising rate of population growth. Without proper treatment and management, uncontrolled sewage generation threatens human health, environment, and society. This chapter briefly introduces sewage management and regulations. The conventional method of sewage and sludge treatment consists of pre-treatment, primary, secondary, and tertiary treatments before the sewage is safe to be discharged. Several sewage treatment technologies are introduced, namely the septic tank, intermittent decanted extended aeration, activated sludge system, membrane sewage treatment system, and finally the anaerobic digester.

INTRODUCTION

Status of Sewage Management in Sabah

Two centuries ago, there were only one billion humans roaming this Earth. Fast forward to today, there is an astounding number of 7.6 billion humans living on our planet. At the beginning of time, the human population grew at a slow rate, but a few centuries ago, we have seen that the growth rate has multiplied radically. In the last two centuries alone, the world population was triple of that the whole history of humanity. The world population changes at an exhilarating rate, and this has great impacts of humanity on the Earth's environment. Still, this also serves as hope for better future, as we have a larger team with advanced thinking who can play a part in contributing solutions to the betterment of global well-being (Roser & Ortiz-Ospina, 2018).

The rising rate of population growth often leads to an increasing amount of waste generation. This trend is happening globally, especially in developing countries, such as Malaysia. Developing countries tend to have higher birth to date ratio, as a result of (1) the high number of poverty, and (2) lack of

DOI: 10.4018/978-1-7998-2645-3.ch010

Figure 1. Population size and annual population growth rate, Sabah, 2014-2018
Source: Department of Statistics Malaysia, 2018

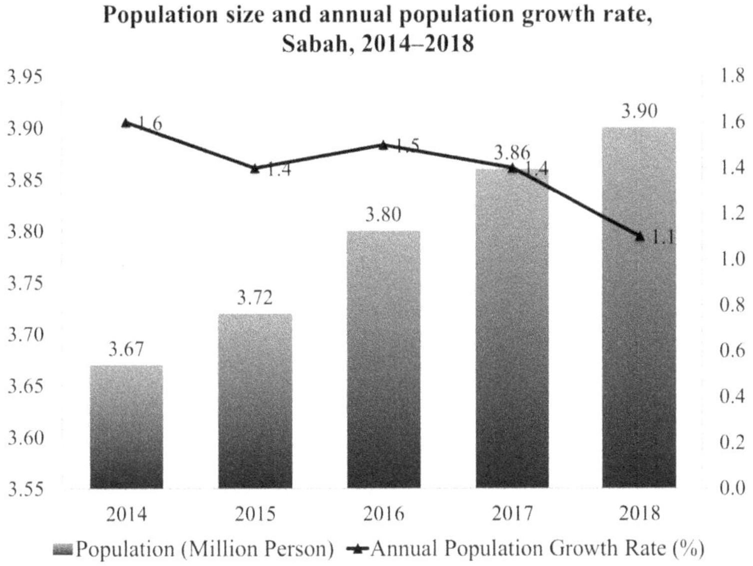

education and family planning. This phenomenon can be seen more clearly in rural states like Sabah. In Sabah, the annual growth rate has been increasing steadily, at an average rate of 1.4% in the last five years. To date, there are approximately 3.90 million (Figure 1) of people residing in Sabah, making it the second most populated state in Malaysia after Selangor (Department of Statistics Malaysia, 2018).

How the populations are spread across the Earth has a substantial effect on the environment. Urbanization of cities like Kota Kinabalu, Sabah allows for better management of the sewage system. This is attributed to the fact that urban cities are packed from the mass of people from rural regions who are migrating to look for employment and opportunities. Denser population allows for the construction of a centralized system that can cater for better management of sewage disposal. Contrarily, it will be difficult to construct a centralized sewage management system for a scattered rural population when looking at the perspective of cost-effectiveness. Urbanization can lead to better resources and waste management; however, if the growth rate outstrips the development rate, environmental problems could arise such as high level of centralized pollution and pressures on resources like water, food, and energy from the ongoing growth.

In the case of tourist arrivals, Sabah received an average of 3.38 million domestic and international tourists annually during the period of 2013 – 2017. The tourist arrivals trend in this five years duration recorded a slight decline from 2013 to 2015 and subsequently increased to the year 2017 (Figure 2). The ratio of Sabah's inhabitants to tourists is almost 1, and it is not impossible that the number of tourists will outnumber the locals. The trend of increasing tourist arrivals to our country is a good mean for our economic and social developments. However, tourism produces large amounts of wastewater and the management has become challenging. Construction of resorts, hotels, and recreational spots could lead to increased sewage generation, and eventually, sewage pollution. Many tourists attraction in Sabah is concentrated at areas surrounding sea, lakes, and rivers, which serves as important water sources to the locals. Wastewater generated from tourism activities has polluted these water bodies surrounding these

Figure 2. Tourist arrivals, Sabah, 2013-2017
Source: Sabah Tourism Board Official, 2018

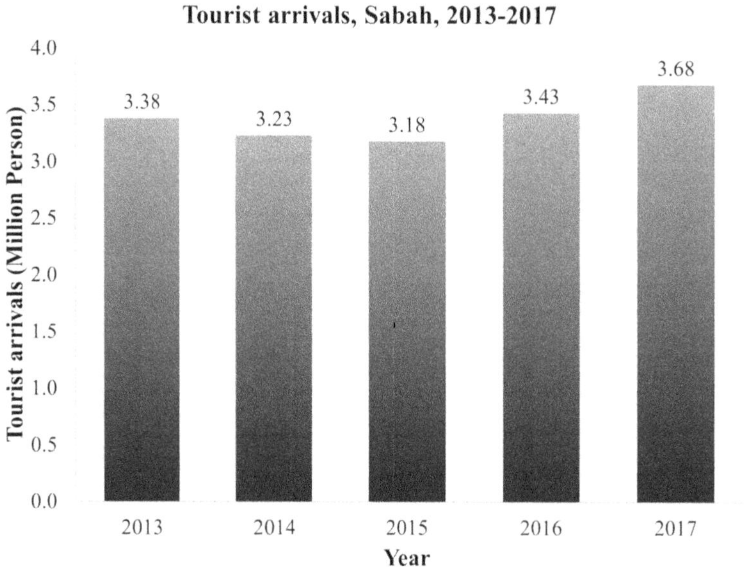

attraction sites, which consequently harms the flora and fauna species and presents a threat to both human health and the environment.

Kota Kinabalu is situated along the West Coast of Sabah with several surrounding islands such as Gaya Island, Sapi Island, and Mamutik Island. Some of these areas are inhabited with 'floating residential/village' or '*kampung air*', which is one of the unique landmarks of Sabah and has been around for many years. The Kampung Air, Tanjung Aru in Kota Kinabalu is one of the biggest 'floating residential/village', with around 3,000 residents living in 240 houses. Other floating settlements also reside around Likas, Sembulan, and Pulau Gaya in Kota Kinabalu, and also in other coastal districts like Semporna, located at the South-East Coast of Sabah.

The settlements are built over water on stilts and are often associated with negative images of garbage pollution, criminality, and an eyesore to the main city. Despite being an eye-catching attraction to some tourists, this type of settlement has no proper sewerage system, apart from being the main source of garbage litters to the neighbouring beaches and hotels. Currently, there are no regulations enforced to prevent the residents from throwing away rubbish into the sea. The absence of sewerage treatment system also perforce the villagers to directly discard their sewage into the sea.

Improper management of sewage that is commonly observed in Sabah is a public concern that needs to be robustly tackled by the government and also the locals. Mismanagement of sewage may lead to sewage runoff that presents hazards to both human and the environment. Sewage contamination of the coastal environment leads to large numbers of transmissible illnesses connected to bathing in beaches and ingestion of seafood. These diseases include diarrhoea, hepatitis, typhoid, and cholera that are caused by pathogens. The disease-causing bacteria can survive for several days to several weeks, while viruses can persist in the marine water and seafood for several months. More severely, hepatitis can live in the sea for more than a year (GESAMP, 2001).

The composition of sewage may contain various types of chemicals and special waste such as medical wastes, nutrients, industrial wastes, and heavy metals, which present additional threats to human and the environment. Nutrient contents from sewage runoff posed as the main source of nutrients to marine life. This often leads to the growth of algae blooms, which consequently cause damage to the coral reefs, as they block the filter-fed corals, hampering their survivability. The toxins from algae bloom also inflict health risks to human exposure.

Succinctly, sewage pollution presents hazards which impose risks to both human health and the environment such as:

- Risks to human health as a result of the increase in total pathogens such as bacteria and viruses, and also toxins secreted by the algae blooms.
- Risks to tourism activities.
- Risks to the coral reef, aquatic flora and fauna, and marine ecosystem.

This chapter discusses the sewage management in Sabah in terms of the laws, regulations, and policy responses on sewage issues. The sewage concerns mainly arise from the population and tourism problems which are discussed early in this chapter. Due to that, proper treatments are required to manage the sewage as many threats may occur from improper management and treatment. The current treatments and management are discussed on the public, private, and individual perspectives. Additionally, the technologies which are significant to the sewage treatment application are introduced in the last section.

Law, Regulation, and Policy Response on Sewerage Services

Malaysia has seen remarkable sewerage development over the last three decades through a public-private partnership. During the Malaysian pre-independent years, sewage was discharged through a pour-flush system with no proper sanitation management. It wasn't until post-independent days before the system was upgraded into the individual septic tank. Both of these systems are seen as primitive/primary in view of their technological capabilities. In the 1970s – 1990s, Imhoff tanks started to be utilized in sewage treatment facilities aside from the activated sludge system and biological filters. Still, the secondary systems are inadequate for full sanitary management which are capable to address the public health and river/sea pollution due to sewage pollution. Thus, fully mechanized tertiary treatment plants using more advanced technologies were commenced around the year 2000s. The mechanical plants include technologies such as extended aeration system, oxidation ditch, rotating biological contactors, sequenced batch reactors and trickling filters (Wan Abdullah, 2011). Generally, there are two acts that mainly regulated the sewage management system, which is (1) Sewerage Services Act 1993 (Act 508), primarily for the enforcement of sewerage services and management, and (2) Environmental Quality Act 1974, Environmental Quality (Sewage) Regulations 2009, which essentially managed the effluent discharge of sewage treatment.

In terms of sewerage management, sewerage services were managed by the local Sanitary Board before independence, and subsequently by Municipals (urban) and Ministry of Health (rural). The Sewerage Services Act 1993 (Act 508) was commenced in June 1993 by the parliament that "amend and consolidate the laws relating to sewerage systems and sewerage services throughout Malaysia for the purpose of improving sanitation and the environment and promoting public health, and to provide for matters connected therewith and incidental thereto" (Laws of Malaysia, 2006). This act covers ten sections as listed below:

- Part I – Preliminary;
- Part II – Responsibility for Sewerage Systems and Sewerage Services;
- Part III – Director General of Sewerage Services;
- Part IV – Public Sewerage Systems;
- Part V – Private Sewerage Systems and Septic Tanks;
- Part VI – Powers of Entry;
- Part VII – Charges;
- Part VIII – Approval of Plans and Specifications of Sewerage System or Septic Tank;
- Part IX – Licensing; and
- Part X – Miscellaneous.

The Sewerage Services Department was successively formed in December 1993 as the regulatory body for sewerage services under Act 508. It is mainly responsible for the enforcement of issues governing the approval to construct sewerage facilities, sewerage operation and maintenance, planning and development of sewerage infrastructure. On the subject of effluent discharge, the Department of Environment is responsible as prescribed in the Environmental Quality Act 1974, Environmental Quality (Sewage) Regulations 2009. The standard limit of discharged effluents is discussed in Section 9-3.

In the case of Sabah state, the state sewerage services are regulated under the Sewerage Services Enactment 2017 (No. 5 of 2017), which contains twelve sections as follows:

- Part I – Preliminary;
- Part II – Responsibility for Sewerage Systems and Sewerage Services;
- Part III – Director of Sewerage Services;
- Part IV – Licensing;
- Part V – Approval of Plans and Specifications of Sewerage Systems or Septic Tanks;
- Part VI – Public Sewerage Systems;
- Part VII – Private Sewerage Systems and Septic Tanks;
- Part VIII – Charges;
- Part IX – General Offences and Penalties;
- Part X – Powers of Entry;
- Part XI – Enforcement; and
- Part XII – Miscellaneous.

Currently, the sewerage services in Sabah is under the control of local authority (*Pihak Berkuasa Tempatan*, PBT) as specified in Part IX of the Public Health Ordinance 1960, excluding the local authorities for districts of Kota Kinabalu and Tawau who have their own by-laws, i.e. Kota Kinabalu Municipal Council (Sewerage) By-Laws 1960 and Tawau Town Board (Sewerage) By-Laws 1960. A Sewerage Branch has been set up under the Public Works Department (*Jabatan Kerja Raya*, JKR), which is responsible for executing the sewerage matters from the local authority. Besides, they are responsible for the planning, design, and implementation of sewerage systems, operation and maintenance, and also providing technical advice on subjects relating to sewerage services to other government bodies.

The planning of sewerage services projects in certain areas, either public or private management would first require the approval from the local authority. The implementation of the projects in terms of design and operation would need to be consented by the Sewerage Branch under JKR before it can proceed. Subsequently, the discharge effluent during operation would be monitored by the Department of Environment and should comply with the standard limit imposed.

Before the Sewerage Services Act, 1993 (Act 508) was approved by the Cabinet, there wasn't a clear law that regulates the sewerage services. As a result, there were a number of rivers contaminated by sewage due to improper management of the service. This issue resulted in the local authority issuing an order that the construction of sewerage system would require for Environmental Impact Assessment (EIA) to be conducted. Once the act was drafted and approved by the Cabinet, the requirement for conducting the EIA is no longer in place as the development of sewerage services would follow the regulations as stated in the act. EIA was ordered before to address the issues at hand and foresee any potential problems at the initial stage of planning and design of the project. The environmental impacts are evaluated in order to protect the environment as well as the wellbeing of humans.

SEWAGE TREATMENT

Introduction to Sewage Treatment

The Malaysian sewage treatment in the past is considerably unhygienic as no proper regulations are in place apart from the primitive technologies available during that time. Improper sanitation management has led to water pollution due to faecal contamination which gives rise to water-borne diseases such as diarrhoea, cholera, and even fatality. Due to these critical circumstances, the Malaysian government has imposed more stringent regulations governing these issues. Better technologies are being utilized in the sewage treatment in order to protect public health, water resources, as well as preserving the environment.

The main objectives of sewage treatment are to reduce the organic contents and kill pathogenic bacteria contained in sewage. It's important to treat sewage as the water reclaimed can be reused which would contribute to the issues of water shortage. This section will discuss the characteristics of sewage and also several types of sewage treatment systems, focusing on the general treatment processes applied in a dwelling unit setting and municipal state.

The American Journal of Public Health (AJPH) in 1924 has compiled a set of terms used in sewerage and sewage disposal practice. Several selected general terms and definitions of sewage treatment are listed as follows:

1. **Sewage:** (1) Wash water and water-carried animal, culinary and in some cases industrial wastes. (2) Liquid waste containing human excreta, ordinarily flowing in or from a house drainage system or sewer. "Excreta" includes faeces, urine, secretions from the skin, expectoration, etc. (3) Liquid wastes from dwellings and institutions, stables and business buildings. It may also contain liquid wastes from industries. (4) A combination of (a) the liquid wastes conducted away from residences, business buildings and institutions and (b) from industrial establishments, with (c) such ground, surface and stormwater as may be admitted to or find its way into the sewers. (5) The ordinary liquid contents of a sewer containing organic wastes, which may or may not include street wash.

2. **Sewage Treatment:** Any artificial process to which sewage is subjected in order to partially remove o so alter its impurities as to render it less offensive or dangerous and more fit to discharge.
3. **Sewer:** A conduit for carrying sewage.
4. **Sewer System:** The collecting system of sewers and appurtenances including sewage lifts where required.
5. **Sewerage System:** Comprehensive term including all constructions for collection, transportation, pumping, treatment and final disposition of sewage.
6. **Sewage Works:** Main pumping station, treatment plant and means of disposal.
7. **Collecting System:** All sewers from the house to the outfall.
8. **Sludge:** Semi-liquid and largely organic suspended sewage solids as deposited in tanks or as subsequently treated.
9. **Public Sewer:** A common sewer controlled by public authority.
10. **Private Sewer:** One privately owned and used by one or more properties.
11. **Septic Tank:** An open or covered continuous horizontal flow tank in which sewage and the deposited sludge are retained long enough to allow considerable anaerobic decomposition. The period of detention of sewage is usually from eight to twelve hours.
12. **Activated Sludge Process:** Sewage treatment in which sewage standing in or flowing through a tank is brought into intimate contact with air and with biologically active sludge, previously produced by the same process. The effluent is subsequently clarified by sedimentation.
13. **Imhoff Tank:** A deep two-storied tank invented by Dr Karl Imhoff, consisting of an upper or continuous sedimentation chamber and a lower or sludge digestion chamber. The floor of the upper chamber slopes steeply to trapped slots through which solids may settle into the lower chamber. The lower chamber receives no fresh sewage but is provided with gas vents and with means for drawing digested sludge from near the bottom.
14. **Trickling Filter:** An artificial bed of coarse material such as broken stone, clinkers, slate, slats or brush over which it trickles to the underdrains giving the opportunity for organic matter to be oxidized by bio-chemical agencies.

Characteristics and Composition of Sewage

Sewage contains a mixture of municipal and industrial wastewaters, composed of organic and inorganic matters in the state of dissolved, suspended or volatile form. Various microorganisms are also found in sewage which could be beneficial or detrimental to human wellbeing. It is important to characterize the sewage prior to assigning a suitable sewage treatment system, in order to make the treatment works efficiently. Sewage is characterized through several properties including the physical, chemical, and biological constituents. These can be summarized as shown in Table 1.

Thomas (2001) described the average composition of contaminants present in domestic or raw sewage as shown in Table 2.

Sewage Treatment System

The sewage treatment system or sewerage system is generally divided into four types, which are (1) Type I – Public Sewerage Systems; (2) Type II – Private Sewerage Systems; (3) Type III – Individual Septic Tank Systems; and (4) Type IV – Primitive Systems/Pour Flush. Each system is categorized into

Table 1. Characteristics of sewage

Property	Description
Physical	The physical property of sewage includes parameters that can be identified using physical means such as colour, odour, temperature, turbidity, total solids. **1. Colour:** Fresh sewage has a yellowish-grey to brownish colouration, while decomposed sewage usually possesses dark grey colour as a result of organic matter oxidation. **2. Odour:** Fresh sewage has soapy or oily odour whereas decayed sewage has an unpleasant odour due to H_2S and CH_4 produced. **3. Temperature:** Sewage generally has a relatively higher temperature than water due to the presence of biological activities. **4. Turbidity:** Sewage contains higher turbidity than water because of the solids contained in it. **5. Total Solids:** The solid content in sewage is identified as all matters except the water. The solid matter can be classified into dissolved solids, suspended solids, and volatile solids. Dissolved solid is the solid matters that dissolved in water. The undissolved solid is classified as suspended solid; suspended solid that floats are referred to as floatable solid or scum, whereas suspended solid that settles is called settleable solids or sludge. Volatile solid that evaporates at 500°C – 600°C is termed as volatile solids.
Chemical	Sewage is composed of both organic (carbohydrates, proteins, fats, and oil & grease) as well as inorganic (chlorides, nitrogen, phosphorus, sulfur) chemicals. Additionally, it contains a range of gases such as CH_4, H_2S, NH_3, and CO_2 that are formed from the decomposition of sewage. Some parameters that are related to the chemical property of sewage are pH, biological oxygen demand (BOD), chemical oxygen demand (COD), and dissolved oxygen (DO). **1. pH:** The pH of sewage is typically high (basic) and becomes low (acidic) after septic and subsequently increased (basic) during the treatment processes. The fluctuation of pH value is generally in the range of pH 5 – 8. **2. BOD:** BOD is a significant parameter that measures the quality of water. It measures the amount of oxygen required by microorganisms to oxidize and break down the organic contents in the sewage. BOD testing is typically carried out after 5 days and measured at standard room temperature. Sewage with high BOD content posed high pollution level if improperly discharged to a water body. This is because it can cause critical depletion of oxygen content in the water which consequently affects the aquatic life which needs oxygen to breathe. Raw sewage may contain BOD count of up to several hundred as compared to BOD of drinking water that should be below 1. It needs to be treated to at least BOD 20 mg/L (Standard A) or 50 mg/L (Standard B) at 20 °C as required by the Malaysian law. **3. COD:** COD measures the amount of oxygen needed to oxidize organic matters in sewage using chemical oxidation method typically utilizing strong oxidizing agents like potassium dichromate ($K_2Cr_2O_7$) and potassium permanganate ($KMnO_4$). COD is closely associated with BOD, the difference being that BOD measures the amount of organic matter being biologically oxidized, whereas COD measures the amount of organic matter being chemically oxidized. The higher the BOD and COD level of sewage, the higher its oxygen depriving capacity when discharged into the watercourse. This is due to oxygen being consumed both biologically and chemically to decompose the organic matters. More organic matters can be chemically oxidized compared to being biologically oxidized, thus COD value, in general, is higher than BOD. COD is crucial in industrial sewage test as toxic chemicals can be tested through COD, but not through BOD test. The effluent standard of COD in Malaysia is 50 mg/L (Standard A) or 100 mg/L (Standard B). **4. DO:** DO is a measure of the amount of oxygen dissolved in sewage. The presence of DO indicated that it is aerobic or fresh, and that oxidation has occurred after treatment.
Biological	Sewage contains various ranges of microorganisms which give the biological characteristics of sewage; both harmless and pathogenic bacteria are present in sewage. The bacteria in raw sewage could range in several hundred thousand to millions per mL. These bacteria are the ones responsible for decomposing the compounds present in sewage aided by some enzymes. They can be classified into three groups, depending on the modes of the process: **1. Aerobic bacteria:** Bacteria which requires the presence of free oxygen to live. **2. Anaerobic bacteria:** Bacteria which can live in the absence of free oxygen. **3. Facultative bacteria:** Bacteria which can live under the presence or absence of free oxygen.

Table 2. Average composition of domestic sewage

Contaminants	Concentration (mg/L)		
	Weak	Medium	Strong
Suspended solids	100	220	350
BOD_5	110	220	400
COD	250	500	1000
Total Phosphorus	4	8	15
Ammoniacal nitrogen	12	25	50
Nitrates	0	0	0

Source: (Thomas, 2001)

Table 3. Types of the sewage treatment system

Treatment System	Category	Mode
Public Sewerage Systems	Centralized/Decentralized	Connected
Private Sewerage Systems	Centralized/Decentralized	Connected
Individual Septic Tank Systems	Decentralized	On-site
Primitive Systems/Pour Flush	Decentralized	On-site

Source: (Abdul Wahab, 2013)

centralized or decentralized system and built as a connected sewerage system or individually on-site as shown in Table 3. Connected sewerage system (Figure 3) refers to sewers from municipal being connected to the centralized sewerage system, while an on-site sewerage system (Figure 4) is a small scale treatment unit that is not connected to the main public sewerage, typically for a single dwelling unit. The connected sewerage system consists of the elements of pump and pipe network, sewage treatment facility (privately or publicly operated), and the discharge point. The on-site sewerage system typically consists of one element of treatment system like a septic tank which would require desludging periodically.

The design of the sewerage system or sewage treatment plant is calculated based upon the population equivalent (pe) parameter. Population equivalent is most used to depict the size of sewage treatment plant, and can be defined as the equivalent of a fixed population of a varying population-based upon predetermined values as tabulated in Table 4. Residential areas population equivalent is set as 5 persons per unit which can be used to calculate population in a certain area. On the other hand, the population equivalent for commercial areas is calculated based on the floor area to approximate the number of people using it daily. The population equivalent size can be converted to volumetric flow rate which is expected to be collected, pumped, and treated.

The calculated flow rate is important to determine the sizes of the piping network, pumps, and sewage treatment plant designs. The flow rates can be calculated using formulas provided in Malaysian Standard MS 1228: 1991 Code of Practice for Design and Installation of Sewerage Systems. The design should take into account both current and future requirements of sewage treatment and management due to the increasing trend of population growth. The designed sewage treatment plants should be able to cater the population served for at least 25 – 30 years down the line and can be calculated by studying the trend of population growth rate in the certain location.

Figure 3. Schematic of a general connected sewerage system

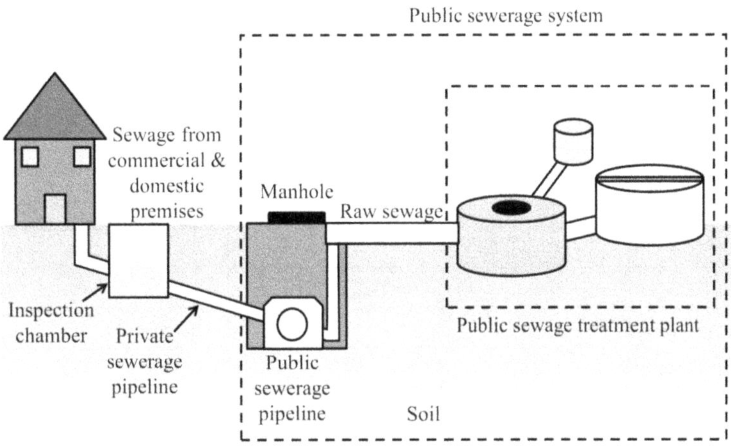

Figure 4. Schematic of a typical on-site sewerage system

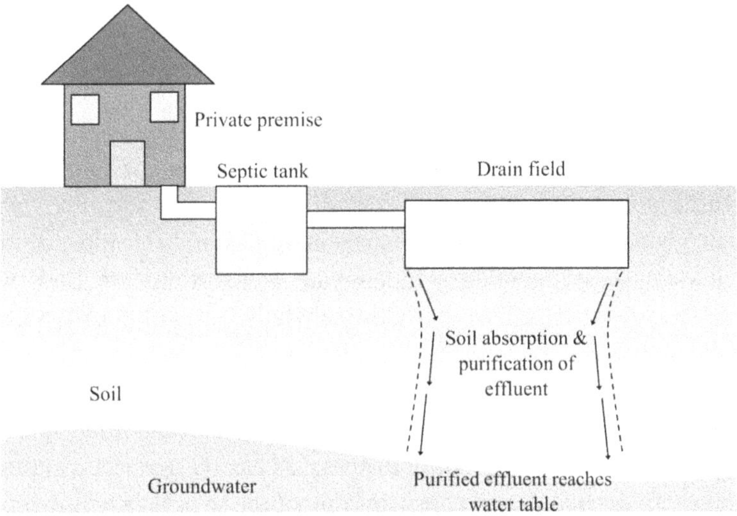

Pumps and Piping Network

Pumps and piping network is constructed connecting homes to the sewage treatment plant in order to transfer the sewage to be treated. As shown in Figure 3, a network of private sewerage pipelines is linked to public sewer pipes which serves as the main pipeline to transmit the sewage to the centralized treatment plant. The sewage treatment plant contractor would be responsible for the installation of public sewerage pipelines, while the private sewerage pipelines would be managed by the homeowner.

Table 4. Population equivalent to type of establishments

Type of Establishment	Population Equivalent
Residential	5 per unit
Commercial: (Includes entertainment/recreational centres, restaurants, cafeteria and theatres)	3 per 100 m gross area
School / Educational Institutions: - Day schools / institutions - Fully residential - Partial residential	0.2 per student1 per student0.2 per non-residential student1 per residential student
Hospitals	4 per bed
Hotel (with dining and laundry facilities)	4 per room
Factories (excluding process water)	0.3 per staff
Market (Wet Type)	3 per stall
Petrol kiosks / Service stations	18 per service bay
Bus Terminal	4 per bus bay

Source: (Malaysian Standard MS 1228, 1991)

Public Sewerage System

Public sewerage system refers to a sewerage system that is regulated by the local authority to treat and process sewage wastewater mainly in the urban areas. This system comprises of the collection, transportation, pumping, treatment and final disposal of sewage. In Malaysia, it can be managed either by the local authority or being handed over to private management. As of 2011, most of the sewage treatment systems in peninsular Malaysia are operated by Indah Water Konsortium Sdn. Bhd., whom services cover 88 out of 144 local authorities (61%). This excludes the whole states of Sabah, Sarawak, and Kelantan, including the Johor districts of Johor Bahru and Pasir Gudang. The total operational area and population served are indicated in Table 5.

In Sabah, the local authorities are the responsible party to deliver urban services such as sewage and wastewater treatment in their corresponding districts, as clearly described in the Local Government Ordinance 1961. The services should cover the provision of the sewerage treatment system, wastewater treatment plant, maintenance and repair works of pipe and pump system, and also individual septic tank desludging. Through mutual agreements between all local authorities (excluding *Dewan Bandaraya Kota Kinabalu*, DBKK) with JKR, the tasks of planning and constructing the common wastewater treatment system will be conducted by JKR and to be handed over once completed for operation and maintenance. However, the most local authorities have also formed another agreement which gives power to JKR on

Table 5. Area and population equivalent covered by Indah Water Konsortium services

	Area (km^2)	%	Population Equivalent (pe)	%
Indah Water Konsortium	68,505.88	51.8	19,333,568*	72.4
Non-Indah Water Konsortium	63,769.54	48.2	7,361,729	27.6

*Exclusive of 2.96 million population using the pour-flush system.
Source: (Wan Abdullah, 2011).

the operation and maintenance of the common wastewater treatment system as well, due to a shortage in labour force, competency and resources.

There are several public sewerage treatment systems established, especially in urban districts like Kota Kinabalu, Penampang, Keningau and Tawau. In the Sabahan context, public sewerage treatment system is more commonly categorized in the common wastewater treatment system, while the private sewerage treatment system is generally classified in the individual wastewater treatment system. The public sewerage treatment system is more preferred in urban cities as it can serve a large population while taking the burden off the local authorities from the responsibilities of controlling loads of individual wastewater treatment system within their authority. Besides, maintaining a single centralized public sewerage treatment system would be easier than much decentralized individual wastewater treatment system. In terms of technology usage, oxidation pond is favoured as the ideal treatment system. Although it is basically just a simple biological process that is less efficient than the mechanical process, oxidation pond is preferred because it requires less maintenance, low cost and easier to operate. More importantly, it is suitable for the tropical climate in Sabah, since high temperature accelerates the treatment efficiency.

Private Sewerage System

The private sewerage system is a sewerage system that is not owned or operated by the local authority or a municipality and is not connected to the public sewerage system. This type of sewerage system is mainly owned and operated by business premises, apartments and condominiums, and housing estates. These property buildings that are not connected to the public sewerage system have their own sewerage treatment plant that treats the sewage and wastewater collected within their service area.

The private treatment plants constructions in these dwellings are completed by the housing estates, and apartments and condominium developers. For housing estates, the authority on operation and maintenance of a private sewerage system would be transferred to the local authorities after 0.5 – 1 year. In the case of apartments and condominium buildings, the operation and maintenance of the treatment plants would be overseen by the developer's own management unit. The sewage treatment plant for a private sewerage system typically employs mechanical treatment to treat the wastewater. The frequently used treatment technologies include extended aeration, oxidation ditch, rotating biological contactor, and Imhoff tank.

Individual Septic Tank System

An individual septic tank is generally the most common type of sewerage system in Malaysia. Especially in Sabah, this form of the sewerage system is broadly utilized in almost every single dwelling, particularly in rural residential areas. The system is installed in areas where there is no centralised sewerage system and if the discharge would not cause a damaging effect on the environment. Moreover, individual septic tank system is a cheap solution to sewage disposal. Nonetheless, it only partially treats sewage and condensed units of individual septic tank in one area can strain the capacity of the receiving surrounding, subsequently leads to unpleasant odour and health risks generally functions as a retention tank, made from concrete or metal, whereby solids from the sewage mainly settles at the bottom forming sludge while the fluids overflows into drain.

An individual septic tank should be periodically maintained to avoid excessive accumulation in the tank which could cause overflow and blockage when it exceeds the capacity. The responsibility falls to the corresponding owner, but the tasks of desludging can be executed by local authorities or private service provider when necessary or as requested by the owner. The sludge collected would subsequently be disposed to the sludge treatment plant or as deemed suitable by the local authority/private contractor.

Discharge Effluent

The discharge effluent from a sewage treatment plant must adhere to the effluent water quality standards as prescribed and regulated by the Department of Environment Malaysia. Since most conventional sewage or wastewater treatment plants operate by using the activated sludge process, most heavy metals and noncompliant organic compounds cannot be effectively removed from the wastewater. However, these compounds need to be removed to meet the identifies effluent water quality standards. Parameters with adverse effects on both human health and the environment are regularly identified by the regulatory body. This includes compounds that cannot be treated effectively by municipal treatment plants. The effluent water quality standards are discussed in Section 9-3.

Conventional Sewage Treatment Process

The conventional sewage treatment process involves a few stages of treatment, namely the pre-, primary, secondary and tertiary (optional) treatment. Typically, the treatment process is designed from pre-treatment until secondary treatment if the secondary treatment can produce effluent within the standard of discharge limit. Figure 5 shows the steps that are performed in the sewage treatment process. The process includes physical, biological, and chemical processes, some of which are covered in Chapter 6 and Chapter 9. The schematic diagram of the overall sewage treatment process is illustrated in Figure 6.

Solid Waste (Sludge) Management

Sludge refers to the solid wastes that settle at the bottom of the septic tank or treatment ponds of a sludge treatment plant. It is an active organic matter that can turn septic if left untreated over a duration of time. Untreated sludge is a serious hazard to public health and the environment. On the other hand, treated sludge is generally stabled, inert, and safe to be utilized. It can be used to improve the soil condition and be used as landfill. The conventional process of sludge treatment is shown in Figure 7.

In Malaysia, the amount of domestic sludge produced yearly are increasing due to the positive population growth. However, there are limited facilities throughout the country to treat and dispose of the sludge waste produced by the sewage treatment plant. The sludge produced by the urban cities are being treated by the sewage treatment plants with additional capacity, but this is not available mostly in rural areas due to the high cost incurred. In the rural state of Sabah, for example, sludge treatment is not carried out due to the absence of sludge treatment plants. More sludge treatment facilities should be developed using appropriate technologies such as sludge lagoon, sludge settling tank, digester, and mechanical dewatering. This is important to ensure sustainable sludge management that will ensure a cleaner and safer environment for the country to thrive in the future. In this regard, construction of sludge treatment plants throughout the country would require large areas of land to produce plants of large scale.

Figure 5. Steps in the conventional treatment process

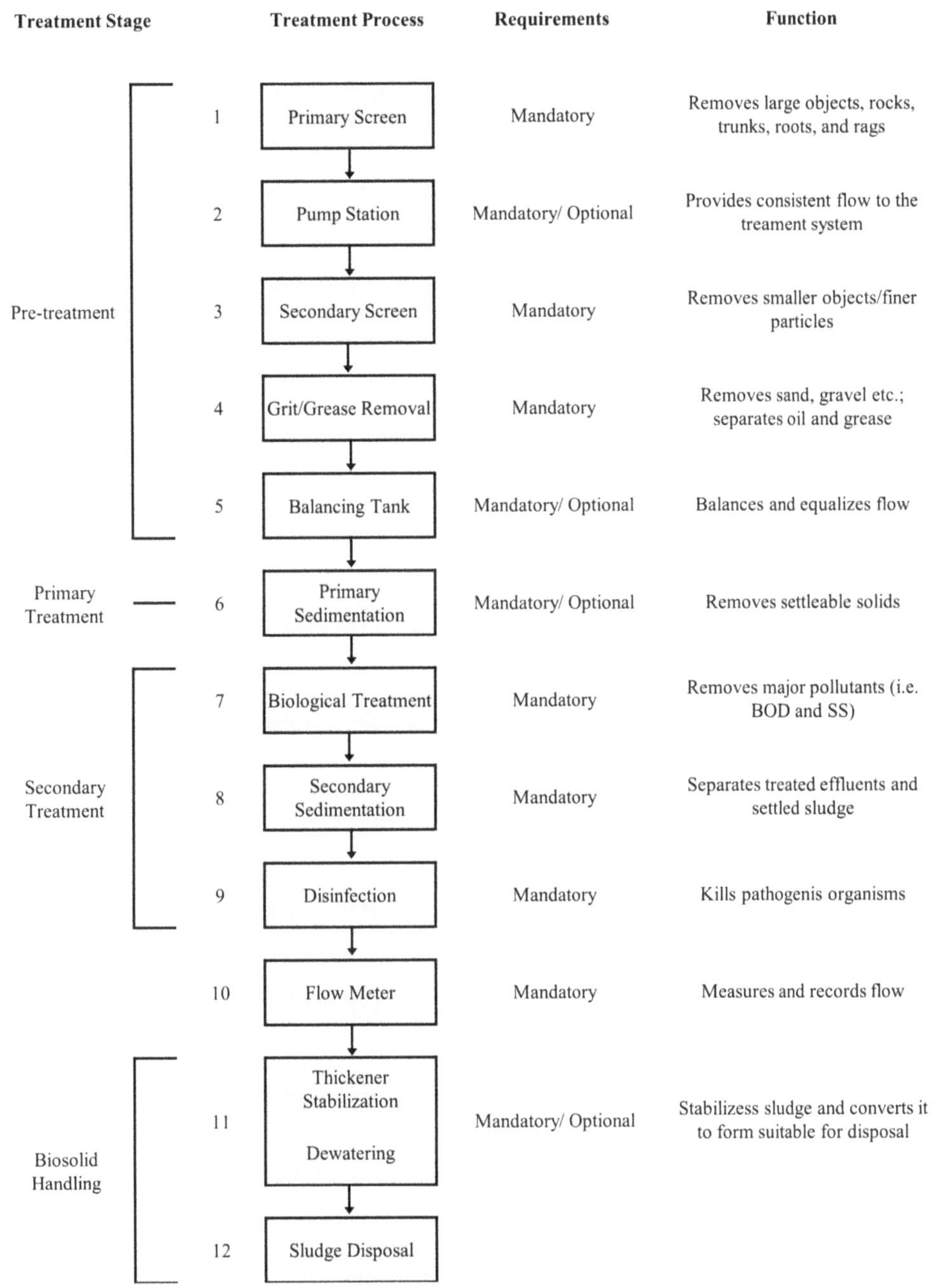

SEWAGE TREATMENT TECHNOLOGIES

Individual Septic Tank

An individual septic tank is commonly used in urban areas and housing estates. The tank is located at a suitable place (usually behind the house) and is made of metal or concrete.

Figure 6. Schematic drawing of the conventional treatment process

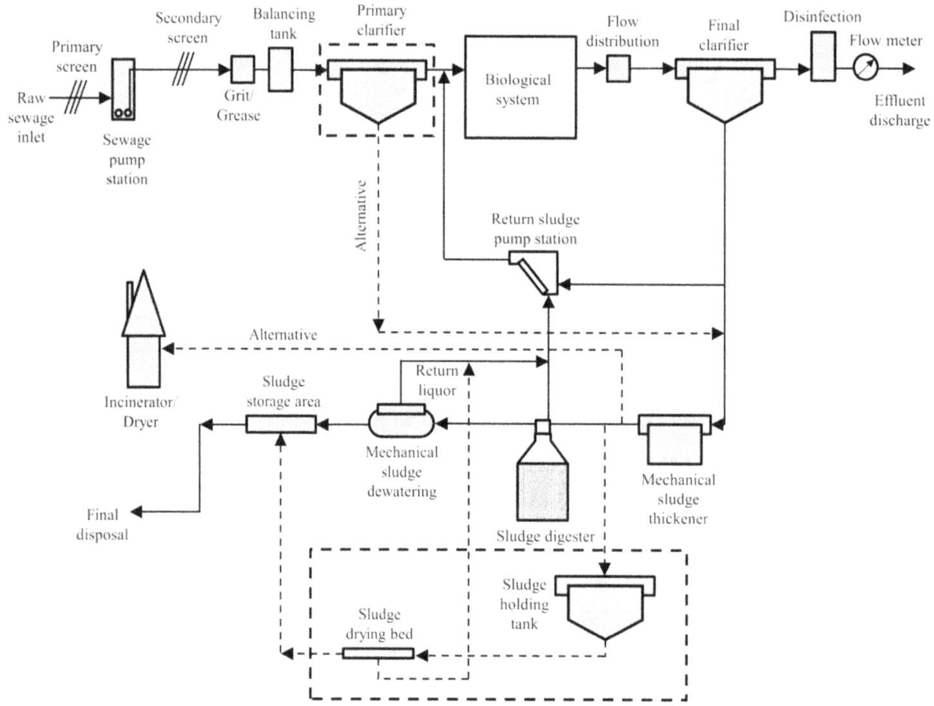

Septic tank is the first stage of biological treatment in which separation of solid and liquid occurs due to the difference in sewage density. In this treatment, household sewage flows by gravity into the septic tank through suitable pipes. Then, the solid matter or sludge will settle down and undergo anaerobic decomposition. The typical retention time of 24 hours is sufficient to allow for the sludge settlement. On top of the tank, oil and grease from the sewage float and form a layer of scum. Meanwhile, between the layers, a clear zone where the water has very few solids is formed. The clear water and some gases then flow out from the system through the underground pipe as affluent as shown in Figure 8.

Septic tanks comprise two chambers where a dividing wall is added to keep sludge and scum on one side of the chamber and clear water is on the other side. The inlet and outlet pipe are designed as an open (both at the top and bottom) T-shaped baffle. The purpose was to direct solids to settle down towards the bottom (for the inlet pipe) and allows clear zone water to exit the tank without coming into contact with the scum layer (for the outlet pipe).

However, these settlement tanks do not fully treat the sewage and require periodic desludging. The maximum volume of sludge that can be stored by a septic tank is approximately a third of its total volume. If desludging is not executed, the sludge level may surpass the maximum level. The sewage retention time decreases which in turns lead to system failure due to the incomplete breakdown of sewage. As a result, untreated sewage is released into rivers, causing the death of aquatic life and arising public health.

Under the Water Services Industry Act 2006, owner of premises with septic tanks is responsible for the maintenance of the septic tank. According to Malaysian Standards (MS 1228), the septic tank needs to be de-sludged regularly at least once every two years. Nevertheless, the frequency of desludging is dependent on the capacity and design of the septic tank.

Figure 7. Sludge treatment process

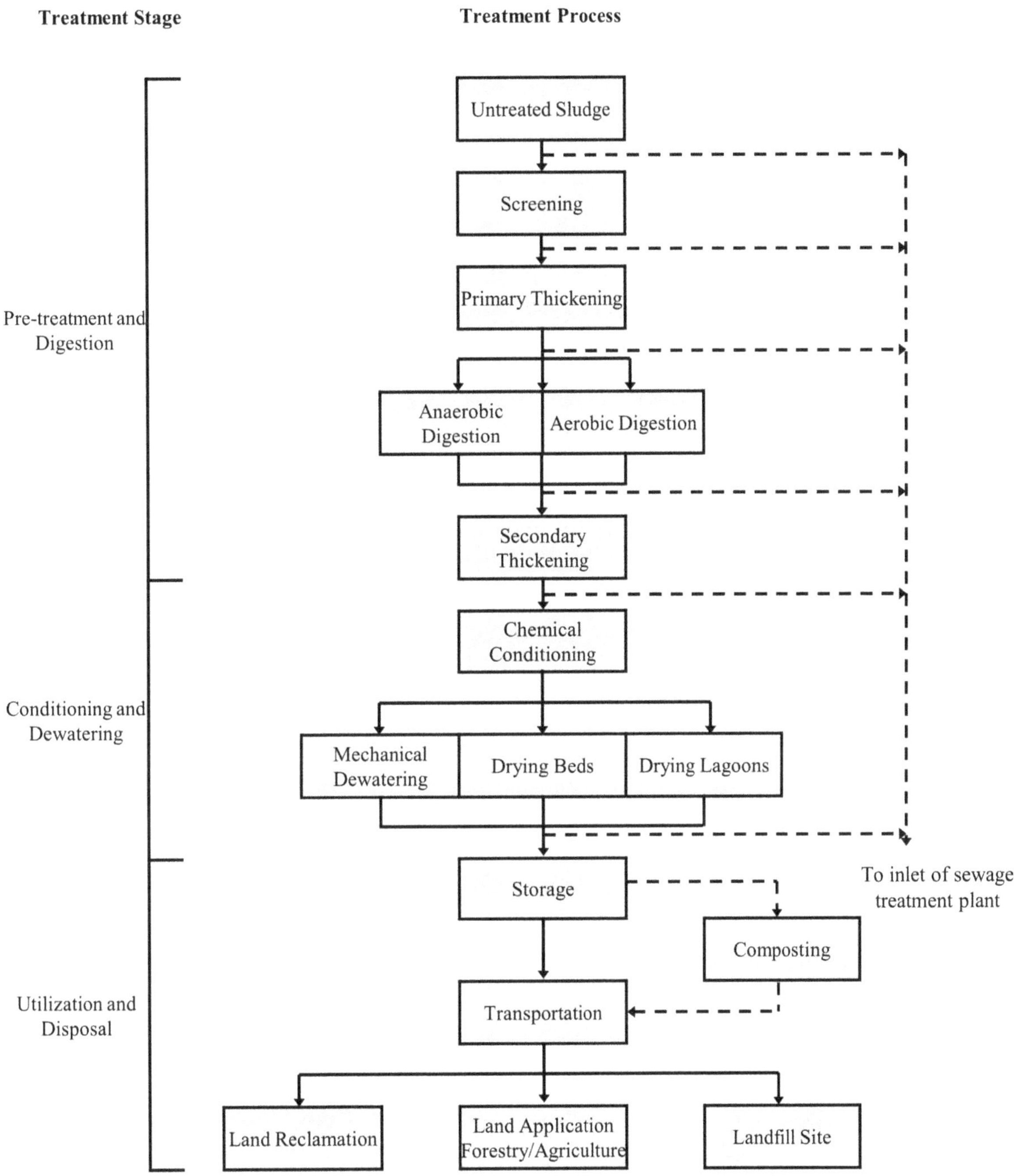

Several cases of overflow septic tank failure have been reported in Kota Kinabalu which has caused the areas to be enclosed with an unbearable smell (The Borneo Post, May 2017). Following that incidents, the National Water Services Commission (SPAN) had enforced the Septic Tank Treatment Regulations in which a failed septic tank could result in fines of not more than RM 50,000 (The Borneo Post, Jul

Figure 8. Schematic of conventional individual septic tank

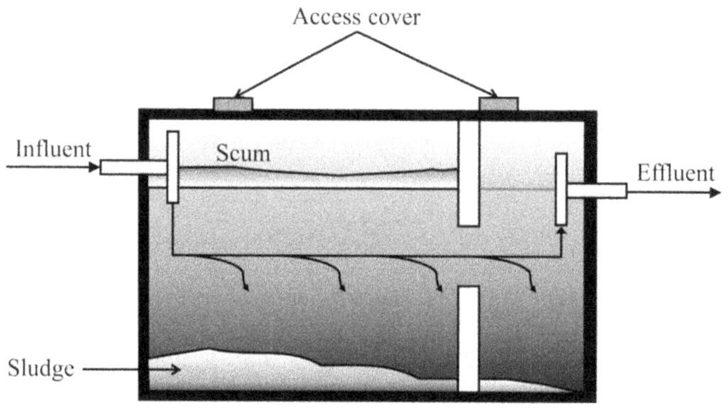

2017). With that, the owner of the premises must maintain their septic tank either by personal desludging or pay a licensed contractor for the desludging service.

Preventive actions of not throwing solid food waste and oil into the draining system can be taken into account as well since it can leads to further clogging and overflowing of the septic tanks. Once clogging and overflowing happen, the system can be regarded as a total failure.

Intermittent Decanted Extended Aeration (IDEA)

The Intermittently Decanted Extended Aeration (IDEA) system is a fusion between the typical sequencing batch reactor (SBR) and extended aeration (EA) systems. Development of IDEA system is based on the recognized advantages of the two hybridized systems with some modification is made onto it. The IDEA treatment process involved a concentrated suspension of microorganisms within an aerated reactor. These reactors provide suitable conditions for enhanced nitrogen removal by nitrifying and denitrifying of micro-organisms. Figure 9 shows a schematic of the IDEA system. This system consists of 3 stages: aeration, settling and decanting.

Figure 9. Schematic of the IDEA system

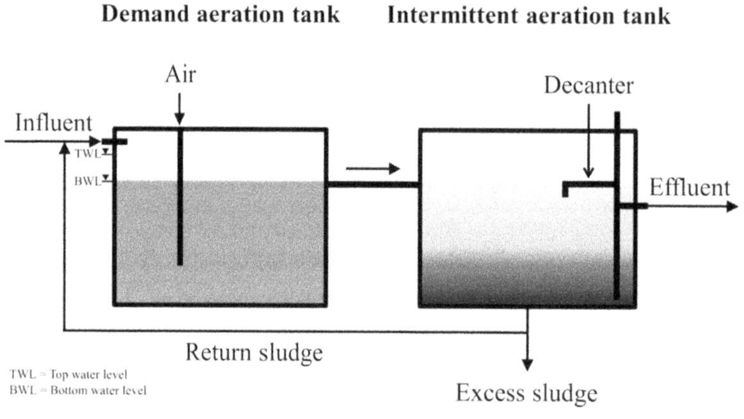

In the first phase of aeration, the configuration was similar to the extended aeration system, which is a "food starved" regime. The mass of micro-organisms within the reactor is significantly higher than the mass of pollutants (BOD) in the influent sewage. In other words, low food/microorganism ratio and has high biomass concentration. Nitrification occurs at this stage in which conversion of ammonia into nitrite takes place. Primarily, rapid surface mixing between the influent and activated sludge (from the recycle stream) are involved in which the mixing action will aid in dissolving oxygen into the water whilst forcing the sludge into suspension from its settled state. As a result, rises of water level and supplied oxygen will significantly remove BOD and ammonia. In some IDEA design, alum is used in the system to bind any present phosphorous in the sewage, and permitting it to settle out.

In the settling phase, the system is characterized by a lack of aeration and mixing. The surface aerators are stopped to allows solids settling to the bottom of the tank and create a sludge blanket within the reactor. The lack of molecular oxygen presents anoxic conditions in the sludge which makes the bacteria in the sludge to take up oxygen bounded in nitrates (NO_3^-) for metabolism. This results in the effusion of nitrogen as a gas, and ammonia denitrification (a process that decreases oxygen requirement).

For the decanting phase, the decanters are opened to allow clear clean discharge from the IDEA system for further processing in downstream processes. This will remove the need for equalization tank and upstream infrastructure.

A regional sewage treatment plant (RSTP) located in Penampang By-Pass has utilized the IDEA system. The RSTP area can withstand until 4 Intermittent Decant Extended Aeration (IDEA) tanks which able to cater up to 300,000 population equivalent (pe). However, for the initial phase only 1 IDEA tanks with tank serving of 75,000 pe are constructed.

The simplicity of IDEA enabled the significant capital economy to be achieved as well as operational simplicity. A higher degree of removal of carbonaceous pollutants (particularly in relation to nitrogen removal) is also achieved. IDEA was a novel enhancement of the conventional activated sludge process since primary and secondary treatment was able to be merged into one, rather than three treatment units. For tank configuration, it was recommended to utilize dual tank process.

Activated Sludge System

Activated sludge system (Figure 10) is a biological treatment technology that uses the aerobic process to treat various types of wastewater. In an activated sludge system, sewage effluent that has been mechanically pre-treated is brought to an aeration tank, whereby aerobic organisms are used to break down organic matter into carbon dioxide and water under an oxygen-rich environment. Activated sludge process can digest roughly 90% of the organic content of the sewage effluent, but it still contains a significant amount of nutrients like nitrate and phosphate. Discharge of the effluent to the water body at this stage can instigate eutrophication. Therefore, tertiary treatment is required to further treat the effluent.

Subsequently, the sewage effluent is channelled to a settling tank (clarifier). From this tank, some of the sewage is continuously recirculated back to the aeration tank while some are usually transferred to the sludge digester for tertiary treatment. The purpose of recycling a portion of the sewage is to return some sludge to maintain the sufficient concentration of activated sludge in the aeration tank to ensure efficient treatment.

Figure 10. Schematic of a conventional activated sludge system

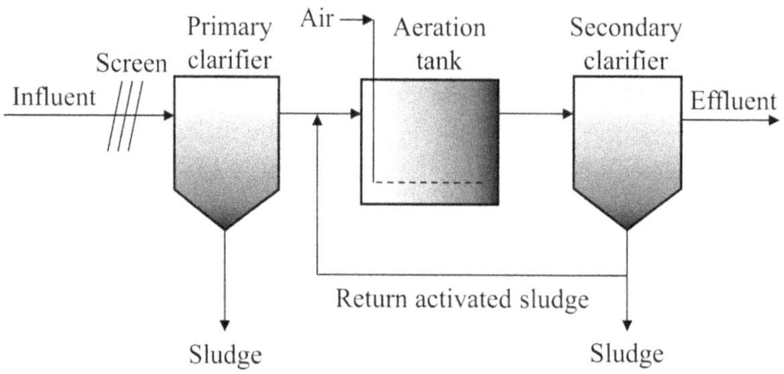

Membrane Sewage Treatment System

Membrane sewage treatment system is a secondary treatment of sewage that generally consists of a single unit of a membrane bioreactor. The basic membrane bioreactor comprises of succeeding units of bioreactor which houses aerobic bacteria that will digest organic matters with the presence of oxygen, and a membrane module that filters the suspension of organic material and bacteria from pure water. In the bioreactor, the effluent is retained for a duration of time required to complete the reaction and subsequently, the membrane is allowed to filter the organic matter from water. Like the activated sludge system, some portion of the sludge is returned to the bioreactor for the same purpose discussed previously. This configuration is known as the side-stream membrane bioreactor. In contrast, the membrane bioreactor can be assembled in a way that the membrane module is immersed inside the bioreactor, whereby excess sludge is occasionally discharged at a rate needed to provide sufficient sludge retention time for efficient treatment. This type of configuration is known as the submerged membrane bioreactor.

The main advantages of using membrane bioreactor as a secondary sewage treatment are smaller footprint, higher efficiency, and high effluent quality. The principle of membrane bioreactor for sewage treatment is by using biological oxidation of organic matter present in the sewage and subsequent separation of the resultant solids from the sewage effluent by the separation by the membrane. The solids created as a by-product of the oxidation process can be easily disposed of, resulting in a cleaner effluent for further tertiary treatment. This process is continuous and can be easily controlled and operated, making it one of the best technology available for sewage treatment.

The membrane bioreactor can also be operated at a significantly higher solid concentration compared to a conventional activated sludge system. This makes the membrane bioreactor system a suitable technology to treat various high strength wastewater. Currently, the cost required to build a membrane bioreactor plant for secondary sewage treatment is relatively higher than the conventional plant, making it an unpopular choice for the rural application. Nonetheless, with more membrane bioreactor plants are being developed, price drops of membrane etc. are expected in the near future. This could lead to the rapid acceptance of the membrane bioreactor system by the wastewater treatment industry.

The small footprint of membrane bioreactor makes it very attractive for development in urban as well as rural areas. Overall, membrane bioreactor can be used as a single main unit operation which can equate a complete sewage treatment plant, provided that pre-treatment processes of large solids and grit

Figure 11. Schematic of a conventional membrane sewage treatment system

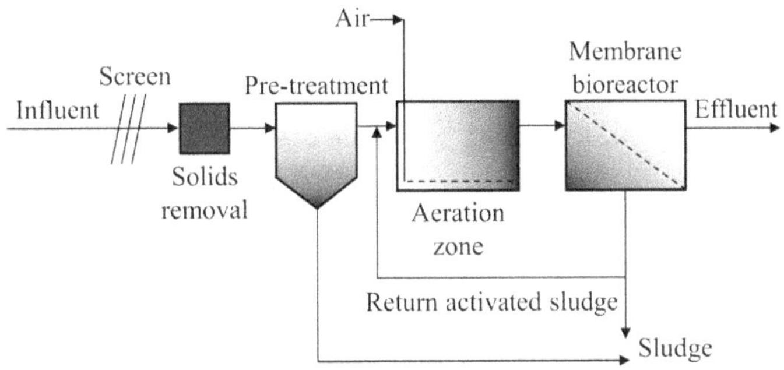

removal are conducted as shown in Figure 11. Consequently, high-efficiency membrane bioreactor can produce effluent of high quality that adheres to the required standard effluent of discharge.

Anaerobic Digester (Sewage to Biogas Production)

An anaerobic digester is mainly employed to biologically digest organic matters like sewage sludge, municipal solid wastes, agricultural effluent, food wastes, and animal wastes with the help of bacteria under an oxygen-deprived environment. The anaerobic digestion of sewage sludge occurs in four consecutive steps: hydrolysis, acidogenic, methanogens, and methanogens. Biogas is produced as a result of the anaerobic digestion process. Figure 12 illustrates a typical anaerobic digester system that captures the biogas produced to generate energy. This type of setting is widely used by the palm oil mills to power their plant, hence reducing the operating costs significantly.

Biogas is an important source of sustainable and renewable energy source for heat and electricity generation. Biogas is composed of two primary components of methane (55 – 70%) and carbon dioxide (30 – 45%). Other components are also present in smaller portion, such as nitrogen (0 – 15%), oxygen (0 – 3%), water (1 – 5%), hydrocarbons (0 – 200 mg/m^3), hydrogen sulfide (0 – 10,000 ppm$_v$), ammonia (0 – 100 ppm$_v$), and siloxanes (0 – 41 mg Si/m^3) (Awe et al., 2017).

Figure 12. Schematic of typical sewage to biogas production system

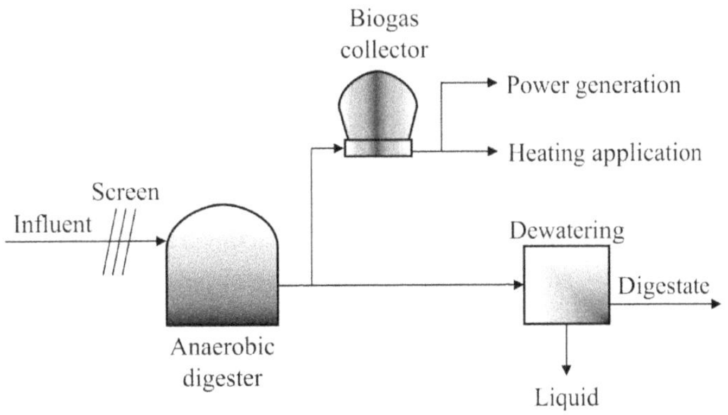

REFERENCES

Abdul Wahab, N. A. (2013). *Decentralized domestic wastewater management in Malaysia* [PowerPoint slides]. WEPA Workshop & 8th Annual Meeting: Siem Reap, Cambodia. Retrieved from http://www.wepa-db.net/activities/2013/20130218-19_4th_wiw/pdf/2_4.pdf

Awa, O. W., Zhao, Y., Nzihou, A., Minh, D. P., & Lyczko, N. (2017). A Review of Biogas Utilisation, Purification and Upgrading Technologies. *Waste and Biomass Valorization*, *8*(2), 267–283. doi:10.100712649-016-9826-4

Department of Statistics Malaysia. (2018). *Sabah @ a Glance*. Published online at www.dosm.gov.my

Dovers, S., & Butler, C. (n.d.). *Population and environment: a global challenge*. Published online at www.science.org.au

GESAMP. (2001). *Protecting the Oceans from Land-based Activities - Land-based Sources and Activities Affecting the Quality and Uses of the Marine, Coastal and Associated Freshwater Environment.* GESAMP Reports and Studies 71.

Malaysia. (2006). *Sewerage Services Act 1993 (Act 508)* (January 1 2006). Kuala Lumpur, Malaysia: Percetakan Nasional Malaysia Berhad.

Painter, H. A., & Viney, M. (1959). Composition of a domestic sewage. *Journal of Biochemical and Microbiological Technology and Engineering*, *1*(2), 143–162. doi:10.1002/jbmte.390010203

Roser, M., & Ortiz-Ospina, E. (2018). *World Population Growth*. Published online at OurWorldInData.org.

Sabah Tourism Board Official. (2018). *Statistics*. Published online at www.sabahtourism.com

Thomas, M. V. (2001). *Municipal wastewater treatment using sequencing batch reactor (SBR) under tropical conditions* (Master's thesis, University of Malaya, Malaysia). Retrieved from http://studentsrepo.um.edu.my/2773/

Wan Abdullah, W. A. R. (2011). *An overview of Malaysia's sewerage management* [PowerPoint slides]. Water Environment Partnership in Asia (WEPA) 3rd International Workshop: Manila, Philippines. Retrieved from http://www.wepa-db.net/pdf/1203forum/20.pdf

ADDITIONAL READING

Bradley, R. M., & Dhanagunan, G. R. (2004). Sewage sludge management in Malaysia. *International Journal of Water*, *2*(4), 267–283. doi:10.1504/IJW.2004.005526

Chen, X., Xu, Z., Yao, L., & Ma, N. (2018). Processing technology selection for municipal sewage treatment based on a multi-objective decision model under uncertainty. *International Journal of Environmental Research and Public Health*, *15*(3), 1–18. doi:10.3390/ijerph15030448 PMID:29510560

Demirbas, A., Edris, G., & Alalayah, W. M. (2017). Sludge production from municipal wastewater treatment in sewage treatment plant. *Energy Sources. Part A, Recovery, Utilization, and Environmental Effects*, *39*(10), 999–1006. doi:10.1080/15567036.2017.1283551

Hanum, F., Yuan, L. C., Kamahara, H., Abdul Aziz, H., Atsuta, Y., Yamada, T., & Daimon, H. (2019). Treatment of sewage sludge using anaerobic digestion in Malaysia: Current state and challenges. *Frontiers in Energy Research*, *7*, 1–19. doi:10.3389/fenrg.2019.00019

Neyens, E., Baeyens, J., De Heyder, B., & Weemaes, M. (2004). The potential of advanced treatment methods for sewage sludge. *Management of Environmental Quality*, *15*(1), 9–16. doi:10.1108/14777830410513559

Pereira, A. J. M., Elvira, P. S. I., Oneto, S. J., Cruz, D. L. R., Portela, J. R., & Nebot, E. (2015). Enhancement of methane production in mesophilic anaerobic digestion of secondary sewage sludge by advanced thermal hydrolysis pretreatment. *Water Research*, *71*, 330–340. doi:10.1016/j.watres.2014.12.027 PMID:25682559

Sabeen, A. H., Ngadi, N., Noor, Z. Z., Raheem, A. B., Agouillal, F., Mohammed, A. A., & Abdulkarim, B. I. (2018). Characteristics of the effluent wastewater in sewage treatment plants of Malaysian urban areas. *Chemical Engineering Transactions*, *63*, 691–696.

Shen, D. S., Huang, B. C., Feng, H. J., Zhao, B., Zhao, J. M., Zhang, H. Y., & Liu, P. Q. (2013). Performance of a novel decentralized sewage treatment reactor. *Journal of Chemistry*, *2013*, 1–6.

Topare, N. S., Attar, S. J., & Manfe, M. M. (2011). Sewage/Wastewater treatment technologies: A review. *Science Reviews & Chemical Communications*, *1*(1), 18–24.

KEY TERMS AND DEFINITIONS

Activated Sludge System: Treatment method which utilizes aerobic microbes to feed on organic contaminants in wastewater to produce a high-quality effluent.

Anaerobic Digester: A system that uses microorganisms to break down biodegradable substances into biomass and biogas under anaerobic condition.

Individual Septic Tank: The simplest form of sewerage system which is commonly installed in individual houses or premises using the principle of biological treatment, and separation of solid and liquid that occurs due to the difference in sewage density.

Intermittently Decanted Extended Aeration (IDEA): System which is the fusion between the typical sequencing batch reactor (SBR) and extended aeration (EA) systems.

Membrane Bioreactor: A system that combines membrane separation process and activated sludge process.

Private Sewerage: Sewerage system that is not owned or operated by the local authority or a municipality, and is not connected to the public sewerage system.

Public Sewerage: Sewerage system that is regulated by the local authority to treat and process sewage wastewater mainly in the urban areas, which comprises of the collection, transportation, pumping, treatment and final disposal of sewage.

Sewage Treatment Process: Process of eliminating the contaminants present in municipal sewage wastewater which includes the pre-treatment, primary, secondary, and tertiary treatment processes.

Sewage Treatment System: Collective term which includes the collection, transportation, pumping, treatment and final discharge of sewage.

Sludge: Semi-solid substances and mostly organic suspended sewage solids that are produced as a by-product from the sewage treatment process.

Chapter 11
Palm Oil Mill Wastewater and Treatment

ABSTRACT

Palm oil is an essential agricultural commodity in Malaysia, as Malaysia is one of the largest producers and exporters globally. The processes of palm oil extraction and purification generate a large amount of wastewater known as palm oil mill effluent (POME). Currently, most treatment processes are carried out using a conventional ponding system which is outdated as it requires large land areas and long retention time. Discharge of poorly treated POME directly to the surroundings leads harms the environment, as it contains high contents of oil and BOD. Therefore, this chapter discusses the treatment method of POME from a newer perspective of membrane technology integrated with biological treatment. Various methods such as clarification system with nanofibre unit, cooling system, activated sludge system, and anaerobic membrane bioreactor, are reviewed on their principles of operation.

INTRODUCTION

Overview of Palm Oil Mill Industry in Sabah

Sabah is one of the states formed Malaysia which is located in the north of Borneo Island. The land size of Sabah is 74,000 km^2, most of which is covered by tropical forest and received high rainfall with an average of 3,000 mm annually. In the past, before the 1980s, the economy of Sabah relied on timber.

From the mid of the 1980s up to today, Sabah has transformed into the largest palm oil producer in Malaysia and the 3rd largest in the world. 2016 statistic shows that about 1.55 million hectares of land in Sabah has been planted with oil palm. The total oil plant planted in Malaysia was 5.74 million hectares. The reason for the fast-growing of Sabah's palm oil plantation compared to the other states in Malaysia is due to the abundant land space and fertile soil.

DOI: 10.4018/978-1-7998-2645-3.ch011

A total count of 126 palm oil mill is registered in Sabah alone as of 2015. The total production of crude palm oil produced is about 5.5 million metric tonnes per year. The production capacity usually ranged from 30 – 120 tonnes of fresh fruit bunch (FFB) per hour, where a conventional palm oil mill here runs at a capacity of 40 – 60 tonnes of fresh fruit bunch per hour.

However, the most significant factor is the slightly higher percentage of oil extract from the fruit bunch compared to that planted in West Malaysia. As a consequence of its large production scale, the palm oil industry has also been identified as the largest contributor to Malaysia's pollution load, by discharging a large volume POME into the environment. The large production of palm oil has exposed Sabah to high generation of wastewater which is produced during the extraction of crude palm oil from fruit bunches.

Due to its large generation and without proper treatment and management, wastewater produced from palm oil industry has been reflected as the most polluting industry in Sabah. This has led to many news reports and has increased public concern. Consequently, the government authority responsible to maintain environmental quality, the Department of Environment (DoE) has decided to make the standard discharge limit more stringent, changing it from Standard B to Standard A. Despite the change of discharge limit, minimal improvement has been seen in wastewater discharge from palm oil mills. One of the possible reason is the conventional wastewater treatment is unable to respond correctly due to the inconsistent weather condition in Sabah, especially during heavy rainfall season, a large volume of wastewater and very high BOD.

The conventional and currently utilized a ponding system to treat wastewater derived from palm oil mill can be considered as an outdated technology as other new technologies are readily available. However, most palm oil mill is reluctant in adopting new technologies mostly due to the high capital cost involved for the installation of new technology and low maintenance cost of the existing technology.

Recently, the palm oil industry has shown a positive development where some of the palm oil industries have started to venture into new technologies such as biofuel production from biomass and wastewater. This technology is attractive not only because it is able to reduce solid waste production and wastewater discharge, but it is also able to reduce operational cost and generate extra income for the company. The biofuel and biogas can be utilized in power generation, eliminating the need for diesel purchase.

If this positive trend proceeds, effluents from palm oil mill will no longer be categorized as wastewater, but rather a source of income for the mills. Therefore new technologies such as microalgae bioremediation and biofuel production as well as membrane utilization are beneficial to the palm oil industry. This will allow the industry to lower operational cost as it will be paid off through the production of by-products in the form of biofuel.

Palm Oil Processing

The key principles in palm oil mill processing have been comparatively unchanged for many years. It mainly constitutes of the sterilization, stripping, digesting, screw press, decanting and clarifying processes to produce the main product that is crude palm oil. The process flow in a typical palm oil mill is illustrated in Figure 1.

The initial procedure in the palm oil processing is the fresh fruit bunch receiving, handling, and transfer. Ripened palm fruit mesocarp hold around 56 – 70% of palm oil (Mba et al., 2015). The oil extraction begins at the reception station where ripe fresh fruit bunch is weighed on the weighbridge and graded. The fresh fruit bunch is then discharged into the loading ramp which acts as a temporary holder before the fresh fruit bunch is being loaded into the fruit cages.

Figure 1. Palm oil mill process flow diagram

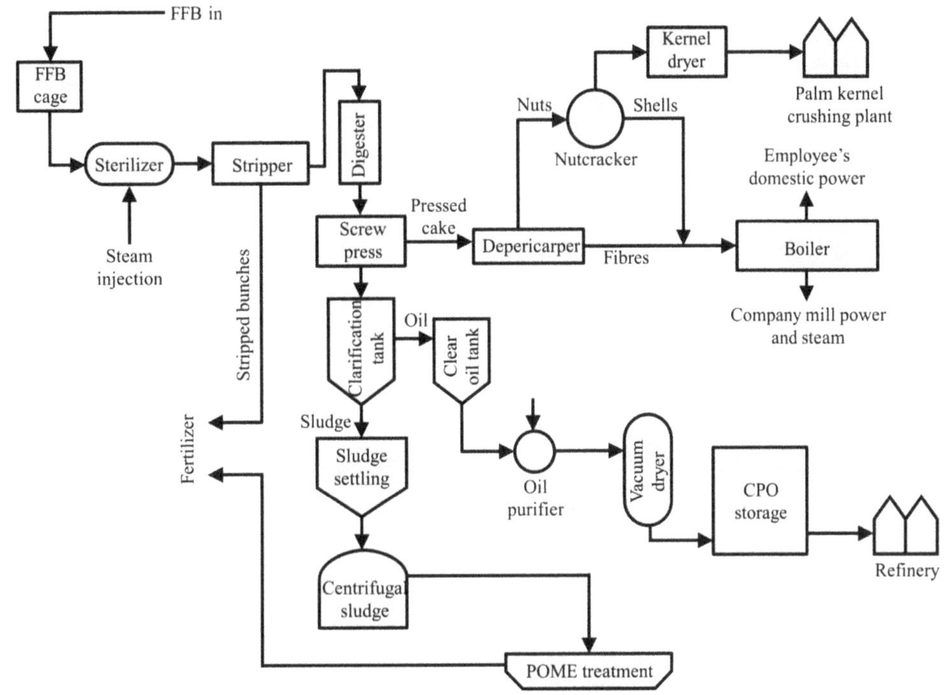

Next, the fresh fruit bunch undergoes the sterilization process. In this process, the palm fruit-laden fruit cage is sent into the sterilizer – a cylindrical pressure vessel, before the process is started. In a batch mode, the sterilization vessel is filled with steam under pressure; while in continuous mode, the process is carried out in a vessel overflowing with steam at atmospheric condition. Sterilization is carried out mainly to hinder the biological activities which contribute to the fresh fruit bunch quality deterioration. It is also performed to promote the fruits to loosen up and freeing it from the fruit bunches for ease of recovery and separation during the subsequent stripping and threshing processes.

The steam sterilization process is the focal point of one of the largest water-consuming in palm oil processing, at around 30 – 60% of total process steam (Mohd. Omar et al., 2018). The type of sterilizer used has a major impact on the steam and energy rate consumption which will have an effect on the energy efficiency of the palm oil extraction process. There are several types of typically used steam sterilizer which are horizontal, vertical, and continuous sterilizer as shown in Table 1.

The sterilized fresh fruit bunches will move to the next operation to extract the oil, which is the stripping or threshing section. This unit is normally a rotating drum machine which functions to thresh the palm fruits from the bunches and separate them. The stripped palm fruits are then transferred to a digester, while the empty fruit bunches are used to provide additional fuel or turned into organic fertilizers through composting. In the digester, the fruits are subjected to hot water or steam treatment to break down the fruits and leach out the oil from ruptured oil-bearing cells of the fruits. The temperature is maintained at 80 – 90 °C as heat is supplied by the steam jacketed digester. This process also facilitates the hydrolyzation of resins, gums, and starch materials, and coagulation of proteins.

The digester content is then channelled to the screw-press machine, a mechanical process which presses out the crude palm oil from the fibrous mesocarp. The press liquor product comprised of a

Table 1. Types of steam sterilizer

Types of Steam Sterilizer	Operation	Steam Usage (kg/tonne FFB)	Operating Parameters
Horizontal vessel sterilizer	Batch process, single-peak, double-peak or triple-peak	110 – 130	Temperature: 121 – 134 °C Pressure:150 kPa Duration: 40 min
Vertical vessel sterilizer	Batch process, single-peak, double-peak or triple-peak	305 – 335	Temperature: 140 °C Pressure:400 kPa Duration: 60 min
Continuous sterilizer	Continuous process at atmospheric condition	300 – 360	Temperature: 98 °C Pressure:101 kPa Duration: 60 – 20 h/d

Source: (Mohd. Omar et al., 2018)

mixture of oil, water, and some fruit debris of various concentrations. A by-product formed from this process is the press cake consisting of fibrous materials, nuts, and moisture, which usually still holds around 5 – 6% of oil.

The press cake will undergo the depericarping process, which is to separate the nuts from the fibres. The fibres can be used as boiler fuel, whereas the nuts are further processed to extract the palm kernel. Subsequently, the nuts are channelled to the nutcracker section before the kernel is separated from the shell using hydro-cyclone. The kernel and shell separation is principally upon the difference of specific gravity.

The resulting digested crude palm oil contains around 35 – 45% of palm oil, 45 – 55% of water, and the remaining of the fibrous compounds (Ahmed et al., 2015). The oil needs to be separated from the rest of the materials in the clarification tank. In this process, it is crucial to maintain the temperature at 80 – 90 °C to prevent the oil from being solidified and to ease the oil separation. Due to its hydrophobic nature, the oil will part from the water and forms a top layer which will be continually skimmed off from the top of the clarifier. The bottom layers still contain some oil which is passed through a sludge separator to recover the oil before returned back to the clarifier. The oil is further dried before being sent to the storage tanks.

PALM OIL MILL WASTEWATER

The palm oil mill is one of the industries which recorded the highest water usage and wastewater generated. In the Sections that follow, the aspects associated with water demand, effluent generated and characteristics of POME will be briefly discussed. Figure 2 shows the process flow chart for palm oil processing showing the wastewater generation points.

Water Consumption

A vast amount of water is needed to extract the crude palm oil from the palm fruits. For each tonne of fresh fruit bunch processing, about 1.5 m^3 (~1.5 tonnes) of water are used, and almost 50% of it is discharged as POME (Ahmed et al., 2015). Majority of this freshwater is used to feed the boiler, and

Figure 2. Process flow chart of palm oil processing

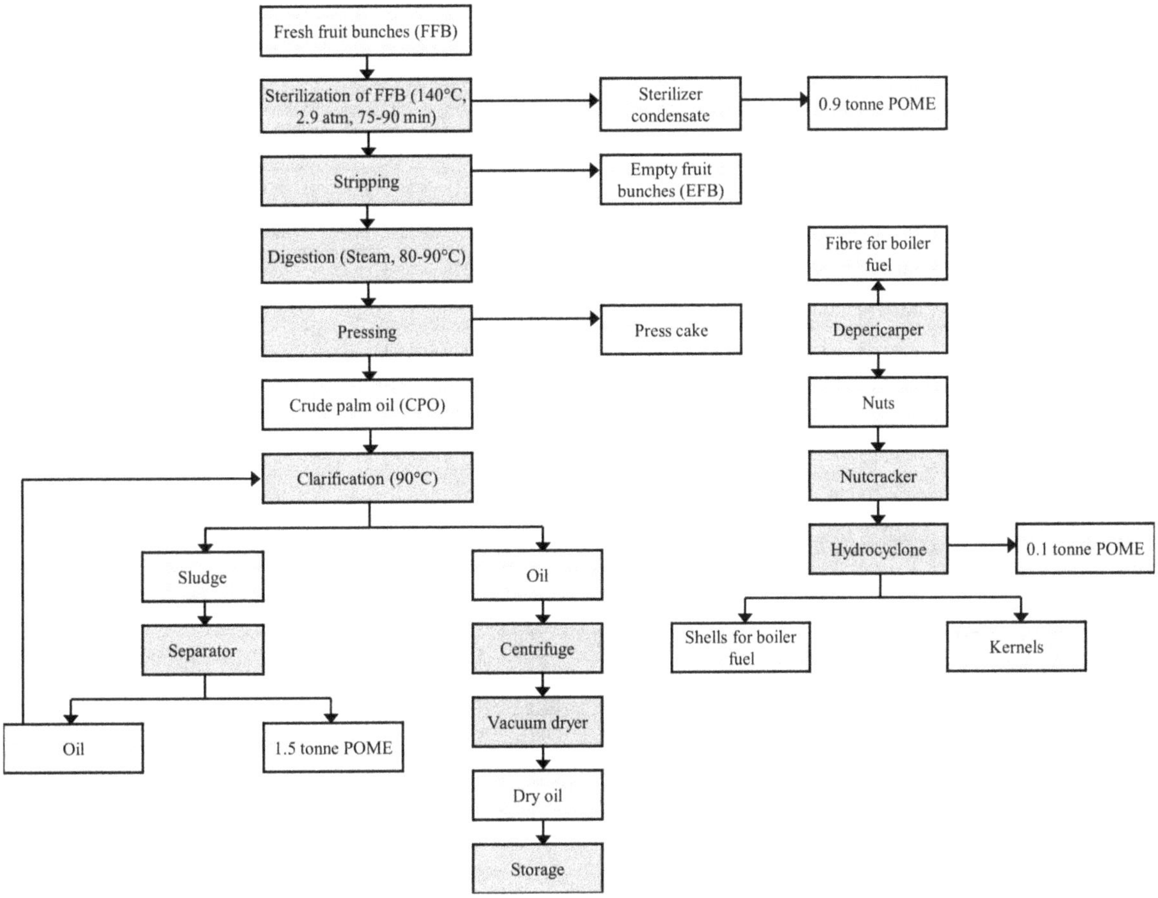

some used during the inline processes while some for cleaning the equipment/oil spill in the plant. The main sections with high water utilization in the palm oil processing are allocated at sterilization, steam jacketed units (digester, clarifier), kernel polishing, and plant clean-up.

Effluent Discharge

Aside from having high water consumption during palm oil processing, it also discharged a huge amount of effluent from the process. It is measured that for every tonne of fresh fruit bunch, around 0.5 – 0.75 tonne of POME is produced. The main contributors to the POME discharged are clarification wastewater (60%), sterilizer condensate (36%), and hydro-cyclone wastewater (4%). Consequently, one tonne of crude palm oil production produced 1.5, 0.9, and 0.1 m^3 of POME from the three main sources respectively (Ahmed et al., 2015).

Table 2. Comparison of POME characteristic

Parameters	Units	Industry 1 [a]		Industry 2 [b]				Industry 3 [c]
		Raw POME	Anaerobically Digested	Raw POME	Anaerobic		Aerobic	Effluent
					Influent	Effluent		
Temperature	^{0}C	84±1	28±1	NM	NM	NM	NM	NM
pH	-	3.5±0.1	7.8 ± 0.2	NM	NM	NM	NM	8.8
BOD_5	mg/L	25,545	25,545	NM	NM	NM	NM	42 [d]
COD	mg/L	55,775	1,372	79,723	32,520	1,959	1,439	946
Total Solids	mg/L	NM	NM	67,200	40,427	26,704	22,579	25
Suspended Solids	mg/L	18,479	512	49,300	11,780	2,456	567	72
Dissolved Solids	mg/L	NM	NM	19,415	17,560	18,381	16,324	NM
Volatile Suspended Solids	mg/L	NM	NM	28,400	8,758	423	25	NM
Total Nitrogen	mg/L	711	134	NM	NM	NM	NM	NM
Ammoniacal Nitrogen	mg/L	36	36	72.8	39.2	64.4	0	17
Kjeldahl Nitrogen	mg/L	NM	NM	672	644	560	238	NM
Oil and Grease	mg/L	8,020	8,020	17,410	11,019	3,856	258	33
Total Alkalinity	mg/L	NM	NM	523	3,818	6,400	6,016	1560
Bicarbonate alkalinity	mg/L	NM	NM	NM	NM	1,975	1,856	NM
Volatile Fatty Acids	Mg/L	NM	NM	NM	NM	NM	NM	252

*NM: Not measured

Source: [a] (Vijayaraghavan et al., 2007); [b] (Zhang et al., 2008); [c] (Borneo Samudera Sdn Bhd, Malaysia, September 2011);
[d]BOD measured for 3 days at $30^{0}C$

Characteristics of Palm Oil Mill Effluent (POME)

POME has the physical characteristic of a yellowish to brown coloured wastewater that originated from the plants constituents such as lignin and phenolics (Ho et al., 1984; Zahrim et al., 2009) as well as re-polymerization of colouring compounds after anaerobic treatments (Kitts et al., 1993; Zahrim et al., 2009). POME is also high in acidity and organic carbon content. Raw POME is rich with colloidal suspension with 95 – 96% water, 0.6 – 0.7% oil, 4 – 5% total solids. About 2 – 4% in the total solids is suspended solids which are mainly developed from debris from palm fruit mesocarp (Borja & Banks, 1994; Khalid and Wan Mustafa, 1992; Ma, 2000). Normally, POME is discharged at a relatively high temperature of 80 – 90 °C.

POME would typically contain a relatively great amount of N, P, K, Mg (Ho et al., 1984; Habib et al., 1997; Muhrizal et al., 2006). Nitrogen originally present in POME in the form of inorganic (protein) nitrogen and will be converted to ammoniacal nitrogen as time progresses (Chow, 1991). The nitrogen content of POME is typically ranged around 200 mg/L as ammonia nitrogen and 500mg/L total nitrogen (Ma et al., 2001).

Other pollutants found in POME is toxic metal, such as Pb (Habib et al., 1997), however, their concentration is normally insignificant, recorded less than 17.5 µg/g (James et al., 1996). The detection of Pb in POME is derived from the contamination from metal and plastic pipes, containers and tanks where Pb

Table 3. Parameter limits for watercourse discharge (second schedule)

Parameter	Unit	Limit
BOD_3 [a]	mg/L	100
COD	mg/L	*
Total Solid	mg/L	*
Suspended Solids	mg/L	400
Oil and Grease	mg/L	50
Ammoniacal Nitrogen	mg/L	150[b]
Total Nitrogen	mg/L	200[b]
pH	-	5-9
Temperature	^{0}C	45

[a]BOD is measured in 3 days at $30^{0}C$
[b]value of the filtered sample

is usually found in paints and glazing materials (James et al., 1996). Table 2 shows comparisons of the general characteristic of raw and treated POME. Based on these characteristics, it is clear that it would present a high deteriorating effect on the environment if POME is discharged without any prior treatment.

CONVENTIONAL TREATMENT OF PALM OIL MILL WASTEWATER

Conventional Treatment Process

With the issues related to wastewater from palm oil mills, the enforcement on the limit of discharges should be made more stringent especially in Sabah, a state that still possesses a lot of natural value to be preserved. In the effort to preserve the environment, the Malaysian government has drafted the Environmental Quality Act 1974 which includes the limits of POME discharge that has to be adhered to. Since then, all palm oil mills are required to treat POME prior its discharge to any watercourse. Table 3 enforced limit of POME discharge, extracted from Environmental Quality (Prescribed Premises) (Crude Palm-Oil) Regulations 1977.

The most common method in treating POME is ponding system. More than 85% of the palm oil mills in Malaysia have adopted this method for POME treatment (Ma et al., 1995). However, the treatment that is based mainly on biological treatments of anaerobic and aerobic systems is not entirely efficient to treat POME, which leads to environmental pollution issues. It was identified that the treated POME using ponding system sometimes does not conform to the discharge standard and the removal of nitrogen from POME is usually insignificant because nitrification process is not common in the ponding system (Chin et al., 1996; John, 1985).

A typical POME treatment plant will consist of a de-oiling tank, acidification, anaerobic and facultative ponds with respective hydraulic retention times (HRT) of 1, 4, 45 and 16 days (Ma & Ong, 1985). Figure 3 depicts the typical treatment system employed by most palm oil mill in Malaysia. The illustration shows that a large area is required for construction of the ponding system and that long period of treatment is also required, which range at 30 – 60 days before discharge.

Figure 3. Typical ponding system applied by the palm oil mills in Malaysia

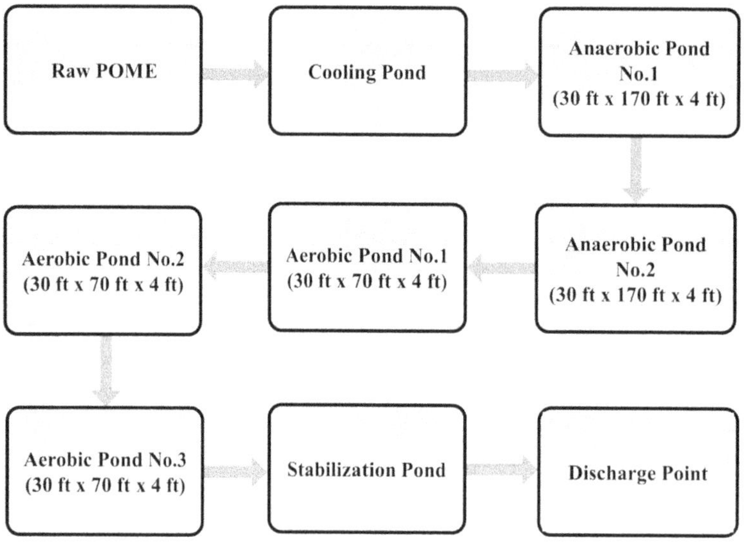

Problems Associated to Conventional Treatment

There are some problems associated with the conventional treatment of POME using ponding system as aforementioned in the preceding section. The problems essentially include the requirement of large land area for construction, treatment efficiency which tediously subjected to the weather, and long treatment period.

Ponding system was developed back in time when lands were abundant and the prices were still low. However, with the high demand for CPO and the increase in land prices, this method is no longer reliable. On the subject of weather, open ponding is continually connected to the problems with the rainy season. For example, during heavy raining, the ponds are subjected to an increase in water content which affects the effectiveness of the treatment mechanism. The improper treatment resulted to have an adverse effect on the quality of the discharged effluent from the treatment system, which may not adhere to the standard regulation imposed. Besides, flooding may occur especially when the ponds constructed have a shallow depth.

A detailed cost calculation done for POME management in Indonesia has shown that conventional system of POME treatment discussed in Section 11-3.1 is not only the system with the highest environmental pollution and the lowest utilization of renewable resources but also the system giving rise to the low profit (Schuchardt et al., 2005).

Potential for By-products Generation

Oil palm is by far one of the most multipurpose crops in Malaysia. There are many benefits of oil palm cultivation which can contribute to the wellbeing of the socio-economics of the surrounding production areas due to its wide range of products generation. However, the main problem in oil palm cultivation, harvesting, processing, and replanting is its abundant volume of biomass wastes and wastewater generated. This includes empty fruit bunch, mesocarp fibre, palm kernel shell, palm oil mill effluent,

Table 4. Potential value-added from palm oil processing waste material

Waste Material	Potential Value-Added Production
Palm oil trunks, leaves and fronds	• Nutrients recycling to the oil palm plantation
Empty fruit bunch	• Fuel for energy production and steam generation • Compost fertilizer through the composting process • Particleboards, pulp and paper production
Mesocarp fibre	• Fuel for energy production and steam generation
Palm kernel shell	• Fuel for energy production and steam generation • Activated carbon production
Palm oil mill effluent/sludge	• Biogas production through anaerobic fermentation for additional energy • Compost fertilizer can be composted by its own or with empty fruit bunch

and palm oil trunks, leaves, and fronds. These biomass wastes and wastewater have great potential for by-products generation which can add value to the economics of the palm oil industry. Table 4 displays some potential value from palm oil processing waste material.

The oil palm tree comprised of about 90% biomass and the rest is oil. The biomass wastes are high in fibre content and these wastes are typically processed through simple mechanical means such as shredding, palletizing, and chipping for size decrement. About 75% of the biomass wastes, mostly from the palm oil trunk and fronds are usually left to rot in the plantation for the objective of mulching and nutrient recycling, whereas the other 25% from the empty fruit bunch, palm kernel shell and mesocarp fibre are used to feed the boiler for steam and electricity generation and simultaneously decrease the dependency on non-renewable fossil fuel.

The POME wastewater also possesses several potential valuable by-products production, for example in the production of biogas. It has been known that the anaerobic ponds in POME treatment release a great amount of biogas in the form of methane and carbon dioxide. Every ton of CPO produced will emit an approximate of 46 m^3 (32.9 kg) methane, corresponds to an equivalent of 384 m^3 (756 kg) CO_2 (Schuchardt et al., 2007). The methane gas is a valuable fuel source which can be used to generate electricity for plant usage.

Another by-product produced during the treatment of POME by using biological treatment is palm oil mill sludge known as Palm Oil Mill Sludge (POMS). POMS has a high moisture content with a pH of 8.4 and enriched with nutrients. A typical concentration of nitrogen, phosphorus and potassium in POMS are 3.6, 0.9 and 2.1 mg/L respectively (Yaser et al., 2007). The high concentrations of nutrients, as well as the bad odours of POMS, are the main reasons it is considered as a highly polluting material. Therefore, most palm oil mill will reserve to manage POMS by drying and using it as fertilizers. This method is known as composting (Parveen et al., 2010).

POTENTIAL FOR MEMBRANE BIOREACTOR TREATMENT SYSTEM

Various methods of POME treatment system are employed in the Malaysian palm oil industry in order to comply with the standard of discharge limit imposed in the Environmental Quality Act 1978. The most commonly used treatment method is a ponding system, consisting of the anaerobic pond, facultative pond, and aerobic pond. Even though membrane technology has been implied in various applications

Figure 4. Schematic diagram of the alternative scheme 1

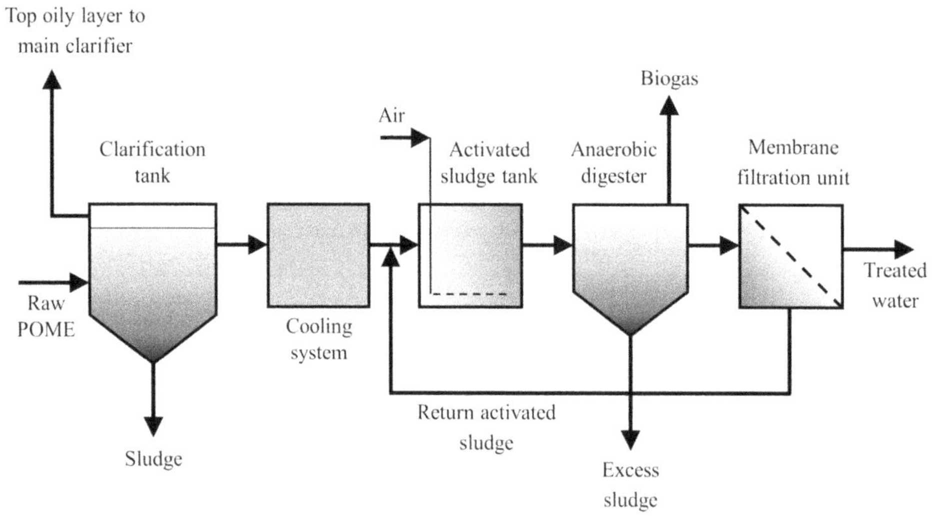

like food, medical and chemical fields, it has yet been applied in large industrial treatment of POME in Sabah. Membrane filtration separation technique is an effective method to treat POME. There are many advantages associated with membrane filtration technology, for instance, it requires less energy, small footprint requiring fewer land areas, easy to manage, produce consistently high purity water, and environmentally friendly.

Membrane process will be a significant tool for advancing the water quality for wastewater treatment. In other regions, the membrane has seen the emergent application in industrial usage concerned with wastewater in need of treatment prior to discharge due to its effectiveness to reject bacteria and contaminants. Although membrane is always associated with the problem of fouling, reduced flux due to impurities, and high membrane cleaning cost, these problems can be prevented by applying stringent membrane operation in accordance to its optimum operating condition.

In this case study, some potential alternative methods of POME treatment technologies are discussed by using membrane technology integrated with biological treatment. The systems will be mostly closed systems to combat the weather issues related to conventional methods discussed in Section 11-3.2.

For the first scheme (Figure 4), the system comprised of pre-treatment such as clarification tank, cooling system, activated sludge system, and anaerobic digestion tank as the first stage; and in the second stage, a membrane filtration unit using microfiltration or ultrafiltration. The pre-treatment ensures that the membrane filtration unit can operate smoothly without compromising the operation and reduce its susceptibility towards fouling of the membrane pores. This is made feasible as the series of pre-treatment can sufficiently reduce the contaminants and remove large solid particles before the wastewater reached the membrane filtration system. Hence, the membrane filtration unit acts more as a polishing process in the POME treatment. Polishing process denotes a treatment process which prepares industrial water to the required specification for discharge or reuse. This system is practical to be implemented for palm oil mill with novice expertise in membrane utilization for POME treatment due to its ease of operation. Moreover, this system can be further upgraded and developed to a more advanced system particularly the membrane filtration unit if desired.

Figure 5. Schematic diagram of the alternative scheme 2

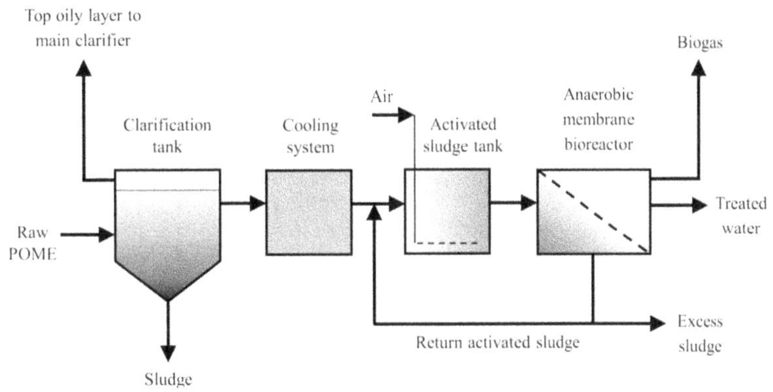

For the second scheme (Figure 5), it includes a pre-treatment system consisting of a clarification tank, cooling system, and activated sludge system for the first stage. Further treatment using anaerobic membrane bioreactor is employed to further treat the wastewater from the activated sludge tank. This system has fewer unit operations than scheme 1 aside from utilizing an anaerobic membrane bioreactor. This is because, since the pre-treatment has fewer unit operations, the membrane unit could be easily subjected to fouling, but stringent membrane operation can prevent this issue. Also, it relies on operators' expertise in handling and controlling the membrane bioreactor unit so as to ensure that the overall operation is within the optimal operating parameters. For instances, the transmembrane pressure and critical flux must be adhered to at a stringent level in order to prevent fouling of the membrane surface. An option would be to use advance sophisticated control system that can safeguard the anaerobic membrane bioreactor unit at a strict operation within the specified operating condition. However, this translates to high investment costs. Thus, this system is suitable for palm oil plant capable of advanced technology installation for POME treatment, particularly in terms of investment cost and workforce expertise.

Clarification System

In this treatment system, a clarification tank is first employed to hold the POME from palm oil mill processing. The primary objectives are to separate any oil residue from the POME and deposit solids present by sedimentation as this can have a negative impact and damage on the membrane. The overall membrane efficiency heavily depends on the clarification, thus the tank must have a sufficient size to ensure maximum oil recovery. Figure 6 demonstrates two possible operations for the clarification tank, whereby in the first method, a conventional way is used which utilizes a skimmer to skim the oily layer from the POME and recycled back to the main clarifier. The settled sludge is removed from the tank bottom, while the wastewater is channelled to the cooling system/pond.

In the second alternative mechanism, the skimmed oily layer is sent to the main clarifier in palm oil processing, while the wastewater from mid-layer (containing less oil residual and sediment) of the secondary clarifier (Figure 6(right)) is channelled through a nanofibre unit to recover any residual oil from the wastewater. This is to ensure maximum oil recovery. The wastewater passing through the nanofibre unit is then channelled to the cooling system. Nanofibre is associated with good oil adsorption in residual

Figure 6. Clarification tank mechanism; conventional method (left) and using nanofibre for direct oil recovery (right)

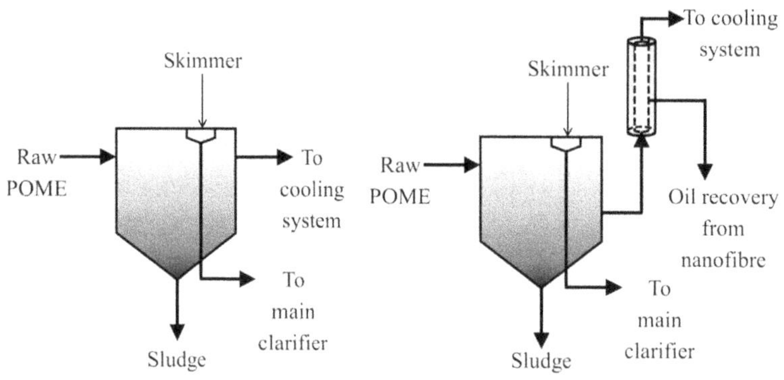

oil recovery from POME. The application of nanofibre unit in POME treatment is further discussed in Section 11-4.2.

Nanofibre Unit

The main problem that needs to be tackled in POME treatment using membrane is the presence of oil and grease. The oil droplets can cause contamination and blockage of the membrane pores and reduce membrane performance. Thus, they must be completely removed from the POME wastewater prior to entering the membrane system. One of the very promising techniques is by using adsorption materials such as polyurethane microfibre or nanofibre adsorbent.

The nanofibre membranes are pictured as films comprising of nanofibres overlap with each other in a completely random manner. The nanofibre membranes possess some characteristics which are close to the ideal membranes used for the membrane distillation processes. The characteristics of the nanofibre membranes include high porosity, pore size ranging from tens of nanometre to several micrometres, interconnected open pore structure and highly permeable to vapours. Saturated nanofibre can be desorbed by mechanical roller press to recover the adsorbed oil. It can be reused up to 20 times without depriving its adsorption capacity (Chung et al., 2017).

The nanofibre unit functions to absorb the oil that escaped the skimmer to further polish the wastewater from any oil residual to safeguard the membrane unit employed at the end of the POME treatment line process. The previous study reports that the optimal size of the nanofibre for this application range between 300 – 700 nm, therefore, allowing higher adsorption of oil. Based on past experience, 1 g of nanofibre is capable of adsorbing 55 g of oil. This renders nanofibre the most suitable material for this application, as the high capacity will benefit the plant by ensuring less replacement of the nanofibre material needed in the nanofibre unit. Besides, its reusability ensures a cost-saving in terms of material as it can be used for multiple times without jeopardizing its adsorption ability, while the ease of operation provides less operability and maintenance costs.

The system's performance will be affected by the flow rate of water, configuration (i.e. horizontal or vertical), and design of the unit. The unit's design is especially critical, and care needs to be taken into consideration in terms of the volume of wastewater and amount of trace oil as it affects the unit's size

Figure 7. Nanofibre unit with single-pass up-flow (left) and double pass up-flow designs (right)

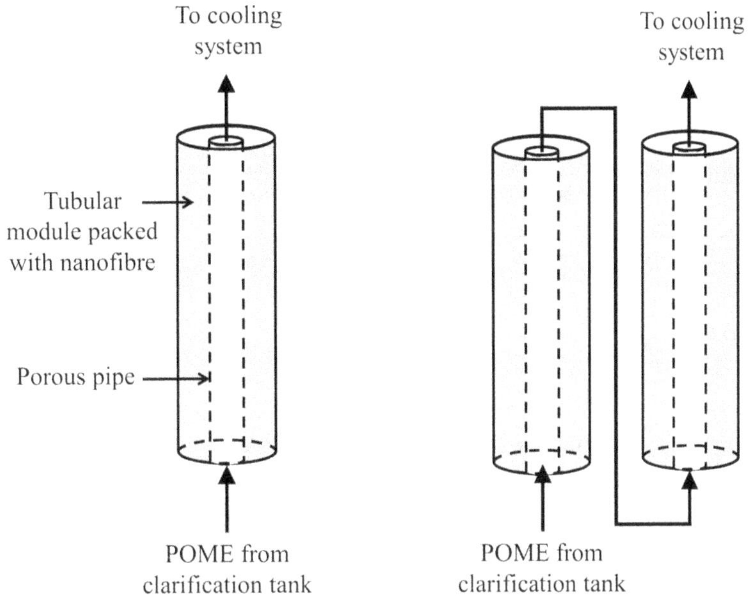

and amount of nanofibre to be utilized. Moreover, it should be designed so as to be able to overcome the flow resistance to ensure that oil droplets are properly adsorbed to the nanofibre material.

Figure 7 illustrates two designs for the nanofibre unit that can be implemented in the POME treatment discussed earlier. The unit provides simple installation showing a tubular module of the nanofibre unit, whereby a tubular tank is filled with a nanofibre bundle while a porous pipeline is installed at the centre of the module. The porous pipeline is employed to provide passage for the POME wastewater and allows oily residual contained in the POME to pass through to the nanofibre bundle. The nanofibre is hydrophobic in nature with high affinity to oil, thus this will ascertain that only the residual oil will be adsorbed to the nanofibre bundle.

The unit can be designed as a single pass or double-pass system. Both units offer some pros and cons which are comparatively justified in Table 5. In order to overcome the flow resistance and allows the oil residual to be efficiently adsorbed to the nanofibre bundle, mild range of turbulent flow with the up-flow regime can ascertain that the resistance can be overcome and operated optimally. This can be attained by providing enough pressure to overcome the gravitational force in order to achieve up-flow performance. Hence, as the wastewater flows through the porous pipe, the oil droplets will be forced through the pores and adsorbed to the nanofibre. Gravity flow is not suitable to be exercised in this unit as it cannot provide enough pressure to overcome the flow resistance and ensure that the oil droplets can be effectively adsorbed to the nanofibre adsorbent. All in all, utilizing the nanofibre unit to the POME wastewater treatment can prevent the membrane system from being affected by the oil and grease which is uniquely found in wastewater from palm oil mill.

Table 5. Comparison between single-pass and double-pass nanofibre units

Single-Pass Nanofibre Unit	Double-Pass Nanofibre Unit
Lower capital cost	Higher capital cost
No secondary unit, thus it needs to be constantly monitored for oil saturation of the nanofibre to ensure the efficiency of the unit	The secondary unit acts as a safeguarding unit to collect oil residual that escapes the first unit
Need to be periodically maintained to supervise the saturation point in case the nanofibre needs immediate oil recovery and/or replacement	The extra unit ensures complete removal of oil and for backup, in case the first unit reached saturation point, less maintenance needed

Cooling System

The cooling system utilized can be closed cooling system technologies or open ponding system. Selection on which suitable cooling system can be based upon several criteria, for instance, the capital cost, operating cost, maintenance, efficiency, and consistency of operation. The cooling process is highly affected by the surface area, conductive material, and operational mechanism to remove heat. There are several technologies that can be used here, such as cooling tower, closed deep ponding, and open ponding which is compared in Table 6. Aside from these aforementioned technologies, other viable technologies that are available such as water cascade system and heat exchangers can also be explored for this application.

In POME treatment, cooling the POME to a certain temperature before subjected to the subsequent process is highly important. This is because the efficiency of the activated sludge process is affected by the temperature of the feed to compensate for the biological condition of the bacteria. Typical activated sludge process operates at a minimum of 19 °C to a maximum 30 °C with optimal solid retention time ranged from 4 – 9 days (Shahzad, Khan, & Paul, 2015). Thus, it is crucial to bring down the POME temperature at this temperature to ensure optimal operation of the activated sludge process.

Open ponding is the most widely used and preferred treatment system of POME in Malaysia, whereby 85% of mill utilized this method as their main treatment. This system operates at a low ambient temperature. The hot discharged POME typically took several days to reduce the temperature in the pond to the desired one (Ismail et al., 2014). The wide application of this method is due to its cheap cost for construction and easy maintenance. However, it is subjected to some disadvantages such as long retention time, requires a large land area, and prone to weather, especially during rainy seasons.

Closed deep ponding is essentially a process using naturally cold water from a deep, cold body of water as a heat sink, such as from the lake, sea etc. in a heat exchange system. The warm water layer is

Table 6. Comparison of closed deep ponding, cooling tower, and open ponding systems

Cooling System	Open Ponding	Closed Deep Ponding	Cooling Tower
Type of system	Open system	Close system	Close system
Maintenance	Easy	Easy	Intermediate
Capital cost	Low	Low	High
Operating cost	Low	Low	High
Efficiency of system	Low	Intermediate	High
Consistency of output	Low	High	High

Figure 8. Closed deep ponding system

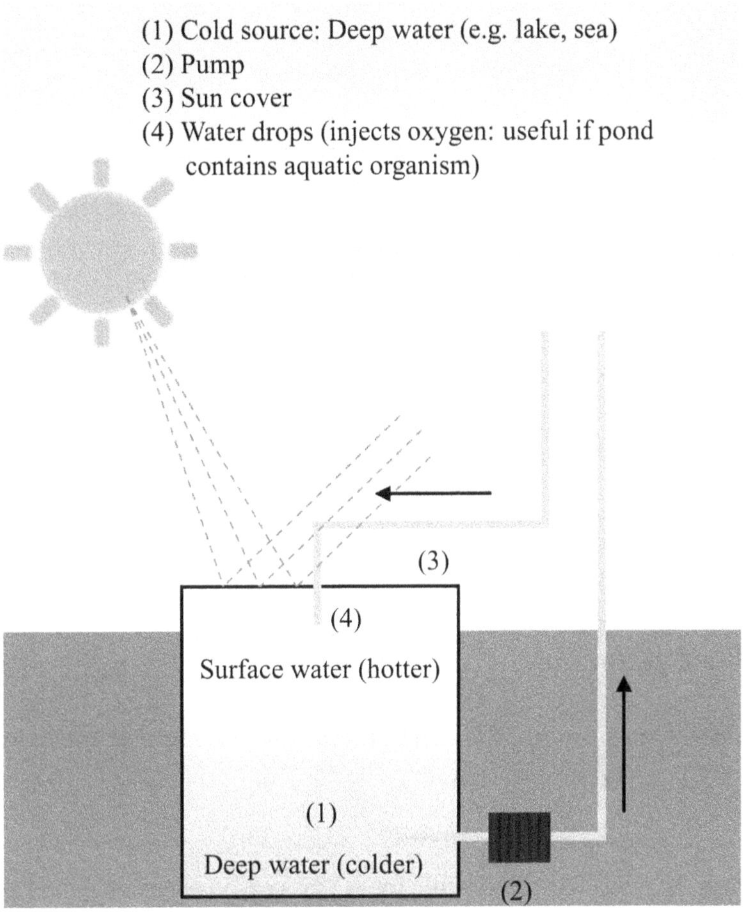

less dense than colder water, thus floats on top. In a deep body of water, there exists a layer of inherently cold water below a particular depth, identified as the hypolimnion. This principle can be applied to the cooling system of POME treatment by having deep well dug and filled with raw POME from the palm oil mill operation. Over time, the void will be filled, reached a steady-state and acts as a heat sink. The successive hot raw POME discharged into the system can be effectively cooled as the mass body of water in the deep pond will absorb the heat. The system (Figure 8) has hotter surface water at the point of discharge, and a colder deep layer which can be used to feed the subsequent process. The sun cover will act as a sun shield to prevent further heating of the surface water and blocks out rainwater from entering into the system, eliminating the problems associated to open ponding.

In cooling tower application, hot water from the heat source is distributed evenly by gravity or pressurized nozzle system to a heat transfer surface (fill material), whilst air is simultaneously forced or drawn through the tower resulting in evaporation of a minor percentage of water as shown in Figure 9. However, this system is associated with high capital and operating costs compared to the two previously discussed cooling systems. But in term of efficiency, the cooling tower can perform better with shorter cooling time.

Figure 9. Cooling tower working principle

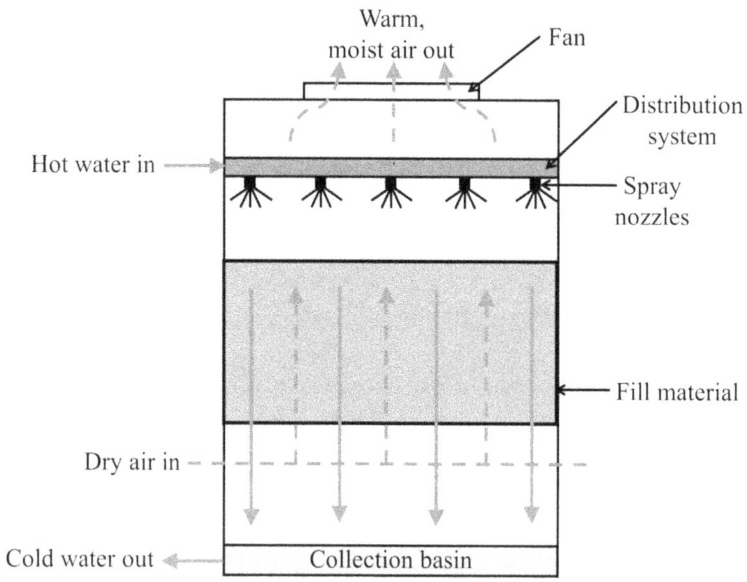

Activated Sludge System

Activated sludge process is an aerobic treatment process that uses aerobic microbes to feed on the organic contaminants of palm oil mill wastewater. The essential principle of activated sludge treatment critically includes:

- Reactor to retain the microbes responsible for the process in suspension and aerated;
- Solid/liquid separation through a settling tank or membrane system
- The recirculating mechanism for the activated sludge back to the activated sludge reactor.

The activated sludge process uses the parameter F/M ratio to classify the various processes used in a water treatment plant. The F/M ratio is expressed in kg of $BOD_5.kg^{-1}$ $VM.d^{-1}$ and can be calculated as shown in Equation 1. This ratio compares the nutrient mass entering with the sludge mass present in the tank. At low F/M ratio, high purification is obtained whereas high F/M ratio is found to produce lower purification levels of the effluent.

$$F / M = \frac{X_{BOD_5} Q_{\inf}}{C_{mlss} V_{t,AS}} \quad (1)$$

where F/M, X_{BOD_5}, $Q_{\inf}$, C_{mlss}, and $V_{t,AS}$ are food to microorganism ratio, the concentration of particulate BOD_5, influent flow rate, the concentration of mixed liquor suspended solids, and activated sludge tank volume respectively.

Sludge bulking formation forms due to the nutrient deficit, especially nitrogen. The primary parameter affecting the sludge settling is the BOD/N ratio, of which ratio 100:4 counts for good sludge settling, whereas ratio lower than 100:3 causes poor settling of the sludge due to the growth of the filamentous organism (Peng et al., 2003).

Another important parameter which ensures the efficiency of the activated sludge process is the return activated sludge. It can be defined as a portion of the settled activated sludge that is accumulated in the secondary clarifier or membrane unit and returned back to the aerated activated sludge tank and mix with the wastewater inside it. This is essential for the activated sludge process in order to maintain the correct concentration of activated sludge in the tank so that the desired extent of treatment can be accomplished within the designated time.

Anaerobic Treatment

The anaerobic digestion is a biological treatment process at which microorganisms present in the wastewater breaks down the biodegradable matters in the absence/lack of oxygen. The process generates biogas (composed mainly of methane and carbon dioxide gases) during the treatment which translates to the excellent potential for recovery of energy. The biogas produced can be collected and used for other purposes such as an alternative source for powering the plant. Nonetheless, it is necessary for the biogas to undergo further separation for carbon dioxide removal in order to enhance its suitability and efficiency to be used as a fuel source.

Anaerobic treatment process performance is strongly affected by the operational temperature in which it operates. The operational temperature can be categorized into three classes which are psychrophilic (lower than 20°C), mesophilic (30 – 45°C) and thermophilic (55 – 65°C) temperature (Lin et al., 2009). Mesophilic temperature is extensively applied in most commercial-scale anaerobic digester as it yields good operational performance. On the other hand, the thermophilic temperature is not considerably utilized because of poorer process stability and high sensitivity towards the dynamic environment (Kim, Ahn, & Speece, 2002).

Membrane Filtration Unit

The membrane filtration unit is a simple unit operation that uses microfiltration, ultrafiltration or reverses osmosis membrane depending on the influent characteristics and quality of water desired. In the treatment line of the first scheme (Figure 4), the filtration unit is the last unit employed in the wastewater treatment process. The POME wastewater which has undergone pre-treatment in the preceding processes enters the second stage treatment which is the membrane filtration to produce clean water. The clean water permeate can be reused back by the palm oil mill for other purposes such as cleaning. Besides, the permeate water may even be further treated and prepared as steam grade water to feed the boiler and used for other applications.

Anaerobic Membrane Bioreactor

Anaerobic membrane bioreactor for POME treatment is an assuring technology due to its capability in separating the biomass from treated water by filtration using a membrane. It simply combines the fundamental of anaerobic digestion and membrane filtration process. The process is able to yield a clean, suspended solid free permeate water and simultaneously produces biogas as a by-product. In the anaerobic membrane bioreactor system, the operating temperature heavily influences the biomass growth rate

Figure 10. Some configurations of anaerobic membrane bioreactor, (a) external cross-flow, (b) submerged, (c) external gas-lift, and (d) external semi dead-end

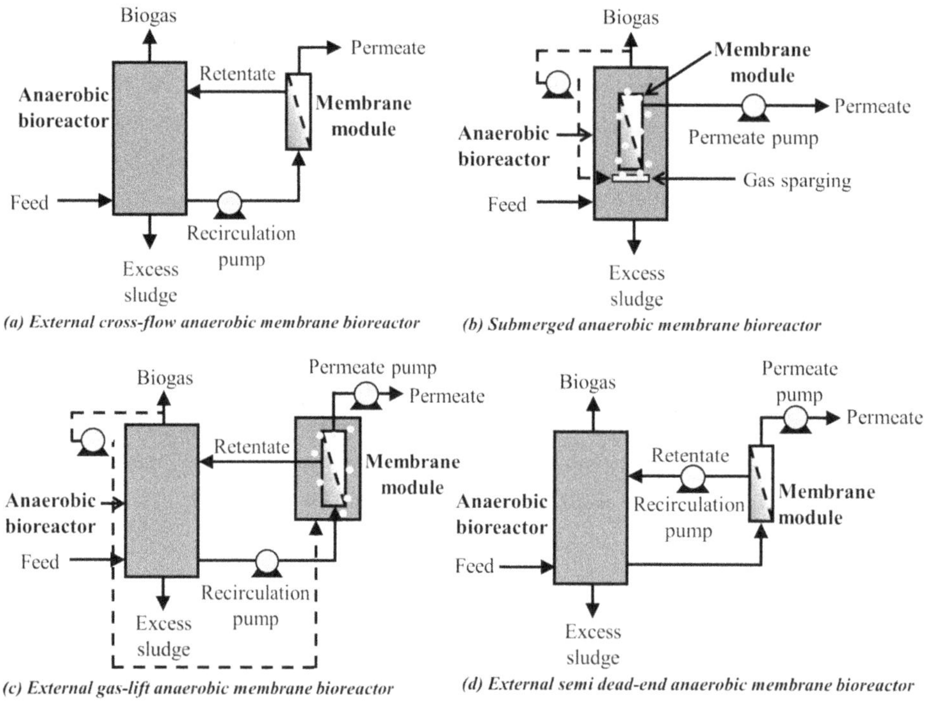

of the microbial population. At lower temperature, the biomass growth rate is observed to be reduced (Martinez-Sosa et al., 2011), whereas, at a higher temperature, the filtration operation is drastically improved. Although, at thermophilic condition, the fouling tendency is significantly higher resulting from the more production of soluble microbial products (SMP) and extracellular polymeric substances (EPS). Besides, the long operational period at thermophilic temperature leads to a considerable reduction in floc size of the sludge which intensified filtration resistance (Lin et al., 2009). However, there are several advantages related to thermophilic temperature operation, like higher methane yield, higher metabolic rates, and efficient elimination of pathogenic microorganisms.

Several configurations are available to be utilized for the anaerobic membrane bioreactor unit for POME wastewater treatment which can be operated under pressure or under vacuum condition as shown in Figure 10. The first configuration is external cross-flow anaerobic membrane bioreactor, whereby the membrane system is detached from the anaerobic bioreactor unit (Figure 10(a)). The pump is used to provide enough transmembrane pressure to enable permeate (clean water) to pass through the membrane, while the retentate is recirculated back to the bioreactor. This system provides cross-flow of the wastewater across the membrane surface which ultimately hinders cake formation of the sludge and act as a fouling preventive mechanism.

The second configuration is a submerged anaerobic membrane bioreactor (Figure 10(b)) with the membrane unit and bioreactor embedded as one unit. This system is typically operated under vacuum pressure, in which the membrane unit is immersed in the bioreactor unit and a pump is used to obtain the permeate from the membrane unit. Since the permeate flow is the controlled variable, the wastewater flow to the membrane unit varies and this could lead to cake formation on the membrane surface which consequently leads to membrane fouling. Thus, the robust bubbling of gas across the surface of the membrane can combat this issue, whereby biogas produced from the anaerobic process can be utilized for this application.

Figure 10(c) illustrates the external gas-lift configuration for an anaerobic membrane bioreactor. The principle operation of this configuration coupled the fundamentals of external cross-flow and submerged anaerobic membrane bioreactor. The membrane system is immersed in an external compartment containing the POME wastewater detached from the main bioreactor and operated under vacuum pressure. As the permeate is withdrawn from the unit, retentate is recirculated back to the main bioreactor. This configuration offers the advantage of the ease of cleaning of the membrane system due to it being easily accessible and physically detached during cleaning or maintenance.

The fourth configuration is shown in Figure 10(d) is an external semi dead-end anaerobic membrane bioreactor. This method occasionally employs the dead-end mode of operation of the membrane, as this can reduce the cost of continuous pumping and lessen the negative effects, for example, the decline in biomass activity of sludge pumping.

REFERENCES

Ahmed, Y., Yaakob, Z., Akhtar, P., & Sopian, K. (2015). Production of biogas and performance evaluation of existing treatment processes in palm oil mill effluent (POME). *Renewable & Sustainable Energy Reviews*, *42*, 1260–1278. doi:10.1016/j.rser.2014.10.073

Borja, R., & Banks, C. J. (1994). Anaerobic digestion of palm oil mill effluent using an up-flow anaerobic sludge blanket (UASB) reactor. *Biomass and Bioenergy*, *6*(5), 381–389. doi:10.1016/0961-9534(94)E0028-Q

Chin, K. K., Lee, S. W., & Mohammad, H. H. (1996). A study of palm oil mill effluent treatment using a pond system. *Water Science and Technology*, *34*(11), 119–123. doi:10.2166/wst.1996.0270

Chow, M. C. (1991). *Palm oil mill effluent analysis* (pp. 11–18). Kuala Lumpur, Malaysia: Palm Oil Research Institute of Malaysia.

Chung, C. H., Janaun, J. A., Semilin, V., & Balan, W. S. (2017). Recovery of residual oil from palm oil mill effluent using polypropylene nanofibre: A field trial. *International Journal of Science and Technology*, *3*(2), 276–294.

Habib, M. A. B., Yusoff, F. M., Phang, S. M., Ang, K. J., & Mohamed, I. S. (1997). Nutritional values of chironomid larvae grown in palm oil mill effluent and algal culture. *Aquaculture (Amsterdam, The Netherlands)*, *158*(1-2), 95–105. doi:10.1016/S0044-8486(97)00176-2

Ho, C. C., Tan, Y. K., & Wang, C. W. (1984). The distribution of chemical constituents between the soluble and the particulate fractions of palm oil mill effluents and its significance on its utilization/treatment. *Agricultural Wastes*, *11*(1), 61–71. doi:10.1016/0141-4607(84)90055-6

Ismail, S., Idris, I., Ng, Y. T., & Ahmad, A. L. (2014). Coagulation of palm oil mill effluent (POME) at high temperature. *Journal of Applied Sciences (Faisalabad)*, *14*(12), 1351–1354. doi:10.3923/jas.2014.1351.1354

James, R., Sampath, K., & Alagurathinam, S. (1996). Effect of lead on respiratory enzyme activity, glycogen and blood sugar levels of the teleost Oreochromisossambicus (Peters) during accumulation and depuration. *Asian Fisheries Science*, *9*, 87–100.

John, C. K. (1985). Treatment of agroindustrial wastes using water hyacinth. *Water Science and Technology*, *17*(4-5), 781–790. doi:10.2166/wst.1985.0179

Khalid, A. R., & Wan Mustafa, W. A. (1992). External benefits of environmental regulation: Resource recovery and the utilization of effluent. *The Environmentalist*, *11*(4), 277–285. doi:10.1007/BF01267698

Kim, M., Ahn, Y., & Speece, R. E. (2002). Comparative process stability and efficiency of anaerobic digestion; mesophilic vs. thermophilic. *Water Research*, *36*(17), 4369–4385. doi:10.1016/S0043-1354(02)00147-1 PMID:12420941

Kitts, D. D., Wu, C. H., Stich, H. F., & Powrie, W. D. (1993). Effects of glucose-glycerine Maillard reaction products on bacterial and mammalian cells mutagenesis. *Journal of Agricultural and Food Chemistry*, *41*(12), 2353–2358. doi:10.1021/jf00036a026

Lin, H. J., Xie, K., Mahendran, B., Bagley, D. M., Leung, K. T., Liss, S. N., & Liao, B. Q. (2009). Sludge properties and their effects on membrane fouling in submerged anaerobic membrane bioreactors (SAnMBRs). *Water Research*, *43*(15), 3827–3837. doi:10.1016/j.watres.2009.05.025 PMID:19555989

Ma, A. N. (1995). A novel treatment for palm oil mill effluent. [PORIM]. *Palm Oil Research Institute of Malaysia*, *29*, 201–212.

Ma, A. N. (2000). Environmental management for the palm oil industry. *Palm Oil Developments*, *3*, 1–9.

Ma, A. N., Chow, C. S., John, C. K., Ibrahim, A., & Isa, Z. (2001). Palm oil mill effluent – a survey. In *Proceedings of the PORIM Regional Workshop on Palm Oil Mill Technology and Effluent Treatment,* pp. 223–269. Palm Oil Research Institute of Malaysia (PORIM). Serdang, Malaysia.

Ma, A. N., & Ong, A. S. H. (1985). Pollution control in palm oil mills in Malaysia. *Journal of the American Oil Chemists' Society*, *62*(2), 261–266. doi:10.1007/BF02541389

Martinez-Sosa, D., Helmreich, B., Netter, T., Paris, S., Bischof, F., & Horn, H. (2011). Anaerobic submerged membrane bioreactor (AnSMBR) for municipal wastewater treatment under mesophilic and psychrophilic temperature conditions. *Bioresource Technology*, *102*(22), 10377–10385. doi:10.1016/j.biortech.2011.09.012 PMID:21962536

Mba, O. I., Dumont, M. F., & Ngadi, M. (2015). Palm oil: Processing, characterization and utilization in the food industry – A review. *Food Bioscience*, *10*, 26–41. doi:10.1016/j.fbio.2015.01.003

Mohd Omar, A. K., Tengku Norsalwani, T. L., Asmah, M. S., Badrulhisham, Z. Y., Mat Easa, A., Mohd Omar, F., Hossain, M. S., Zuknik, M. H., & Nik Norulaini, N. A. (2018). Implementation of the supercritical carbon dioxide technology in oil palm fresh fruit bunch sterilization: A review. *Journal of CO2 Utilization, 25*, 205–215.

Muhrizal, S., Shamshuddin, J., Fauziah, I., & Husni, M. A. H. (2006). Changes in iron-poor acid sulphate soil upon submergence. *Geoderma*, *131*(1-2), 110–122. doi:10.1016/j.geoderma.2005.03.006

Parveen, F. R., Rajeev, P. S., Ibrahim, M. H., & Norizan, E. (2010). Review of Current Palm Oil Mill Effluent (POME) Treatment Methods: Vermicomposting as a Sustainable Practice. *World Applied Sciences Journal*, *10*(10), 1190–1201.

Peng, Y., Gao, C., Wang, S., Ozaki, M., & Takigawa, A. (2003). Nonfilamentous sludge bulking caused by a deficiency of nitrogen in industrial wastewater treatment. *Water Science and Technology*, *47*(11), 289–295. doi:10.2166/wst.2003.0617 PMID:12906302

Schuchardt, F., Wulfert, K., & Damoko, D. (2005). New process for combined treatment of waste (EFB) and wastewater (POME) from palm oil mills — technical, economical and ecological aspects. *Landbauforschung Völkenrode*, *55*, 47–60.

Schuchardt, F., Wulfert, K., Darnoko, D., & Herawan, T. (2007). Effect of new palm oil mill processes on the EFB and POME utilization. *Proceedings of Chemistry and Technology Conference PIPOC 2007*, Kuala Lumpur, August 26-30, 2007. Academic Press.

Shahzad, M., Khan, S. J., & Paul, P. (2015). Influence of temperature on the performance of a full-scale activated sludge process operated at varying solids retention times whilst treating municipal sewage. *Water (Basel)*, *7*(12), 855–867. doi:10.3390/w7030855

Viyayaraghavan, K., Desa, A., & Abdul Aziz, M. E. (2007). Aerobic treatment of palm oil mill effluent. *Journal of Environmental Management*, *82*(1), 24–31. doi:10.1016/j.jenvman.2005.11.016 PMID:16584834

Yaser, A. Z., Rahman, R. A., & Kalil, M. S. (2007). Co-composting of palm oil mill sludge-sawdust. *Pakistan Journal of Biological Sciences*, *10*(24), 4473–4478. doi:10.3923/pjbs.2007.4473.4478 PMID:19093514

Zahrim, A. Y., Rachel, F. M., Menaka, S., Su, S. Y., Melvin, F., & Chan, E. S. (2009). Decolourisation of anaerobic palm oil mill effluent via activated sludge-granular activated carbon. *World Applied Science Journal, 5*(Special Issue for Environment), 126–129.

Zhang, Y., Yan, L., Qiao, X., Chi, L., Niu, X., Mei, Z., & Zhang, Z. (2008). Integration of biological method and membrane technology in treating palm oil mill effluent. *Journal of Environmental Sciences (China)*, *20*(5), 558–564. doi:10.1016/S1001-0742(08)62094-X PMID:18575108

ADDITIONAL READING

Abdulsalam, M., Che Man, H., Idris, A. I., Yunos, K. F., & Zainal Abidin, Z. (2018). Treatment of palm oil mill effluent using membrane bioreactor: Novel processes and their major drawbacks. *Water (Basel)*, *10*(9), 1–46. doi:10.3390/w10091165 PMID:30079254

Ahmad, Z., Ujang, Z., Olsson, G., & Abdul Latiff, A. A. (2009). Evaluation of hybrid membrane bioreactor (MBR) for palm oil mill effluent (POME) treatment. *International Journal of Integrated Engineering*, *1*(2), 17–25.

Igwe, J. C., & Onyegbado, C. C. (2007). A review of palm oil mill effluent (POME) water treatment. *Global Journal of Environmental Research*, *1*(2), 54–62.

Loh, S. K., Lai, M. E., Ngatiman, M., Lim, W. S., Choo, Y. M., Zhang, Z., & Salimon, J. (2013). Zero discharge treatment technology of palm oil mill effluent. *Journal of Oil Palm Research*, *25*, 273–281.

Mba, O. I., Dumont, M. J., & Ngadi, M. (2015). Palm oil: Processing, characterization and utilization in the food industry – a review. *Food Bioscience*, *10*, 26–41. doi:10.1016/j.fbio.2015.01.003

Mohd Azoddein, A. A., Haris, H. B., & Azli, F. A. M. (2015). Treatment of palm oil mill effluent (POME) using membrane bioreactor. *The Malaysian Journal of Analytical Sciences*, *19*(3), 463–471.

Odia, O. J., Ofori, S., & Maduka, O. (2015). Palm oil and the heart: A review. *World Journal of Cardiology*, *7*(3), 144–149. doi:10.4330/wjc.v7.i3.144 PMID:25810814

Poh, P. E., Yong, W. J., & Chong, M. F. (2010). Palm oil mill effluent (POME) characteristic in high crop season and the applicability of high-rate anaerobic bioreactors for the treatment of POME. *Industrial & Engineering Chemistry Research*, *49*(22), 11732–11740. doi:10.1021/ie101486w

Tan, S. P., Kong, H. F., Bashir, M. J. K., Lo, P. K., Ho, C. D., & Ng, C. A. (2017). Treatment of palm oil mill effluent using combination system of microbial fuel cell and anaerobic membrane bioreactor. *Bioresource Technology, 245*(Part A), 916–924.

Zahrim, A. Y. (2014). Palm oil mill biogas producing process effluent treatment: A short review. *Journal of Applied Sciences (Faisalabad)*, *14*(23), 3149–3155. doi:10.3923/jas.2014.3149.3155

KEY TERMS AND DEFINITIONS

Activated Sludge System: Aerobic treatment process that uses aerobic microbes to feed on the organic contaminants of palm oil mill wastewater.

Anaerobic Membrane Bioreactor: Technology that combines the fundamental of anaerobic digestion and membrane filtration process, which works by breaking down the biodegradable matters by anaerobic microorganism and separating the biomass from treated water respectively.

Clarification System: System that separates oil residuals using mechanical means such as skimmer and removes solids present through sedimentation from the POME.

Cooling System: System which cools down the hot POME to a certain temperature before subjected to the subsequent process that is affected by high-temperature fluid, i.e. activated sludge process.

Membrane Filtration: System that is used to separate the biomass from treated POME using either microfiltration, ultrafiltration, or reverse osmosis membrane depending on the influent characteristics and quality of water desired.

Nanofibre: Films comprising of nanofibres overlap with each other in a completely random manner with pore size ranging from tens of nanometre to several micrometres which are used to adsorb the residual palm oil.

Palm Oil Mill Effluent (POME): Wastewater generated by palm oil processing mills which contain oil, grease, and suspended solids.

Ponding System: Conventional POME treatment system which uses the natural biological treatments of anaerobic and aerobic systems.

Chapter 12
Membranes and Microalgae in Wastewater Treatment

ABSTRACT

The application of microalgae-based wastewater treatment was first introduced in the 1940s to treat municipal wastewater. Microalgae have been studied for its various potentials such as for nutrients removal, carbon dioxide (CO2) removal, biofuel production from biomass, etc. This chapter focuses on the potential of microalgae membrane bioreactors for wastewater treatment, microalgae cultivation, and harvesting. Furthermore, the selection of microalgae species is covered by comparison of nitrogen, phosphorus, COD, and BOD removal from various studies. Microalgae membrane bioreactors combine the biological treatment of microalgae with the conventional membrane bioreactor. Still, membrane fouling phenomenon is a challenge in microalgae membrane technology. Thus, several other technologies of immobilized microalgae are introduced which can potentially reduce the membrane fouling occurrence and concurrently remove the need for microalgae harvesting process.

INTRODUCTION

The potential of wastewater treatment by using algal remediation was discovered as early as the 1940s where microalgae were proposed to be used for the treatment of municipal wastewater (Caldwell, 1946). Development of microalgae-based treatment process has contributed significantly to the reduction of pollutant especially biological oxygen demand (BOD), chemical oxygen demand (COD) as well as nutrients, mainly on nitrogen and phosphorus.

Microalgae as a photosynthetic microorganism require only a few basic necessities to thrive, including CO_2 supply, light and nutrient contents such as nitrogen and phosphorus. Given a suited environment, microalgae can be easily cultivated even in wastewaters. A number of researches have reported that microalgae are able to be cultivated in wastewaters occurring from various sources ranging from human

DOI: 10.4018/978-1-7998-2645-3.ch012

sewage, municipal, industrial or even from agricultural wastewater. This makes wastewater treatment using microalgae to uptake nutrients a very attractive method in addition to many other of its uses.

Microalgae-based wastewater treatment has several benefits over conventional wastewater treatments, namely chemical and biological treatment. In biological treatment, the aeration process requires a high costing, contributed by the process of mechanical aeration as well as the management of sludge produced at the end of the treatment. As for chemical treatment, the costing of the chemical itself is already in great amount. On the other hand, the microalgae bioremediation offers a more cost-effective approach where the biomass produced at the end of the treatment can be used for various purposes such as biofuel production for other industries. Furthermore, the microalgae treatment can serve a dual purpose of wastewater treatment and CO_2 sequestration at the same time. The nitrogen removed during the treatment process is also being able to be recovered in the form of biomass and can be further reused.

The utilization of membrane for solid-liquid separation that is widespread can be implemented in the microalgae for wastewater treatment since it provides (1) filtration media for water-microalgae/contaminants separation, and (2) barrier for retaining the nutrients for uptake by the microalgae. Thus, this chapter will cover the concept of wastewater treatment using microalgae and membrane technologies, which covers the prospects of microalgae and microalgae membrane bioreactor for wastewater treatment, and the technologies that are available for this application.

MICROALGAE FOR WASTEWATER TREATMENT

Many researchers have been exploring the utilization of nutrient removal in wastewater through the cultivation of microalgae in the wastewater generated from municipal, agro-industrial or industrial. This has opened up the opportunity to utilize the available unlimited resources of wastewater in order to cultivate microalgae for the dual purpose of removing nutrients and producing biofuels.

Research in this field was conducted to utilize microalgae nutrient uptake to treat municipal wastewater. The samples were obtained from the effluent of primary settling tank (PS), the effluent from an anaerobic digestion tank (AD) and the conflux of wastewaters rejected from sludge-concentrate tanks and dewatering facilities (CR) in a municipal wastewater treatment plant in Busan, Korea. The study was conducted to investigate the feasibility of utilizing the readily available nutrients in sewage for microalgae cultivation as well to treat the wastewater. From the experiment, it was found that the value of total nitrogen and total phosphorus was able to be totally removed from 250 mg/L and 17 mg/L to respectively in 5 days. However, the Chemical Oxygen Demand (COD) content increased over time caused by the increased biomass within the system (Cho et al., 2013). This shows that a reliable separation technique is important at the end of the treatment to ensure the effluent is well under the specified limit.

A combination of microalgae treatment with membrane bioreactor has been experimented to polish effluent sample from an anaerobic membrane bioreactor (MBR) fed with domestic wastewater. The treatment system consisted of two parts, the MBR system and a subsequent microalgae membrane reactor (mMR). Both of the treatment stages utilized a commercial membrane of pore size 0.45 µm (PES, Millipore). The experiment which lasted for 23 days was able to remove on average 50% of NH_4, 75% of NO_2, 35% of NO_3 and 60% of PO_4 consistently from the MBR effluent under the conditions tested.

As for the application of microalgae wastewater treatment in the industrial sector, the researchers are more focusing on nitrogen and phosphorus-rich wastewater such as effluent derived from animal farms and organic industries such as soybean and textile factories. Although in general, most industrial wastewaters contain more heavy metal pollutants and less nitrogen or phosphorus than other types of wastewater. Microalgae treatment is still possible by selecting microalgae strains with high metal sorption capacity. However, only a few algal species have been studied for metal sorption ability (Cai et al., 2013). There are several reports evaluating the nitrogen, phosphorus and heavy metal removal from industrial wastewater as an algae growth medium, such as those from the carpet industry (Chinnasamy et al., 2010).

Selection of Microalgae Species

Microalgae are present in all existing earth ecosystem, not just aquatic but also terrestrial, representing a big variety of species living in a wide range of environmental conditions. It is estimated that more than 50,000 species exist, but only a limited number, of 30,000 have been studied and analyzed (Richmond, 2004). For nutrient removal by microalgae in wastewater, it is important to consider the species that will be utilized in the system for maximum performance, depending on the nature of wastewater. Various species and strains of microalgae have been documented in previous researches worldwide. However, frequently documented microalgae researched for wastewater treatment as well for biofuel production are of *Chlorella* sp. and *Scenedesmus* sp. among other species.

Chlorella sp. microalgae have the ability to remove nitrogen (in the form of ammonia and ammonium ion) while being cultivated using wastewater effluent and reported to be able to uptake nutrients (nitrogen and phosphorus) through the cell membrane (Kim *et al.,* 2010). Cultivation of *Chlorella sp.* as an individual species has been documented capable to treat wastewater or coupled with other microalgae species.

Scenedesmus sp., on the other hand, is usually coupled with other species and has shown a fairly high removal rate. Combination of *Chlorella vulgaris* and *Scenedesmus* sp. in batch and continuous system were able to completely remove the nitrogen, ammonia and phosphorus content by the end of the treatment of 36 and 48 days respectively (Hammouda *et al.,* 1995). However, a lower value of 40% removal of nitrogen has been obtained by Gonzalez-Fernandez *et al.,* (2011) in the system combining *Chlorella Vulgaris* and *Scnedesmus obliquus* with microorganism in treating pig slurry. In research by Gonzalez *et al.,* (1997), a higher initial removal of nitrogen by *Scenedesmus dimorphus* was shown compared to *Chlorella vulgaris*. However, the final efficiency was equal for both species. Previous researches show that microalgae wastewater treatment are high potential, as presented in Table 1.

Limitation of Microalgae in Wastewater Treatment

Nutrients Sources and Limiting Nutrients

The most important essentials needed by microalgae to multiple other than CO_2 is the availability of nutrients in its medium. Nitrogen and phosphorus are the two major nutrients needed by microalgae to conduct photosynthesis. Most microalgae cultivation plant provides these nutrients by using chemical fertilizers to optimize the production capacity. Based on a review conducted by Lam and Lee, (2012), microalgae are the number one consumers of chemical fertilizers than other biofuel sources followed by jatropha, rapeseed, sunflower and palm oil being at the bottom. The article stated that microalgae require an approximate 0.032 kg of N-fertilizer to produce 1 kg of oil in a photobioreactor whereas palm oil only requires 7.79 kg of N-fertilizer to produce 0.163 tonnes of oil. Fortunately, the requirement of microalgae for nutrients in its medium to conduct the photosynthesis in order for its growth is being compensated by using nutrient-rich wastewater.

Table 1. List of researched microalgae for wastewater treatment

Algae Species [a]	Wastewater Characteristic	Removal Rate [b] (%)				Retention Time	Source
		N [c]	P [d]	COD	BOD		
Chlorella sp.	Municipal	82	83	50	*NA	9 days	Wang *et al.*, 2009
		>90	>90	NA	NA	5 days	Cho *et al.*, 2013
C. vulgaris		50	NA	NA	NA	2 days	Kim *et al.*, 2010
	Steel-making plant	100	NA	NA	NA	120 h	Yun *et al.*, 1997
	Synthetic	100	78	NA	NA	10 days	Aslan & Kapdan, 2006
	Textile	45	30	60	NA	12 days	Lim *et al.*, 2010
Chlorella sp. + *Scenedesmus* sp.	Municipal	100	75	97	93	42 days	Hammouda *et al.*, 1995
Chlorella sp., *C. vulgaris, S. quadricauda, S. dimorphus*		100	>70	NA	NA	3 days	Singh & Thomas, 2012
C.vulgaris, S. abundance, S.quadricauda		50	85	NA	NA	18 days	Kassim, 2002
C.vulgaris, S. dimorphus	Piggery	>90	60	NA	NA	220 h	Gonzalez *et al.*, 1997
Chlorella pyrenoidosa	Soybean processing	89	70	78	NA	120 h	Hongyang *et al.*, 2011
C.vulgaris + Cynabacterium *Planktothrix isothrix*		80	100	NA	NA	48 – 96 h	Silva-Benavides & Torzillo, 2012
Oosystis sp., *Chlorella* sp., *Chlamydomonas* sp.	Municipal	>65	>80	NA	NA	14 days	Rasoul-Amini *et al.*, 2013
Neichloris vigensis,Chlorococcum spec., S. rubescens		NA	>10	NA	NA	30 days	Aravantinou et al., 2013
Chlorococcales (order of the *Chlorophyceae*)		67	97	NA	NA	42 days	Ruiz-Martinez *et al.*, 2012
Scenedesmus sp. AMDD		90	90	NA	NA	6 days	McGinn *et al.*, 2012
Chlorella sp. 227		100	100	NA	NA	5 days	Cho *et al.*, 2013
Rhodobacter sphaeroides + *C. sorokiniana*	Synthetic	100	100	NA	NA	1 day	Ogbonna *et al.*, 2000
Scenedesmus sp. LX1	Electronic device factory	46	100	NA	NA	15 days	Zhen-Feng *et al.*, 2011
C. minutissima, C. sorokiniana, S. bijuga	Poultry litter	>15	>25	NA	NA	4 days	Singh *et al.*, 2011
Ooscystis sp.	Fish processing	95	>70	>55	NA	10 days	Riaño *et al.*, 2011

[a] Blank row refers to the previous species, (+) indicates co-cultures, (,) indicates individual cultures.

[b] an approximate value of the overall performance of nutrient removal.

[c] sometimes in the form of nitrogen derivatives (NH_4^+- N, NO_3 - N, NO_2 - N).

[d] in the form its derivatives (PO_4^{3-} – P)

*NA not applicable

Since the source of wastewater is abundance and freely available, this has benefited tremendously in cost reduction especially for the dual application of wastewater treatment and microalgae cultivation purposes. Nonetheless, with the nature of wastewater either from the industries or municipal sources, the content is usually uncontrollable, depending on the origin of the wastewater, it may contain too high, too low, if not none at all of the essentials microalgae growth nutrient requirement, nitrogen and phosphorus. Therefore, it is imperative that careful consideration of the choices in wastewater input is conducted before proceeding with the treatment process. A pre-addition of a certain nutrient into the wastewater or a pre-treatment of the wastewater may have to be conducted in order to prepare a more suitable environment for the microalgae to work efficiently.

Another important aspect of nutrient uptake by microalgae is the limiting nutrient. It is very likely that the limiting nutrient will affect the efficiency of nutrient removal by the microalgae. In a microalgae culture, the limiting nutrient will be exhausted first than the other nutrients in the culture. For example, in a culture with phosphorus as the limiting nutrients, the concentration of nitrogen will still be high even though the phosphorus has been uptake by the microalgae. Over the years, several patterns of limiting nutrient of microalgae have been discovered by researches. In the research done by Anand, (2010), phosphorus was detected to be limiting nutrients for algae growth. Yun *et al.,* (1997), also reported that the removal of ammonia was very slow as a result of the phosphate limitation in the culture medium. However, in some literature nitrogen was reported to be the limiting factor (Gonzalez *et al.,* 1997). This implies that the limiting nutrient for microalgae culture could be affected by other factors. Richmond, (2004) has mentioned that the competition of which nutrient is limiting is highly dependent on the N:P ratio of the medium.

The limiting nutrient is mainly affected by the ratio of nitrogen and phosphorus in the medium, namely the N:P ratio, where medium with high N:P ratio will be phosphorus limited and medium with low N:P will be nitrogen-limited as display in Figure 1. By that note, it is important to consider the outcome of the experiment based on the concentration of wastewater used.

Cost Optimization

The operational cost of the microalgae wastewater treatment is closely related to the choice of system design, either it is conducted in a photobioreactor or in an open pond system as well as the choice of the harvesting process. The total cost of operation can be generally broken down into the system capital cost, operational cost such as the supply of lighting and CO_2 supply, the cost of the harvesting process and finally the overall maintenance cost of the system.

The overall cost of a current microalgae related operation, especially the biofuel production depends on the design of subsystems, including algae cultivation, harvesting, and lipid extraction. Most of the operative plants have chosen to build large microalgae cultivation operations along with coastal areas in order to utilize seawater as the major water source, thus reducing the cost of water usage. This is because, the microalgae cultivation will consume a large quantity of water during its cultivation, which occupies 10 – 20% of the total cost of algae production. Hence, combining algae biomass production with organic wastewater treatment can mitigate costs in the algae-oriented biofuel industry.

As previously discussed, the application of microalgae wastewater treatment can immensely reduce the operating cost for the cultivation of microalgae. This dual application is further advantageous if coupled with CO_2 sequestration process which is much needed basic requirement for the microalgae's photosynthetic activity. Flue gases generated from power plants containing 10 – 15% (v/v) CO_2 can be

Figure 1. The specific growth rate dependence for hypothetical microalgae for the optimal N:P ratio
Source: Richmond, 2004

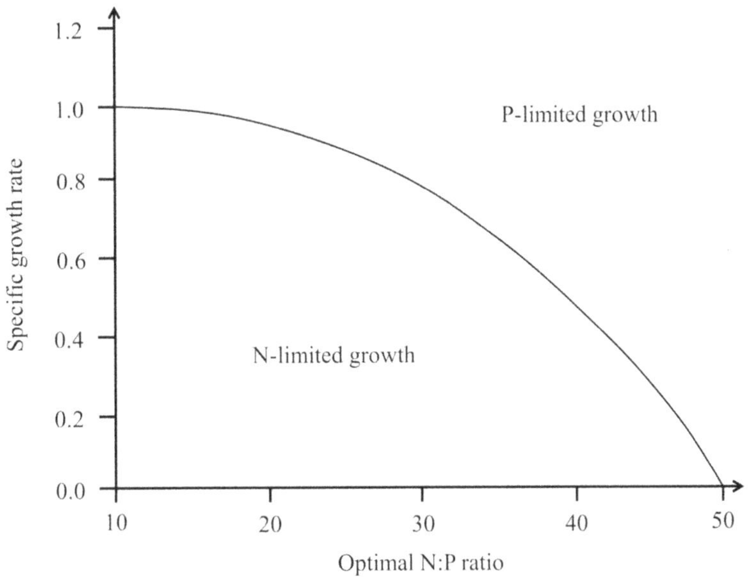

used for microalgae cultivation. Successful cultivation of *Scenedesmus* sp., *Chlorella* sp. and *Nannochloropsis* sp. has been reported with a relatively high microalgae growth rates using 10 – 15% flue or synthetic CO_2 by other researchers (Lee et al., 2002; Jin et al., 2006; Jiang et al., 2011). Reduction of cost in production at the same time doubling the microalgae production is obtained by combining the nutrients removal and carbon dioxide from the waste stream.

High-rate algal pond (HRAP) is by far the most cost-effective available system design for wastewater treatment. It was estimated that the energy required for an open pond system is around 1 W/m^3 whereas 50 – 300 W/m^3 for photobioreactor operation (Acien et al., 2008). It is especially suitable for tropical and subtropical countries where solar radiation is abundant throughout the year. Cost elimination is possible for an open pond treatment because the aeration and lighting are acquired naturally from the environment. Nevertheless, to provide an even light exposure to the microalgae cell, mixing of the culture is required. The most common mixing method is by using a paddle wheel mixed, which will also provide aeration to the culture. The depth of water in the pond should be as shallow as possible ranging from 10 cm and it should not exceed a maximum level of 30 cm to allow sufficient light penetration (Rawat et al., 2011). Even though some research has suggested that an increased pond depth can increase microalgae productivity, the range of depth is still very small at 40 cm (Sutherland et al., 2014). For that reason, a larger area is required for a bigger plant capacity, which contributes to a larger capital cost for land purchase.

Even though most of the cost in wastewater treatment using microalgae are able to be compensated with the usage of waste materials, specifically CO_2 and the wastewater, the issue of capital, operation maintenance cost still arises. This matter is especially significant if the process is conducted in a photobioreactor system. In an area with limited land and wastewater treatment that can optimize the available land has to be constructed, a better choice of the system design is photobioreactor. The photobioreactor presents its own advantages such as it eliminates the risks of contaminations of the microalgae culture, the CO_2 utilization efficiency is higher and it can provide a higher microalgae growth rate. The light supply of the microalgae is also optimized as the photobioreactor is made from transparent material and can provide a more even and higher light penetration than an open pond system. However, the consumption of energy by a photobioreactor is higher in terms of mixing and aeration of the microalgae culture. The capital and maintenance cost is also high if compared to an open pond system as the photobioreactor will consist of mechanical units that need regular maintenance compared to the open pond system.

Culture Contamination

Even though microalgae can strive in almost any water with various conditions, the microalgae culture can be affected by the presence of other organisms in the culture. This case is particularly severe in an open pond system where control of contamination is almost impossible. The presence of predatory zooplankton and protozoa that graze on the microalgae can significantly reduce the growth of microalgae within a few days period. Other than that, fungal parasitism and viral infection are also found to be able to trigger changes in algal cell structure, reduction of algal chlorophyll a, diversity and microalgae succession.

On the other hand, the presence of other microorganisms beside microalgae can slightly inhibit their growth but does not significantly change the treatment effectiveness in term of COD and phosphorus reduction. However, the presence of these organisms has slightly delayed the nitrogen reduction in wastewater treatment. Therefore, sterilization may be necessary in order to minimize the negative effects of contamination from bacteria as well as predatory organism. However, this process also increases the capital cost of the algae cultivation system.

Environmental Factor

Algal growth is influenced by various factors either abiotic, biotic or the operational factor of the cultivation process. Abiotic factors include the quality of intensity of light, the environmental temperature, amount of nutrients available, the salinity of water, supply of oxygen and carbon dioxide, pH level as well as the existence of toxic chemicals in the culture. The abiotic factor is the presence of contaminant or pathogens such as bacteria, fungi and viruses. Finally, the operational factor is such as shear produced by mixing, dilution rate, harvest frequency and the addition of bicarbonate.

As a photosynthesis microorganism, one of the most important elements that affect the biomass yield in microalgae cultivation is light exposure, both in term of cycle and intensity. Light may be supplied either from a natural source which is by sunlight or artificial light source such as fluorescent tubes. Low irradiance exposure will decrease the rate of microalgae biomass production, while too high irradiance may be fatal due to overheating and photo-inhibition where the photosynthesis process declined. Literature has reported that microalgae biomass growth is at an optimum rate with a light illumination range of 20 to 200 µmol. photons m-2s-1 while higher biomass production is observed at higher light intensity (Mata et al., 2012; Sandnes et al., 2005).

In the account of the light cycle, some microalgae could show preference with respect to the duration of the light periods, resulting from the environmental conditions in which they were isolated. In the research done by Jacob-Lopes et al., (2009) the uni-algal culture of *Aphanothece microscopic nägeli* was reported to have a higher biomass yield and CO_2 reduction rate at the was higher for the light cycle of 12/12 (dark/light) of which it was acclimatized in its natural environment. This result is supported by the result found by Mata et al., (2012) that shows the best culture condition for *Scenedesmus* obliquus in biomass growth and nutrient removal is when the culture is aerated, exposed to a 12 h period of daylight, at a higher intensity at 12000 Lux.

Additionally, temperature also plays a significant role in optimizing biomass productivity. The optimal range of temperature may vary for each algal species. However, in general, the most suitable temperature range reported for algal growth is between 15- 25 °C and above the temperature of 35 °C, biomass growth will decline.

Microalgae *Phormidium bohneri* was found to remove more nutrients (NH_3 and PO_4^{-3}) at a higher temperature, as removal rate at 30 °C and 10 °C was compared (Tablot & De La Noue, 1993). Researcher Sandnes et al., (2005) has found that the optimum growth culture of *Nannochloropsis oceanica* was at 28 °C while the growth noticeably reduces at temperature 30 °C.

The controlling of light and temperature are although necessary, it may be difficult in nutrient removal by microalgae as every species of microalgae will respond in a different manner. Moreover, the addition of artificial lighting and heating mechanism will further increase the operating cost. In a tropical climate country such as Malaysia, with the abundance lighting supply and a relatively uniform ambient temperature throughout the year (ranging at 24-32°C), controlling of these two parameters are not be the main concern.

pH level will affect the algae mass yield as it influences the carbon availability. pH also affects the metabolism and biochemical composition of the microalgae cells. Therefore, controlling of pH is essential algae in cultivation to ensure the carbon source is at its optimum availability. In both photoautotrophic and heterotrophic cultures, pH is controlled by the addition of strong acids, alkalis and CO_2. The pH level will increase as a result of CO_2 fixation during the photosynthesis as OH- accumulates in the growth solution. Sparging of carbon dioxide into the culture media is the most convenient method of pH control and can also increase yield in mass algal cultures. However, Yun et al., 1997 found that the CO_2 fixation and ammonia removal in the cultivation of microalgae without pH control is better than culture with pH control.

Separation of Microalgae and Treated Water

Microalgae harvesting is a downstream process which may involve one or more stage of separation between microalgae biomass and its medium. While the purpose of the process differs depending on the field of application, it is performed extensively in wastewater treatment and biofuel production. In wastewater treatment, microalgae harvesting is carried out in order to separate the microalgae biomass that otherwise will contribute to a high chemical oxygen demand, lowering its discharge below the allowable limit. Whereas, the resulting concentrated biomass can be utilized in other application for further processing. Therefore, it is important that the method used will not cause contamination to both the biomass and treated water.

There are a few methods of microalgae harvesting that is sedimentation, flotation, centrifugation or filtration, which is sometimes, required a prior flocculation step. Even with various options of harvesting techniques, separation of microalgae biomass from its medium remains a major hurdle to industrial-scale processing mostly because of the small size of the algal cells, with unicellular eukaryotic algae typically 3–30 μm, and cyanobacteria as small as 0.2–2 μm. As the requirement of microalgae differs in every situation and due to the unique characteristic of microalgae such as size, shape and motility in every species, it is impossible to name one method to be superior to the other. This area remains as an active area of research to this date. Although a universal harvesting method does not exist, experience has demonstrated that for all algal species it is possible to develop an appropriate, economical harvesting system.

Sedimentation technique is based on gravity settling where it separates the feed into the concentrated slurry at the bottom and a clear liquid at the top. This method can be accomplished in two ways which are lamella separator and sedimentation tank. However, this technique is highly dependent on the density difference between the microalgae and its growth medium. Therefore, flocculants are to be added prior to the sedimentation process to increase the microalgae separation and the rate of sedimentation. With the absence of the addition of flocculants, this technique is, however unreliable.

Microalgae harvesting by floatation is described as a physiochemical type of gravity separation technique, in which the separation is performed by bubbling air gas through a solid-liquid suspension. The suspended solids will be attached to the gaseous molecules and carried to the liquid surface as floats, which can be easily skimmed off. However, this technique is dependent on the instability of suspended particle, where a lower instability will correspond to a higher air-particle contact. The size of particles is also very important in floatation, the smaller the particle sizes, the more likely the particle can be levitated by the bubbles. There are three main floatation techniques that entail bubble generation, which is dispersed air floatation, dissolved air floatation and electrolytic flotation.

Flocculation is a method of microalgae harvesting, which is considered advantageous by facilitating cell broth separation. The process flocculation is carried out by adding flocculants (inorganic flocculants or organic polymer/polyelectrolyte flocculants) into the aqueous resulting in the solute particles to collide and adhere to each other. Microalgae cells can form stable suspensions with a chemically reactive cellular surface that has a net negative surface charge due to the ionization of functional groups. The stability of these microalgae suspensions is dependent on the forces that interact between the particles themselves and the particles and water. Even though flocculation presents to be more advantageous than other conventional methods as it targets a larger quantity and wide range of microalgae species, the chemical for flocculation is, however expensive, being only marginally cheaper than centrifugation There are also a few other drawbacks to it as the techniques alone is not enough as it needs to be coupled with floatation or centrifugation and flocculant may be toxic to the treated water or algal cell.

Another microalgae harvesting method which is widely used is by centrifugation. Centrifugation is a separation process that implements the centrifugal force to separate solid and liquid based on the difference in size and density of the materials. The behaviour of the smallest particles in the system has the largest effect on separation efficiency. Centrifugation technique has the efficiency to recover 80–90% of the algal biomass. Although centrifugation is an effective method of microalgae recovery, the main disadvantage is the high investment and operating costs required when processing a large volume of cultures as well as time-consuming. With the high costing of centrifugation, it may be only suitable to be utilized to produce the high-value product but not for suited for a low-cost harvesting method. Although centrifugation is a successful dewatering method, the exposure of microalgae cells to high gravitational and shear forces can damage cell structure.

Membrane filtration method of separation utilizes a permeable medium with a specific pore size through which a suspension is passed. It is a strictly physical, absolute separation technique whose quality performance depends on the membrane pore size. The permeable medium retains the solids that are larger than the pore size and allows the liquid to pass through. Two commonly employed type of filter in harvesting large amount of microalgae is a rotary vacuum drum filter and chamber filter press. A rotary vacuum filter consists of a drum frame covered with filter cloth made of either canvas, nylon, dacron, metal or glass fibre. The internal volume of the drum is divided into the radial chamber to which vacuum

Table 2. Ammonia-rich synthetic wastewater properties

Properties	Concentrations	
	High Concentration	Low Concentration
NH_3 - N (mg/L)	150 ± 10	50 ± 10
PO_4^{3-} - P (mg/L)	20 ± 5	10 ± 5
COD (mg/L)	110 ± 10	60 ± 10
BOD (mg/L)	50 ± 5	30 ± 5
pH	7	7
Turbidity (FAU)	50 ± 5	50 ± 5

is applied. The drum rotates with part of it is submerged in the agitated algal slurry to be filtered. The filtered algal cake is to be scraped off the filter aid and therefore exposing fresh filtration surface. Another method in microalgae harvesting, which utilizes the technique of filtration is membrane filtration using either microfiltration or ultrafiltration.

MICROALGAE MEMBRANE BIOREACTOR FOR WASTEWATER TREATMENT

In this topic, the utilization of microalgae membrane bioreactor will be discussed specifically on wastewater treatment based on one research by the author. An alternative treatment for ammonia-rich wastewater was designed and constructed in a laboratory scale. The wastewater utilized microalgae's ability in up-taking nutrients from its media. A locally isolated microalgae species, *Chlorella* sp. was used in the project because of its track record to have higher nutrient removal performance as well as it being recorded to have many post-harvest applications.

The project used synthetic ammonia-rich wastewater which was prepared daily. Two different concentrations were used for the project to evaluate the system efficiency in a different wastewater condition. The properties of synthetic wastewater feed is tabulated in Table 2.

Microalgae Membrane Bioreactor Setup

The microalgae membrane bioreactor treatment system was fabricated in a laboratory size, to treat wastewater in semi-batch mode. Schematic of the microalgae membrane bioreactor is as shown in Figure 2 and

Figure 3 shows the system set up in the laboratory. The system consisted of one photobioreactor and three other tanks, an influent tank where the untreated wastewater was stored, the excess algae holding the tank and the treated water tank. Two water pumps were used to feed wastewater into the photobioreactor and to flow the mixture of treated wastewater and algae through the membrane. Another pump was used for the membrane backwashing process. The pressure gauge was installed at three of the membrane inlet and outlet to monitor the operating pressure. Gate valve was used to make any necessary adjustment of the water flow. Air diffuser made from PVC pipes were installed at the bottom of the photobioreactor which is connected to an air pump for aeration as well to act as mixing of the wastewater and microalgae to maximize algae-wastewater contact as well contact with light supply. Then, the UF membrane is connected downstream to the photobioreactor before the treated water and algae holding tank. Transparent

Figure 2. Schematic of microalgae membrane bioreactor treatment

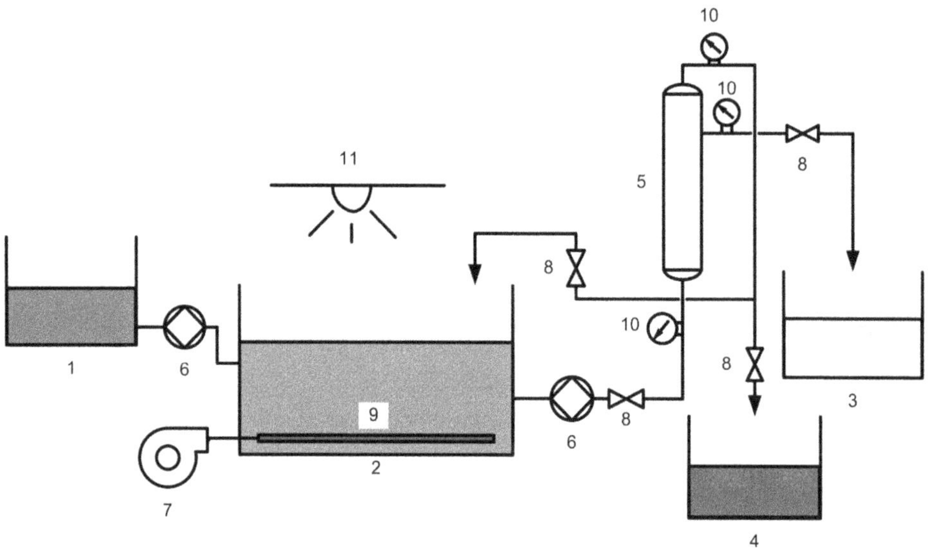

1) Untreated Synthetic Wastewater Tank, 2) Microalgae Photobioreactor, 3) Treated Water Tank, 4) Excess Algae Holding Tank, 5) Ultrafiltration Membrane, 6) Pump, 7) Air Pump, 8) Valve, 9) Air Diffuser, 10) Pressure Gauge, 11) Florescent Lamps

glass (prospect) was used as the photobioreactor material so that light supplies are even throughout the whole reactor.

Microalgae Membrane Bioreactor Operation

The wastewater sample prepared was initially stored in the influent tank. Upon the beginning of the system operation, 30 L of the wastewater was pumped into the photobioreactor where 20 L of microalgae has been cultured beforehand and was prepared in the specified concentration OD_{680} reading at approximately 0.150. The ratio of microalgae and synthetic wastewater volume was 2:3. The microalgae treatment was operated in two different short retention time (RT) inside the photobioreactor. During this operation, carbon dioxide and the light were supplied to the microalgae by aeration and artificial illumination respectively.

After each treatment cycle in the specified RT, the mixture of wastewater and microalgae were pumped using a gear pump into the UF membrane for separation of the treated water and microalgae biomass. The membrane permeate will flow directly into the effluent water, and the concentrated microalgae biomass will be pumped back into the photobioreactor. The filtration process was done until 2/3 of the photobioreactor volume is filtered to maintain the microalgae to synthetic wastewater volume ratio. Membrane cleaning by forwarding flushing and backwashing was conducted after every 25 minutes of the membrane filter operation. 30 L of untreated synthetic wastewater sample feed was pumped into the

Figure 3. Laboratory scale microalgae membrane bioreactor setup

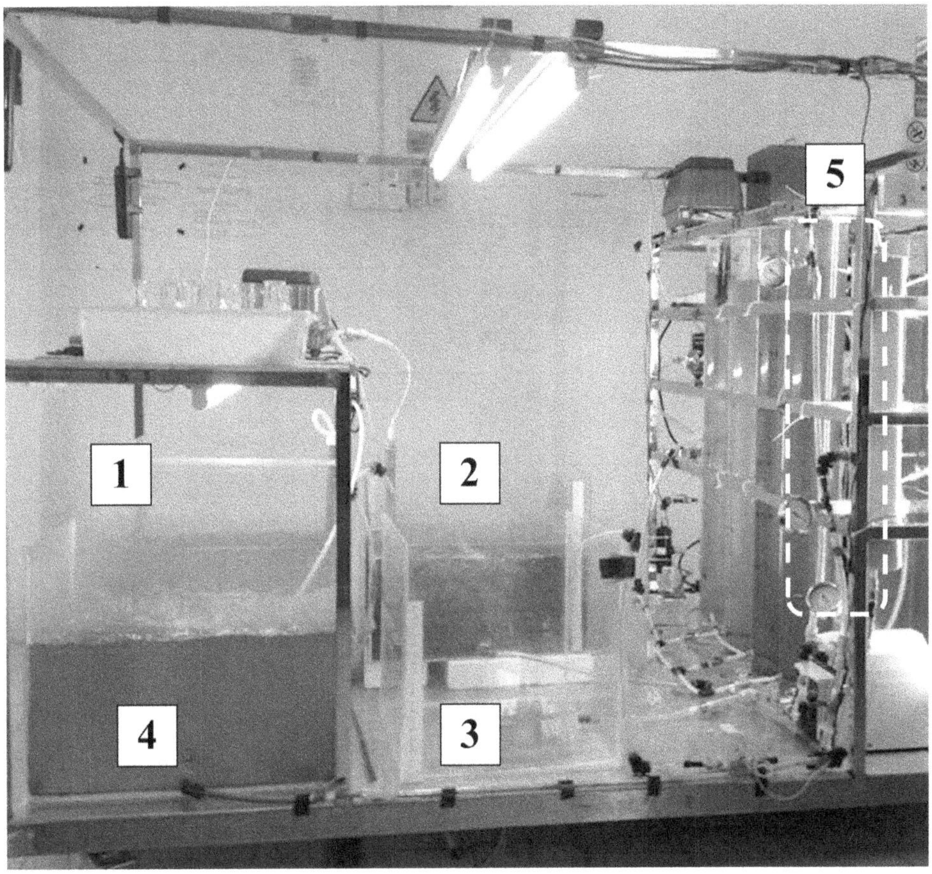

1) Untreated Synthetic Wastewater Tank, 2) Microalgae Photobioreactor,

3) Treated Water Tank, 4) Excess Algae Holding Tank,

5) Ultrafiltration Membrane

photobioreactor for the next treatment operation. The operating parameters for the microalgae membrane bioreactor are summarized in Table 3.

Temperature

The room temperature was monitored daily using a room thermometer and the recorded room temperature was between 20 – 24 °C. The room temperature was not altered during the operation to maintain the original condition of the microalgae environment. The synthetic wastewater feed was kept at a room temperature level which is 22 – 23 °C. The wastewater temperature was monitored using the Prima Long Digital Thermometer.

Table 3. Operating parameter of microalgae membrane bioreactor

Operating Parameter	Specifications
Membrane Filter Unit	Commercial Ultrafiltration Membrane Molecular Weight cut-off: 50 kDa
Ambient Temperature	20 – 24°C
Synthetic Wastewater Temperature	22 – 23°C
CO_2 Supply Flow rate	15 L/min
Synthetic Wastewater pH	7 – 7.5
Light Supply	Intensity: 4000 ± 20 Lux Light/Dark Cycle: 12/12 h
Photobioreactor Working Volume	70 L
Membrane Cleaning	Forward Flushing: 30 s Backwashing: 4 Min TMP: 0.2 bar

Carbon Dioxide Supply

The carbon dioxide is supplied by diffusing ambient air continuously into the photobioreactor using an LP-60 air pump. Air diffuser system was made from 4 parallel 1.5-inch diameter PVC pipes and was horizontally placed at the bottom of the tank. 1 mm holes were evenly drilled along the pipes for air diffusion. Air was pumped with an approximate flow rate of 15 L/min.

pH

The culture pH was monitored daily using the pH meter (Eutech Instrument, pH 700). The pH of wastewater feed was maintained in the range of 7 – 7.5 using the buffering agent salt of Na_2HPO_4 (Mata et al, 2012). However, the pH of microalgae culture was not altered during the treatment.

Light Intensity and Cycle

The light exposure for the experiment was provided artificially with the illumination of two fluorescent lamps (Philips 36W, 2500 Lux). The intensity was measured by using the LX-101 Lux Meter. The intensity of light supplied in the experiment was 4000 ± 20 Lux. Lighting source was supplied in 12/12 light/dark to simulate day and night.

Working Volume

The photobioreactor was constructed with 0.8 m length, 0.39 m width and 0.4 m height. However, the operative volume for the photobioreactor was only 70 L. The volume of water in the tank was daily monitored and any water loss due to evaporation was corrected by the addition of distilled water.

Table 4. Membrane characteristics

Performance of Hollow Fibre	Specifications
Maximum feed Pressure (MPa)	0.6
Admission Transmembrane Pressure (MPa)	0.3
Normal operation pressure (MPa)	0.1 – 0.2
Backwash Pressure (Mpa)	0.1
pH value of feed water	2 – 10
Max thickness of continuous remaining chlorine o feed water (mg/L)	50
Operation mode	Dead end or Cross-flow
Max operating temperature	<45 ° C
Membrane Characteristic	Modified Hydrophilic Double Skin
Hollow-fibre membrane material	Polyacrylonitrile (PAN) AAA+
Molecular Weight Cut Off (MWCO)	50 000 Dalton
Membrane Area	2.5 m^2
Inside fibre diameter	1.0 mm
Outside fibre diameter	1.6 mm
Number of hollow fibre	1500
The geometry size of the membrane module	106 x 590
In & out water calibre of the membrane module	G ¾"
Max Air Pressure for integrity test (MPa)	0.2
Flux (0.1 MPa, 25° C)	1, 100 Litres / Hour
Material of cartridge	ABS
Weight of cartridge with hollow fibre	<10kg
Warranty period of the cartridge (0.2 MPa)	2 years
Time to change the cartridge	3 years
Materials of the cylinder housing	Stainless steel 304
Thickness of housing	1.5 mm

Membrane Filter Unit

UF membrane was used in the experimental setup with the molecular weight cut-off 50 kDa. The membrane was soaked in distilled water for approximately 24 hours stabilized by pressurized filtration with distilled water before any actual microalgae filtration. Specification of the membrane utilized in the treatment system is listed in Table 4.

Membrane Cleaning

To ensure the efficiency of the membrane is not compromised, membrane cleaning was done after every 25 minutes of the membrane operation. It was done by combining the two modes of cleaning, 30 seconds of forwarding flushing, followed by 4 minutes of backwashing. Two flashing modes were done

Table 5. Average nutrient removal for microalgae membrane bioreactor operation

Average Value	High Concentration		Low Concentration	
	3 Days RT	2 Days RT	3 Days RT	2 Days RT
NH_3 Removal (%)	29.6	19.1	80.5	81.9
Effluent NH_3 (mg/L)	107	122	11	9
PO_4^{3-} Removal (%)	19	15.2	21.7	25.5
Effluent PO_4^{3-} (mg/L)	18	17	7.9	7.5

for higher recovery of membrane flux. The operated backwashing TMP was higher than the normal operation, at 0.2 bar.

Microalgae Nutrient Removal

The membrane bioreactor was able to remove the ammonia content in the wastewater feed. However, the microalgae are more preferable to low concentration feed where more than 80% of NH_3 and more than 20% of PO_4^{3-} - was removed when feeding with lower concentration wastewater. Less than 30% of NH_3 and less than 20% of PO_4^{3-} were removed when feeding with high concentration wastewater for both RT. The highest removed nutrient was achieved in operation with low concentration wastewater sample in 2 days RT where 81.9% of NH_3 and 25.5% of PO_4^{3-} was removed from the wastewater.

From the result of NH_3 and PO_4^{3-} removal shown in Table 5, microalgae nutrient uptakes tend to be higher value when the initial nutrient concentration in the wastewater is lower. This result was seen in all the operations, both batch and semi-batch. This can be seen as the higher removal was achieved when low concentration wastewater was used as the feed for the microalgae membrane bioreactor. This result was also achieved by Li et al., (2013). The nutrient uptake is also influenced by the initial concentration of microalgae where a higher microalgae concentration will lead to higher nutrient uptake.

Based on the result analysis of the project and data comparison from other literature, it is recognized that the nutrient uptake by microalgae is a very complex process as it can depend on many parameters that may include nutrient concentration, light intensity, the nitrogen/phosphorus ratio, light/dark cycle. However, the most obvious parameter that affects the efficiency behaviour of the system other than the tested variables was the N:P ratio.

Throughout the whole operation, the recorded PO_4^{3-} removal was lower than the NH3 removal. This implies that in the nature of the synthetic wastewater used, the limiting nutrient is nitrogen. The N:P ratio of used wastewater feed is 5:1 which suggested by Richmond, (2004) to be nitrogen limited based on the optimal N: P ratio for microalgae growth graph. This result was also obtained by Gonzalez et al., (1997) and Yun et al., (1997) where phosphorus removal was low even with the exhaustion of nitrogen content where the microalgae of *Chlorella vulgaris* were used. Both of the research was done using medium containing low N:P ratio. From other literature, Ruiz-Martinez et al., (2012), Hultberg et al., (2013) and Li et al., (2013) that used a medium with very high N:P ratio, the phosphorus removal was reported to be higher than nitrogen removal.

This study has shown that wastewater treatment using microalgae is able to achieve desired nutrient removal efficiency with less required time compared to the biological treatment. Even though the

Table 6. Average turbidity, COD, and BOD reading in an effluent sample of microalgae membrane bioreactor operation

Average Value	High Concentration		Low Concentration	
	3 Days RT	2 Days RT	3 Days RT	2 Days RT
Turbidity (FAU)	< 4	< 4	< 5	< 4
COD Removal %	72.1	72.4	83.1	91.1
Effluent COD (mg/L)	31	30	11	7
BOD Removal %	72.2	73.6	87.8	92.6
Effluent BOD (mg/L)	14	12	4	3

microalgae membrane bioreactor system does not allow recovery of ammonia and urea, this can be compensated by the various possible applications of the resulting biomass from the treatment system.

Ultrafiltration Membrane Performance

The performance of the membrane was evaluated based on its capability on removing the Biological and Chemical Oxygen Demand (BOD and COD) as well as its ability to retain microalgae biomass which was assessed by the turbidity reading of the membrane permeate. Turbidity was recorded to be below 5 Fau for all of the microalgae membrane bioreactor operations. This has shown that the membrane unit was able to retain microalgae biomass that will otherwise contribute to COD and BOD contamination in the permeate water (Table 6).

All of the COD and BOD removal were achieved 70% removal. Average COD removal of 72.1% (3 RT) and 72.4% (2 RT) was recorded for the high concentration wastewater feed and 83.1% (3 RT) and 91.1% (2 RT) for low concentration wastewater feed. In the other hand, BOD removal of 72.2% (3 RT) 73.6% (2 RT) for high concentration wastewater feed was obtained. A higher reading of 87.8% (3 RT) and 92.6% (2 RT) was achieved for the low concentration wastewater sample. This shows that the UF membrane has successfully removed the biomass in the effluent of a microalgae membrane bioreactor.

Physical membrane cleaning was conducted in every 25 minutes by holding off the filtration process and conducting cleaning by forward-flushing and backwashing. The transmembrane pressure (TMP) used during filtration was at 0.1 bar and backwashing TMP was higher at 0.2 bar. Over 90% flux recovery was achieved after membrane the membrane cleaning was conducted.

From the overall result, it can be established that microalgae membrane bioreactor system that incorporates UF membrane as the separation unit of microalgae biomass is highly efficient in producing a good quality effluent water, by removing the NH_3, PO_4^{3-}, turbidity, BOD and COD as well as retaining the microalgae biomass. However, the operating parameters should be further studied to avoid any unnecessary extra cost for membrane maintenance and damages.

Microalgae Harvesting

Microalgae harvesting is a process of collecting microalgae that have been grown and cultivated. The harvesting process of microalgae can be difficult because of their micro size. Thus, harvesting can be costly, particularly those methods that involve a further downstream drying process. Approximately 25%

Figure 4. Summary of processes involved in the conversion of microalgae to biofuel

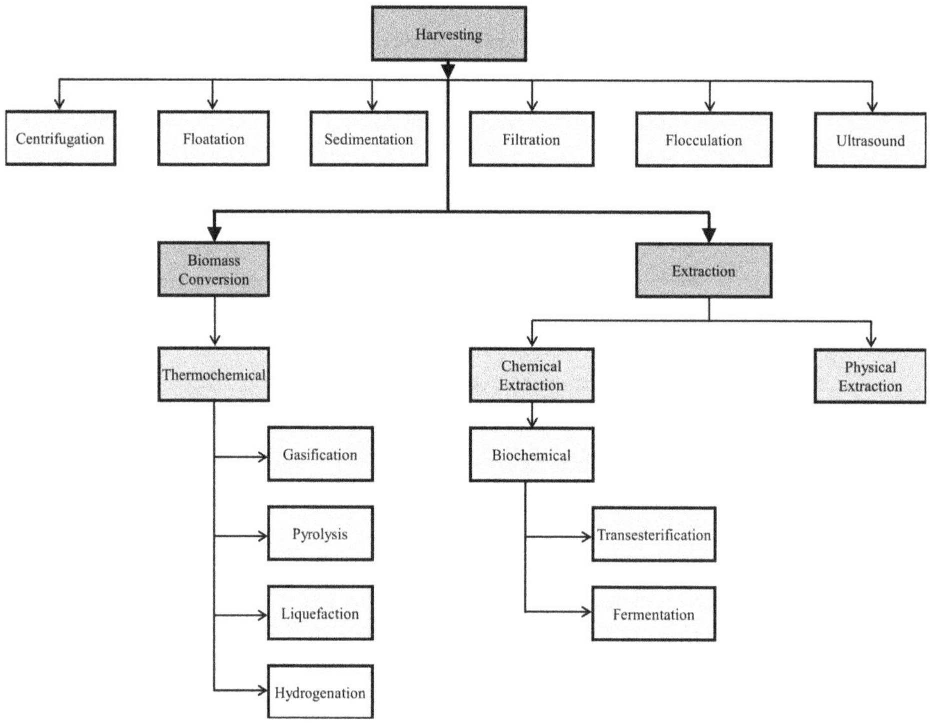

of the production cost is incurred during the harvesting process, which includes the cost of electricity, reagents and the maintenance of the separation equipment. Microalgae harvesting usually involves flocculation followed by filtration, sedimentation, centrifugation, flotation, using electromagnetic or ultrasound techniques. Summary of microalgae conversion to biofuel is shown in Figure 4.

The varieties of technique to harvest microalgae have resulted from technology development of the harvesting process. In the beginning, microalgae were only harvested by natural flocculation or filtration using tools such as fabric or membrane. Although this method can be considered as cheap, it is time-consuming to obtain extractable biomass for further conversion. Thus, the microalgae were first flocculated to shorten the required time for harvesting.

Microalgae biomass harvested has various potential processing applications into other products. Primarily, the biomass is used to produce biofuels, biogas, and other bio-products. Prior to bio-products production, microalgae biomass will undergo the bio-refinery process to extract its macromolecules of carbohydrates, proteins, and lipids, and also the biomolecules. These molecules obtained from microalgae can further be transformed into sugars, polymers, bioethanol, amino acids, fatty acids, biodiesel, pigments, pharmaceuticals, etc.

Flocculation

Harvesting microalgae at the commercial scale usually involves a flocculant to reduce the time required to separate the medium from the algal cells. Flocculant agents are materials that have the ability to support the so-called bridging phenomena between two molecules, leading to the coagulation process.

Flocculation is a process of aggregating the microalgae cells to promote their separation, beginning with the addition of a material (a flocculant) into the medium, which disturbs the stability of the particles in suspension, including microscopic cells, causing them to aggregate. Flocculants with higher molecular weights are generally more effective. High molecular weight flocculants can adsorb several particles at once, forming a three-dimensional matrix. When this occurs, the aggregated cells become easier to harvest. This is why the most effective flocculants are polymers, either natural or synthetic.

Flocculation is not critical in separating algal cells. However, selecting inefficient or inappropriate flocculants can be costly. Organic flocculant can be obtained naturally or synthetically. Natural organic is based on natural polymers like starch and mucilage while synthetic organic is based on various monomers. The okra mucilage that was developed for drag reducing agent probably could be used as a flocculant (Abdul-Bari et al., 2010). The reason behind this idea is the mucilage use to reduce drag in pipelines where the formation of the so-called as bridge phenomena 'disturb' the flow channel whereas this 'bridge phenomena' could be used to flocculate microalgae cell in media.

Organic flocculant has an advantage over the dosage used to flocculate the particle compared to non-organic flocculant. Chitosan, for instance, has reported capable to harvest microalgae up to 98% while the compatibility dosage was reported varied from 20 to 40 mg/L (Lertsutthiwong et al., 2009; Harith et al., 2009). However, as low as 0.5mg/L was also reported suitable to culture freshwater algae (Divakaran & Pillai, 2002) and up to 200mg/L for salty water, which resulted in separation efficiency of more than 98% (Gualteiri et al., 1988). Chitosan efficiency as a flocculant agent affected by the pH level where pH ranges from 5 to 8 is suitable for the flocculation process. The modified cationic chitosan-polyacrylamide was reported efficient to apply as flocculant at pH 5.5 when the dosage is 8mg/L (Wang et al., 2011). However, chitosan is expensive and its efficiency is reduced in salt-water.

Another natural cationic polymer that commercially available: Greenfloc120, which was made from starch, was reported efficient as a flocculant to harvest freshwater microalgae (Vandamme et al., 2009). In addition, this flocculant is also excellent for wastewater treatment. According to You et al. (2009), the synthesis of these cationic polymers: chitosan and starch can perform much better compared to chitosan and starch alone. However, this work so far was only tested to flocculate the algae for wastewater treatment. Organic flocculant was reported to affect the efficiency of the filtration process. According to Ji et al. (2010), the dosage of the organic flocculant had a significant effect on the mitigation of membrane fouling during the filtration process where they suggested the optimal dosage of chitosan flocculant for sustainable filtration time is equivalent to 150 mg/L.

Another technique that was discussed and still under development is the so-called microbial flocculation. This technique works by adding a minute quantity of microbe culture as low as 1g/L into the microalgae culture that needs to be separated. The selected microbes as flocculant agent must capable to release an extracellular polymeric substance in depletion of nutrients. This microbe was feed with an organic substrate such as crude glycerol, which make this microbe less expensive compared to other flocculant agents. In addition, this technique will not damage the microalgae cell, thus allowing media culture to be reused without further treatment. Usually, this technique was applied to remove the taxonomic group of algae in the reservoir for drinking water, which has as high as 90% removal efficiency.

Combination of Harvesting Techniques

After the flocculation process, the separated algal cells then continue to filtration, centrifugation, floatation or sedimentation before a further drying process. Combining filtration, centrifugation, floatation and

Figure 5. Combination of flocculation, floatation, and filtration

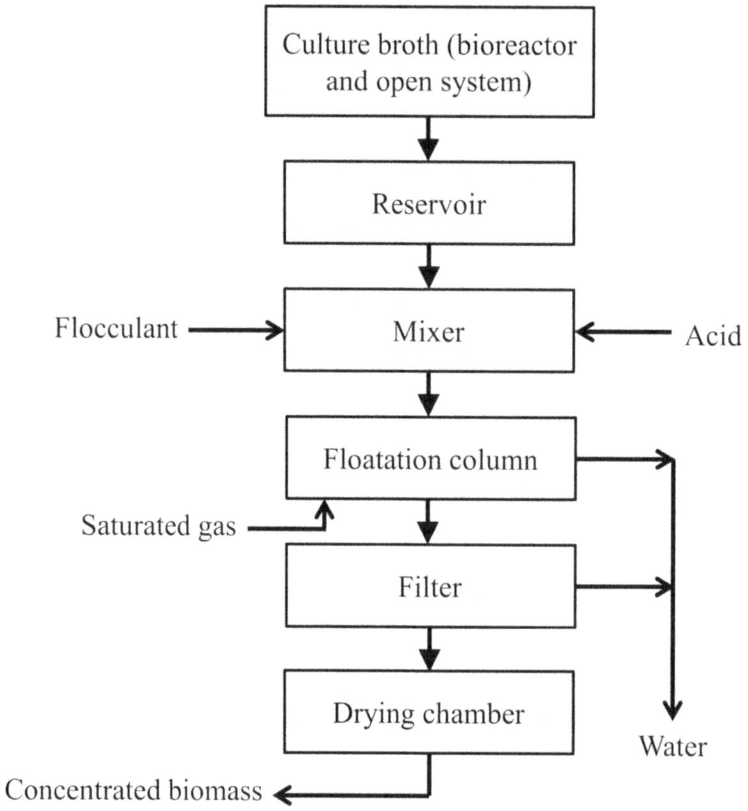

sedimentation can also be applied to harvest microalgae. Figure 5 shows an example of a combination technique for harvesting microalgae. This process is based on harvesting by a combined flotation and filtration technique. In Figure 5, the biomass is first treated with flocculant and acid in mixer reactor before pumped into a flotation column. Here, the flocculated algal biomass is transferred into a filtration unit by an overflow system, and the clarified medium is recycled to the culture units, either open pond or photobioreactor. The harvested biomass is pumped into a drying chamber before advancing to the extraction unit where the algal oil is removed.

Centrifugation is the most preferable method to harvest microalgae for laboratory study. This is because this technique does not require additional chemicals however needs more electrical energy compared to flocculation. In large-scale harvesting process, centrifugation gave good recovery and thickened slurry but the currently available equipment for centrifugation process is too expensive. This is a reason that hinders this technique for commercial purpose. The concentration of biomass could improve centrifugation efficiency. However, to concentrate microalgae with 30% of dry particle needs additional energy which results in additional cost.

The direct filtration process, on the other hand, is a process by which microalgae biomass is harvested directly by using microbe membrane which blocks only algal cell to pass through. This technique seems like the cheapest technique to harvest microalgae. However, this technique requires backwash to maintain the efficiency of the membrane filter and time-consuming.

Microalgae harvesting using an ultrasound technique is under development, on the other hand, is first mentioned in Bosma et al (2003). Here, microalgae cells experience a force that drives them into the planes of pressure nodes when they are exposed to an ultrasonic standing wave. When the field is switched off, the aggregated cells experience rapid gravitational sedimentation because of the gravitational forces. This technique needs further study before can be applied for large-scale especially in open pond where contaminants are high because of its ability not only coagulate microalgae cell but other sediments such as mercury.

MICROALGAE WASTEWATER TREATMENT TECHNOLOGIES

Immobilized Microalgae Wastewater Treatment

Even though microalgae are able to uptake most of the nutrients in the water based on the numbers of researches recorded, the treatment may not be successful to meet the effluent limit such as the chemical oxygen demand (COD) without a proper downstream process, which is the separation of microalgae biomass from treated water. Studies have shown that with the growth of microalgae, the COD reading will increase, in fact, every gram of microalgae will contribute to around 1.25 g of COD. There are several means of microalgae biomass separation technique that are available including floatation, sedimentation, flocculation, filtration and centrifugation. Even though these techniques are established separation processes and has been used for centuries, there could be drawbacks in terms of microalgae separations, especially for wastewater treatment.

The cost of microalgae separation or harvesting either in wastewater treatment or biofuel production can sometimes be the main contributor in the total operational cost. For instance, algae harvest and dewatering contributed up to 30% of the total costs of the whole system in microalgae producing the plant. In a recent report on the techno-economic analysis of autotrophic microalgae for fuel production, it was estimated that biomass harvesting costs will be 21% of the total capital cost of an open pond system (Davis et al., 2011). Therefore, finding a process that can eliminate the need for post cultivation harvesting process can be an advantage, especially in term of operational cost.

Microalgae immobilization has become an option in microalgae utilization, especially for wastewater treatment, owing to the challenge of microalgae separation, especially in a large volume of water. The definition of cell immobilization is by naturally or artificially preventing the cell from moving independently from its original location to all parts of an aqueous phase of a system. Immobilization of microalgae is conducted by attachment of the cell onto materials to eliminate the need for the harvesting of the cell later on. Due to the entrapment of cell, this technique is able to provide several advantages to the treatment operation such as increasing the cell retention time within the bioreactor, making the cell better adapted to the substrate. The risk of the microalgae cell to be washed out is also eliminated and the metabolic activity of microalgae remains constant for a long period, thus it can serve as an alternative process for solving the harvesting problem. Immobilization of microalgae can be conducted either by "passive" or "active" immobilization technique.

Figure 6. Example of an algal turf scrubber with attached growth application

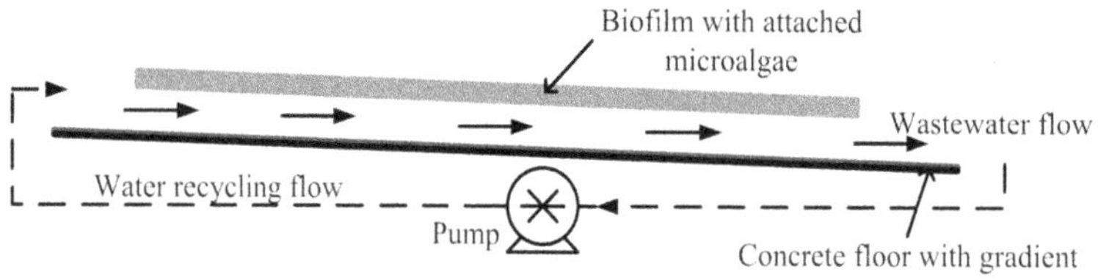

Passive Immobilized Microalgae Wastewater Setups

Similar to most microorganisms, microalgae possess the tendency to naturally be attached to surface and thrive. This tendency can be utilized in order to immobilize microalgae cells on surfaces called carriers. This process is a passive immobilization technique, where the attachment of cells is not facilitated by any chemicals. The immobilizing materials used can be either natural or artificial sources.

Bio-film Microalgae-Reactor with Turf Scrubber

Attachment of microalgae in biofilm has been reported tremendously for wastewater treatment from primary and secondary treated municipal as well as industrial wastewater. Various systems have been developed such as Algal turf scrubber (ATS) as in Figure 6 and Figure 7, a twin- layer photobioreactor, rocker system using polystyrene foam as well as a rotating rope-based reactor and spool harvester for removing nutrients from either the ecosystems or wastewaters.

Bio-film microalgae immobilization technique has become more popular for microalgae cell harvesting in biomass cultivation as well for wastewater treatment. Since its first introduction in 1893 in England by the construction of a trickling filter, the microalgae biofilm has tremendously improved and numerous types of fixed film configuration systems are designed, including rotating biological contactors, submerged bed bioreactors, fluidized biofilm reactors, and combinations of fixed film and rotating algal biofilm (RAB) cultivation. As such, numbers of researches have reported that microalgae immobilized in bio-film are able to uptake the pollutants in wastewaters such as nitrogen, phosphorus and carbon.

Other than its use in wastewater treatment, microalgae immobilization also has been extensively studied for its combination for biofuel production, to solve the difficulty of a harvesting issue. For instance, a flat plate parallels horizontal photobioreactor with glass plate as attachment substrate, the reactor was

Figure 7. Example of biofilm with attached growth of microalgae setup (Top view of the system in Figure 6)

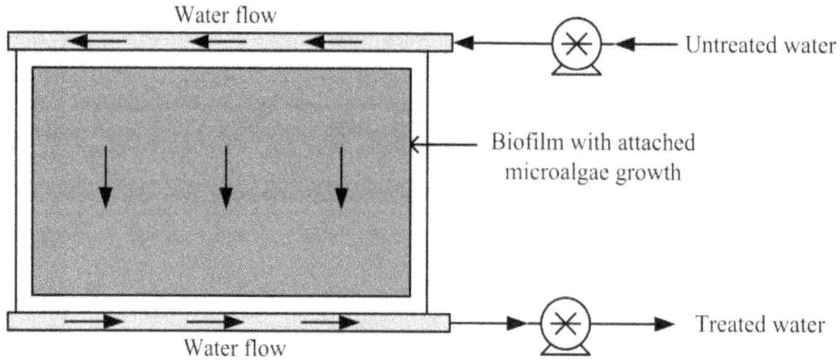

conditioned with municipal wastewater for biofilm formation but switched to synthetic medium for microalgae growth and oil production (Schnurr et al., 2013). A similar concept but using filter paper as attachment material was also reported for growing the various pure strains for algal oil production (Liu et al., 2013; Cheng et al., 2013).

Microalgae biofilm bioreactor application has experimented vigorously on variations of wastewater types such as secondary domestic wastewaters and both digested and raw manure effluents. The results obtained are also promising, for example, uptake rates of nitrate and phosphate from municipal secondary effluents of up to 1.0 g NO-3 m-2 d-1 and 0.34 g P m-2 d-1, respectively, have been reported in an open microalgae biofilm photobioreactor (Boelee et al., 2011). Likewise, up to 83 $\pm$ 25% and 91 $\pm$ 12% of the total nitrogen and phosphorus removed from dairy manure wastewater, respectively, were recovered in the harvested microalgae biomass in an ATS (Mulbry et al., 2008). Posadas et al., (2013) made a solid conclusion that a symbiosis of microalgae and bacteria operation in a biofilm bioreactor is superior to a bacteria bio-film bioreactor. Carbon, nitrogen and phosphorus removals of 91 $\pm$ 3%, 70 $\pm$ 8% and 85 $\pm$ 9%, respectively, were recorded in the algal-bacterial bioreactor at 10 d of hydraulic retention time when treating domestic wastewater.

In the early '90s an ATS system designed by Adey et al., (1993), consisting of a plastic mesh for filamentous algae attachment with intermittent wave surges were reported to have a biomass production of 15–27 g.m-2day-1. Years later, several other studies with this design have shown good nutrient uptake and biomass productivity that typically ranges from 5–20 g.m^{-2}day^{-1}. The filamentous algae grown on the ATS has low fatty acid content, however, reducing its value as a biofuel feedstock.

In the ATS system, microalgae biomass in the form of a filamentous will be formed on a plastic mesh by intermittently passing water over the surface. However, while this form of microalgae biomass has been useful in wastewater treatment and has been utilized in a full-scale treatment system, it may not be as useful for biofuels as other species.

Rotating Drum Reactor System

The rotating biofilm bioreactors are based on the sole biofilm attachment to the photobioreactor in the form of rotating part or attached a fixed part and have shown promising carbon and nutrient removal capacities during the treatment. The microalgae cells will be grown on the surface of a material, and alternation between nutrient-rich liquid and CO_2 rich gaseous phase will occur due to either rotation of the material or wastewater flow through the material. Rotating biological contactors which is common in secondary wastewater treatment can be helped to increase in both carbon dioxide transfer and oxygen release through the use of gas-liquid contactor reactors. Harvesting of the biomass is often conducted by scrapping off the biomass from the attached surface. Figure 8 shows the general setup of the rotating drum reactor (RDR) for microalgae immobilization application.

The RDR is able to simultaneously facilitate algal growth and dewatering while achieving high biomass concentration. Biofilm reactors can also reduce the water and energy requirements for biomass and photo-production compared to traditional suspended culturing strategies. In the early years of microalgae immobilization development, few rotating discs made of different material experimented for the removal of nitrogen and phosphorus in wastewater, for example, aluminium rotating disc and styrofoam rotating disc. More recent studies on microalgae RDR have been focusing on the potential to concurrently remediate wastewater and produce biofuel precursor molecules. This technology can facilitate efficient biomass harvesting via the reported spool harvesting technique.

Figure 8. Example of a microalgae rotating drum reactor system

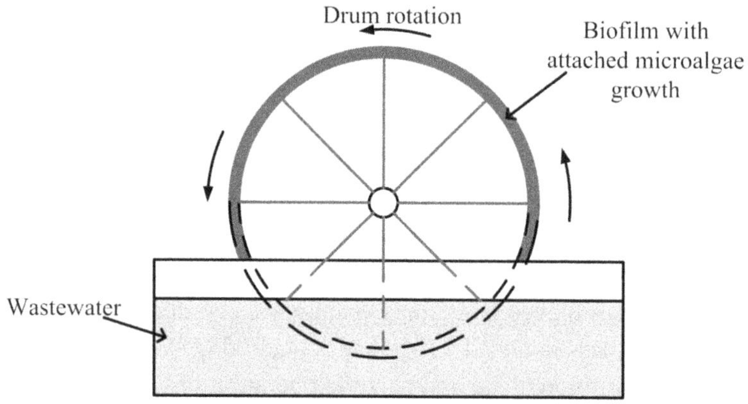

However, optimal biomass harvesting practices need to be determined in the context of biofilm-specific physiology, such as optimal biomass areal density and biofilm thickness as it relates to active photo-production and photosynthesis zones. Reducing photoinhibition and photo limitation is another major design criterion for any photobioreactor, and care must be taken to ensure that cells do not reside too long in either the illuminated or dark zones. The time span of light/dark exposure for each cell has to be tested as it has shown that maximum growth can continue even with a considerable dark period. Light/dark cycling inherent in the rotation of the RDR can make use of the dark period, allowing the cylindrical construction of the rotating drum to support more growth area per illuminated/aperture surface area than flat growth surface designs.

Gross et al., (2013) has developed a RAB system that implemented the attached algal growth concept with the feature of in situ biomass harvest through scrapping. In their system, the algae growing on the surface was able to be harvested by scraping, and thus a stand-alone operation (such as sedimentation or centrifuge) for harvesting biomass from suspended culture systems are avoided.

Active Immobilization by Gel Entrapment of Microalgae

Likewise, the active microalgae immobilization is a technique that involves the application of chemical agents in order to facilitate the attachment of the microalgae cell into the carriers. Such agent may include flocculants, chemical attachment or alginate for encapsulation of cells.

Gel Entrapment Technique

Another widely used immobilization technique for microalgae is natural polysaccharides. These include carrageenan, agar and alginate. However, the highest concern of the natural polysaccharides would be the stability of the materials that could affect the cells storing capability. Some researchers found that ca-alginate could be more stable than carrageenan.

Alginate, which is extracted from brown microalgae, is the most commonly used active immobilization materials for microalgae, specifically sodium alginate. There are a quite number of recorded researches on the application of alginate microalgae immobilization. Quite numbers of benefits can be gained by utilizing sodium alginate in microalgae immobilization, majorly because of its non-toxicity, permeability

Figure 9. Example of alginate encapsulation reactor system

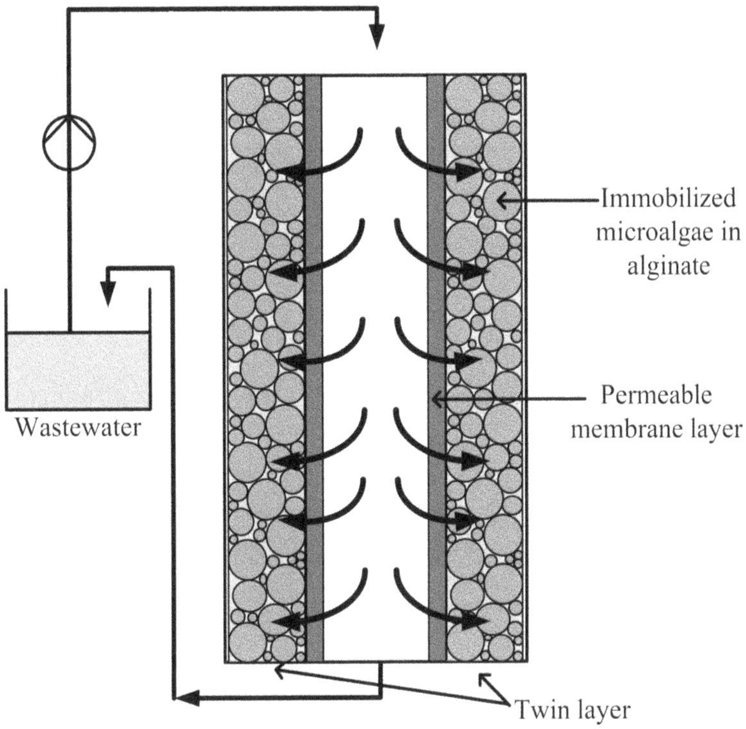

and transparency. This will ensure that the entrapped cells do not suffer from extreme physical-chemical condition changes during the immobilization process. The transparency of small calcium-alginate beads is enough to permit the growth of immobilized microalgae. The formed matrix implies a very gentle environment for immobilized cells. An example of a schematic of alginate encapsulation reactor system to treat wastewater is shown in Figure 9.

The physical properties of alginate solutions at various concentrations are shown in Table 7. As the alginate concentration increased, the solution density increase slightly whereas the solution apparent viscosity at zero shear rate exhibited a typical exponential increment. The surface tension at low alginate concentration (5–20 g/L) was about 68–72 mN/m and it showed a decreasing trend as the concentration increased.

As the alginate concentration increased, the surface tension will decrease, it will cause increase stability of the droplets. However, it will cause a decreased permeability of nutrient and carbon dioxide supply to microalgae. Therefore, 15g/l alginate solution and calcium chloride were used in this experiment. Ohnesorge number (Oh) in a concentration of 15g/l sodium alginate solution is in the range 0.24 – 0.3. Mathematically, Oh can be calculated by using the equation:

$$Oh = \frac{\nu\rho}{\rho d_d \gamma} \tag{1}$$

Table 7. Physical properties of alginate solutions

Alginate Concentration (g/l)	Density (kg/m³)	Viscosity (mPa.s)	Surface Tension (mN/m)	Oh
5	999	38	71	0.075 - 0.093
15	1004	130	70	0.24 - 0.3
25	1008	560	69	1.1 - 1.3
40	1017	2700	57	5.8 - 6.5
50	1023	4700	47	11 - 13

Source: Chan, 2009

where ν is the kinematic viscosity, ρ is the density, d_d is the diameter of a liquid drop, and γ is the liquid surface tension.

The size of the sodium alginate can be predicted using the size prediction model for sodium alginate particle which was developed by taking into account the liquid lost factor (k_{LF}) and the shrinkage factor (k_{SF}) (Chan et al., 2009). Modified Tate's law could be used for size analysis of the alginate drops prediction as described by Equation 2.

$$d_b = k_{LF}k_{SF}\left(0.006d_T\gamma / \rho g\right)^{\frac{1}{3}} \tag{2}$$

where d_b is the diameter of the liquid bead, d_T is the outer diameter tip, and g is the gravitational force (9.81 m/s^2).

A drop was considered spherical if the sphericity factor <0.05. However, the extent of deformation could not be obviously differentiated by human vision. It was found that the minimum alginate concentration required to enable the formation of spherical bead was 15 g/L. The viscosity of alginate solution must be above a certain range around 60–50 mPa s, to form spherical beads.

The minimum and maximum collecting distances are strongly dependent on the viscosity of the solution. Since the solution viscosity increased exponentially with the alginate concentration, Equation 3 and Equation 4 explain the exponential and natural-logarithmic relationship between the collecting distances and Oh.

$$D_{\min} = 1.63e^{0.12Oh} \tag{3}$$

$$D_{\max} = 62.35\ln\left(Oh\right) + 111 \tag{4}$$

where D_{min} is the minimum collecting distance in cm, and D_{max} is the maximum collecting distance in cm.

REFERENCES

Abdul Bari, H. A., Ahmad, M. A., & Yunus, R. B. M. (2010). Formulation of okra-natural mucilage as drag reducing agent in different size of the iron pipes in turbulent water flowing system. *Journal of Applied Sciences (Faisalabad)*, *10*(23), 3105–3110. doi:10.3923/jas.2010.3105.3110

Acien, F. G., González, C. V., Fernández, J. M., García-González, M., Moreno, J., Sierra, E., . . . Molina, E. (2008). Removal of CO2 from flue gases coupled to the photosynthetic generation of biomass and exopolysaccharide by cyanobacteria. In *Proceedings of 11th International Conference on Applied Phycology*. Galway, Ireland. Academic Press.

Adey, W., Luckett, C., & Jensen, K. (1993). Phosphorus removal from natural waters using controlled production. *Restoration Ecology*, *1*(1), 29–39. doi:10.1111/j.1526-100X.1993.tb00006.x

Aravantinou, A. F., Theodorakopoulos, M. A., & Manariotis, I. D. (2013). Selection of microalgae for wastewater treatment and potential lipids production. *Bioresource Technology*, *147*, 130–134. doi:10.1016/j.biortech.2013.08.024 PMID:23994695

Aslan, S., & Kapdan, I. K. (2006). Batch kinetics of nitrogen and phosphorus removal from synthetic wastewater by algae. *Ecological Engineering*, *28*(1), 64–70. doi:10.1016/j.ecoleng.2006.04.003

Boelee, N. C., Temmink, H., Janssen, M., Buisman, C. J., & Wijffels, R. H. (2011). Nitrogen and phosphorus removal from municipal wastewater effluent using microalgal biofilms. *Water Research*, *45*(18), 5923–5933. doi:10.1016/j.watres.2011.08.044 PMID:21940029

Bosma, R., vanSpronsen, W. A., Tramper, J., & Wijffels, R. H. (2003). Ultrasound a new separation technique to harvest microalgae. *Journal of Applied Phycology*, *15*(2/3), 143–153. doi:10.1023/A:1023807011027

Cai, T., Park, S. Y., & Li, Y. (2013). Nutrient recovery from wastewater streams by microalgae: Status and prospects. *Renewable & Sustainable Energy Reviews*, *19*, 360–369. doi:10.1016/j.rser.2012.11.030

Caldwell, D. H. (1946). Sewage oxidation ponds - performance, operation and design. *Sewage Works Journal*, *18*, 433–458.

Chan, E. S., Lee, B. B., Ravindra, R., & Poncelet, D. (2009). Prediction models for shape and size of ca-alginate macrobeads produced through extrusion-dripping method. *Journal of Colloid and Interface Science*, *338*(1), 63–72. doi:10.1016/j.jcis.2009.05.027 PMID:19604515

Cheng, P. F., Ji, B., Gao, L. L., Zhang, W., Wang, J. F., & Liu, T. Z. (2013). The growth, lipid and hydrocarbon production of *Botryococcus braunii* with attached cultivation. *Bioresource Technology*, *138*, 95–100. doi:10.1016/j.biortech.2013.03.150 PMID:23612166

Chinnasamy, S., Bhatnagar, A., Hunt, R. W., & Das, K. C. (2010). Microalgae cultivation in a wastewater dominated by carpet mill effluents for biofuel applications. *Bioresource Technology*, *101*(9), 3097–3105. doi:10.1016/j.biortech.2009.12.026 PMID:20053551

Cho, S., Lee, N., Park, S., Yu, J., Luong, T. T., Oh, Y. K., & Lee, T. (2013). Microalgae cultivation for bioenergy production using wastewaters from a municipal WWTP as nutritional sources. *Bioresource Technology*, *131*, 515–520. doi:10.1016/j.biortech.2012.12.176 PMID:23453233

Davis, R., Aden, A., & Pienkos, P. (2011). Techno-economic analysis of autotrophic microalgae for fuel production. *Applied Energy*, *88*(10), 3524–3531. doi:10.1016/j.apenergy.2011.04.018

Divakaran, R., & Pillai, V. N. S. (2002). Flocculation of river silt using chitosan. *Water Research*, *36*(9), 2414–2418. doi:10.1016/S0043-1354(01)00436-5 PMID:12108734

Gonzalez, L. E., Canizares, R. O., & Baena, S. (1997). Efficiency of ammonia and phosphorus removal from a Colombian agroindustrial wastewater by the microalgae *Chlorella vulgaris* and *Scnedesmus dimorphus*. *Bioresource Technology*, *60*(3), 259–265. doi:10.1016/S0960-8524(97)00029-1

Gonzalez-Fernandez, C., Molinuevo-Salces, B., & Garcia-Gonzallez, M. C. (2011). Nitrogen transformations under different conditions in open ponds by means of microalgae-bacteria consortium treating pig slurry. *Bioresource Technology*, *102*(2), 960–966. doi:10.1016/j.biortech.2010.09.052 PMID:20943377

Gross, M., Henry, W., Michael, C., & Wen, Z. (2013). Development of rotating algal biofilm growth system for attached microalgae with in situ biomass harvest. *Bioresource Technology*, *150*, 195–201. doi:10.1016/j.biortech.2013.10.016 PMID:24161650

Gualteiri, P., Barsanti, L., & Passarelli, V. (1988). Chitosan as flocculant for concentrating *Euglena gracilis* cultures. *Annales de l'Institut Pasteur. Microbiology*, *139*(6), 717–726. doi:10.1016/0769-2609(88)90076-2 PMID:3150942

Harith, Z. T., Yusoff, F. M., Mohamed, M. S., Din, M. S. M., & Ariff, A. B. (2009). Effect of different flocculants on the flocculation performance of microalgae, *Chaetoceros calcitrans*, cells. *African Journal of Biotechnology*, *8*(21), 5971–5978. doi:10.5897/AJB09.569

Hongyang, S., Yalei, Z., Chunmin, Z., Xuefei, Z., & Jinpeng, L. (2011). Cultivation of *Chlorella pyrenoidosa* in soybean processing wastewater. *Bioresource Technology*, *102*(21), 9884–9890. doi:10.1016/j.biortech.2011.08.016 PMID:21911289

Hultberg, M., Carlsson, A. S., & Gustafsson, S. (2013). Treatment of drainage solution from hydroponic greenhouse production with microalgae. *Bioresource Technology*, *136*, 401–406. doi:10.1016/j.biortech.2013.03.019 PMID:23567708

Jacob-Lopes, E., Scoparo, C. H. G., Lacerda, L. M. C. F., & Franco, T. T. (2009). Effect of light cycles (night/day) on CO_2 fixation and biomass production by microalgae in photobioreactors. *Chemical Engineering and Processing*, *48*(1), 306–310. doi:10.1016/j.cep.2008.04.007

Ji, J., Qiu, J., Wai, N., Wong, F. S., & Li, Y. (2010). Influence of organic and inorganic flocculants on physical-chemical properties of biomass and membrane-fouling rate. *Water Research*, *44*(5), 1627–1635. doi:10.1016/j.watres.2009.11.013 PMID:19954810

Jiang, L., Luo, S., Fan, X., Yang, Z., & Guo, R. (2011). Biomass and lipid production of marine microalgae using municipal wastewater and high concentration of CO_2. *Applied Energy*, *88*(10), 3336–3341. doi:10.1016/j.apenergy.2011.03.043

Jin, H. F., Lim, B. R., & Lee, K. (2006). Influence of nitrate feeding on carbon dioxide fixation by microalgae. *Journal of Environmental Science and Health. Part A*, *41*, 2813–2824.

Kassim, T. I. (2002). Possible use of microgreen algae to remove phosphate and nitrate from wastewater. In *Proceedings of International Symposium on Environmental Pollution Control and Waste Management* (pp. 628–632). Tunis, Tunesia. Academic Press.

Kim, J., Lingaraju, B. P., Rheaume, R., Lee, J. Y., & Siddiqui, K. F. (2010). Removal of Ammonia from Wastewater Effluent by *Chlorella vulgaris. Tsinghua Science and Technology*, *15*(4), 391–396. doi:10.1016/S1007-0214(10)70078-X

Lam, M. K., & Lee, K. T. (2012). Potential of using organic fertilizer to cultivate *Chlorella vulgaris* for biodiesel production. *Applied Energy*, *94*, 303–308. doi:10.1016/j.apenergy.2012.01.075

Lee, J. S., & Lee, J. P. (2003). Review of advances in biological CO2 mitigation technology. *Biotechnology and Bioprocess Engineering; BBE*, *8*(6), 354–359. doi:10.1007/BF02949279

Lertsutthiwong, P., & Rojsitthisak, P. (2011). Chitosan-alginate nanocapsules for encapsulation of turmeric oil. *Die Pharmazie*, *66*(12), 911–915. PMID:22312692

Li, S., Luo, S., & Guo, R. (2013). Efficiency of CO_2 fixation by microalgae in a closed raceway pond. *Bioresource Technology*, *136*, 267–272. doi:10.1016/j.biortech.2013.03.025 PMID:23567690

Lim, S. L., Chu, W. L., & Phang, S. W. (2010). Use of *Chlorella vulgaris* for bioremediation of textile wastewater. *Bioresource Technology*, *101*(19), 7314–7322. doi:10.1016/j.biortech.2010.04.092 PMID:20547057

Liu, T., Wang, J., Hu, Q., Cheng, P., Ji, B., Liu, J., ... Wang, H. (2013). Attached cultivation technology of microalgae for efficient biomass feedstock production. *Bioresource Technology*, *127*, 216–222. doi:10.1016/j.biortech.2012.09.100 PMID:23131644

Mata, T. M., Melo, A. C., Manuel, S., & Nídia, S. C. (2012). Parametric study of a brewery effluent treatment by microalgae *Scenedesmus obliquus. Bioresource Technology*, *107*, 151–158. doi:10.1016/j.biortech.2011.12.109 PMID:22244957

McGinn, P. J., Dickinson, K. E., Park, K. C., Whitney, C. G., MacQuarrie, S. P., Black, F. J., ... O'Leary, S. J. B. (2012). Assessment of the bioenergy and bioremediation potentials of the microalga *Scenedesmus sp.* AMDD cultivated in municipal wastewater effluent in batch and continuous mode. *Algal Research*, *1*(2), 155–165. doi:10.1016/j.algal.2012.05.001

Mulbry, W., Kondrad, S., Pizarro, C., & Kebede-Westhead, E. (2008). Treatment of dairy manure effluent using freshwater algae: Algal productivity and recovery of manure nutrients using pilot-scale algal turf scrubbers. *Bioresource Technology*, *99*(17), 8137–8142. doi:10.1016/j.biortech.2008.03.073 PMID:18487042

Narasimhan, A. M. (2010). *Microalgal bioremediation of nutrients in wastewater and carbon dioxide in flue gas* (Master dissertation). Missouri University of Science and Technology Database.

Ogbonna, J. C., Yoshizawa, H., & Tanak, H. (2000). Treatment of high strength organic wastewater by a mixed culture of photosynthetic microorganisms. *Journal of Applied Phycology*, *12*(3/5), 277–284. doi:10.1023/A:1008188311681

Posadas, E., Garcia-Encina, P. A., Soltau, A., Domínguez, A., Diaz, I., & Munoz, R. (2013). Carbon and nutrient removal from centrates and domestic wastewater using algal–bacterial biofilm bioreactors. *Bioresource Technology*, *139*, 50–58. doi:10.1016/j.biortech.2013.04.008 PMID:23644070

Rasoul-Amini, S., Montazeri-Najafabady, S., Shaker, S., Safari, A., Kazemi, A., Mousavi, P., ... Ghasemi, Y. (2013). Removal of nitrogen and phosphorus from wastewater using microalgae free cells in bath culture system. *Biocatalysis and Agricultural Biotechnology*, *3*(2), 126–131. doi:10.1016/j.bcab.2013.09.003

Rawat, I., Kumar, R. R., Mutanda, T., & Bux, F. (2011). Dual role of microalgae: Phycoremediation of domestic wastewater and biomass production for sustainable biofuels production. *Applied Energy*, *88*(10), 3411–3424. doi:10.1016/j.apenergy.2010.11.025

Riano, B., Molinuevo, B., & García-González, M. C. (2011). Treatment of fish processing wastewater with microalgae-containing microbiota. *Bioresource Technology*, *102*(23), 10829–10833. doi:10.1016/j.biortech.2011.09.022 PMID:21983413

Richmond, A. (2004). *Handbook of Microalgae Culturing* (1st ed.). UK: Blackwell Science Ltd.

Ruiz-Martinez, A., Garcia, N. M., Romero, I., Seco, A., & Ferrer, J. (2012). Microalgae cultivation in wastewater: Nutrient removal from anaerobic membrane bioreactor effluent. *Bioresource Technology*, *126*, 247–253. doi:10.1016/j.biortech.2012.09.022 PMID:23073115

Sandnes, J. M., Kallqvist, T., Wenner, D., & Gislerød, H. R. (2005). Combined influence of light and temperature on growth rates of *Nannochloropsis oceanica*: Linking cellular responses to large-scale biomass production. *Journal of Applied Phycology*, *17*(6), 515–525. doi:10.100710811-005-9002-x

Schnurr, P., Espie, G., & Allen, G. (2013). Algae biofilm growth and the potential to stimulate lipid accumulation through nutrient starvation. *Bioresource Technology*, *136*, 337–344. doi:10.1016/j.biortech.2013.03.036 PMID:23567700

Silva-Benavides, A. M., & Torzillo, G. (2012). Nitrogen and phosphorus removal through laboratory batch cultures of microalgae *Chlorella vulgaris* and cyanobacterium *Planktothrix isothrix* grown as a mono-algal and as co-cultures. *Journal of Applied Phycology*, *24*(2), 267–276. doi:10.100710811-011-9675-2

Singh, A., Nigam, P. S., & Murphy, J. D. (2011). Renewable fuels from algae: An answer to debatable land based fuels. *Bioresource Technology*, *102*(1), 10–16. doi:10.1016/j.biortech.2010.06.032 PMID:20615690

Singh, G., & Thomas, P. B. (2012). Nutrient removal from membrane bioreactor permeate using microalgae and in a microalgae membrane photoreactor. *Bioresource Technology*, *117*, 80–85. doi:10.1016/j.biortech.2012.03.125 PMID:22609717

Sutherland, D. L., Turnbull, M. H., & Craggs, R. J. (2014). Increased pond depth improves algal productivity and nutrient removal in wastewater treatment high rate algal ponds. *Water Research*, *53*, 271–281. doi:10.1016/j.watres.2014.01.025 PMID:24530547

Tablot, P., & De La Noue, J. (1993). Tertiary treatment of wastewater with *Phormidium bohneri* (Schmidle) under various light and temperature conditions. *Water Research*, *27*(1), 153–159. doi:10.1016/0043-1354(93)90206-W

Vandamme, D., Foubert, I., Meesschaert, B., & Muylaert, K. (2009). Flocculation of microalgae using cationic starch. *Journal of Applied Phycology*, *22*(4), 525–530. doi:10.100710811-009-9488-8

Wang, B., Zhang, Y., & Miao, C. (2011). Preparation of cationic chitosan–polyacrylamide flocculant and its properties in wastewater treatment. [Oceanic and Coastal Sea Research]. *Journal of Ocean University of China*, *10*(1), 42–46. doi:10.100711802-011-1741-5

Wang, L., Min, M., Li, Y., Chen, P., Chen, Y., Liu, Y., ... Ruan, R. (2009). Cultivation of green algae *Chlorella sp.* in different wastewaters from municipal wastewater treatment plant. *Applied Biochemistry and Biotechnology*, *162*(4), 1174–1186. doi:10.100712010-009-8866-7 PMID:19937154

You, L., Lu, F., Li, D., Qiao, Z., & Yin, Y. (2009). Preparation and flocculation properties of cationic starch/chitosan crosslinking-copolymer. *Journal of Hazardous Materials*, *172*(1), 38–45. doi:10.1016/j.jhazmat.2009.06.120 PMID:19631460

Yun, Y. S., Lee, S. B., Park, J. M., Lee, C. I., & Yang, J. W. (1997). Carbon dioxide fixation by algal cultivation using wastewater nutrients. *Journal of Chemical Technology and Biotechnology (Oxford, Oxfordshire)*, *69*(4), 451–455. doi:10.1002/(SICI)1097-4660(199708)69:4<451::AID-JCTB733>3.0.CO;2-M

Zhen-Feng, S., Xin, L., Hong-Ying, H., Yin-Hu, W., & Tsutomu, N. (2011). Culture of *Scenedesmus sp.* LX1 in the modified effluent of a wastewater treatment plant of an electric factory by photo-membrane bioreactor. *Bioresource Technology*, *102*(17), 7627–7632. doi:10.1016/j.biortech.2011.05.009 PMID:21684736

ADDITIONAL READING

Abdel-Raouf, N., Al-Homaidan, A. A., & Ibraheem, I. B. M. (2012). Microalgae and wastewater treatment. *Saudi Journal of Biological Sciences*, *19*(3), 257–275. doi:10.1016/j.sjbs.2012.04.005 PMID:24936135

Goncalves, A. L., Pires, J. C. M., & Simoes, M. (2017). A review on the use of microalgal consortia for wastewater treatment. *Algal Research, 24*(Part B), 403–415.

Hwang, J. H., Church, J., Lee, S. J., Park, J., & Lee, W. H. (2016). Use of microalgae for advanced wastewater treatment and sustainable bioenergy generation. *Environmental Engineering Science*, *33*(11), 882–897. doi:10.1089/ees.2016.0132

Larronde-Larretche, M. (2018). Development of a novel membrane bioreactor for cost-effective wastewater treatment and microalgae harvesting (Doctoral dissertation). Retrieved from http://theses.gla.ac.uk/30805/

Lavrinovics, A., & Juhna, T. (2018). Review on challenges and limitations for algae-based wastewater treatment. *Construction Science*, *20*(1), 17–25.

Luo, Y., Le-Clech, P., & Henderson, R. K. (2017). Simultaneous microalgae cultivation and wastewater treatment in submerged membrane photobioreactor: a review. *Algal Research, 24*(Part B), 425–437.

Marbelia, L., Bilad, M. R., Passaris, I., Discart, V., Vandamme, D., Beuckels, A., ... Vankelecom, I. F. (2014). Membrane photobioreactors for integrated microalgae cultivation and nutrient remediation of membrane bioreactors effluent. *Bioresource Technology*, *163*, 228–235. doi:10.1016/j.biortech.2014.04.012 PMID:24814249

Paches, M., Martinez-Guijarro, R., Gonzalez-Camejo, J., Seco, A., & Barat, R. (2018). Selecting the most suitable microalgae species to treat effluent from an anaerobic membrane bioreactor. *Environmental Technology*, *2018*, 1–10. PMID:29963975

Praveen, P., Heng, J. Y. P., & Loh, K. C. (2016). Tertiary wastewater treatment in membrane photobioreactor using microalgae: Comparison of forward osmosis & microfiltration. *Bioresource Technology*, *222*, 448–457. doi:10.1016/j.biortech.2016.09.124 PMID:27756022

Shchegolkova, N., Shurshin, K., Pogosyan, S., Voronova, E., Matorin, D., & Karyakin, D. (2018). Microalgae cultivation for wastewater treatment and biogas production at Moscow wastewater treatment plant. *Water Science and Technology*, *78*(1), 69–80. PMID:30101790

Wang, Y., Ho, S. H., Cheng, C. L., Guo, W. Q., Nagarajan, D., Ren, N. Q., ... Chang, J. S. (2016). Perspectives on the feasibility of using microalgae for industrial wastewater treatment. *Bioresource Technology*, *222*, 485–497. doi:10.1016/j.biortech.2016.09.106 PMID:27765375

KEY TERMS AND DEFINITIONS

Microalgae: Photosynthetic microscopic-sized algae mostly found in freshwater and marine system which requires only a few basic necessities to thrive, including CO_2 supply, light and nutrients.

Microalgae Biomass: Microalgae material that has high values as a renewable energy resource.

Microalgae Cultivation: Process of growing microalgae to produce biofuels and bioproducts.

Microalgae Harvesting: Process of collecting microalgae that have been grown and cultivated.

Microalgae Immobilization: Process by naturally or artificially preventing the microalgae from moving independently from its original location to all parts of an aqueous phase of a system by attachment of the microalgae onto materials to eliminate the need for the harvesting of the microalgae later on.

Microalgae Membrane Bioreactor: Membrane technology that combines the photobioreactor and membrane bioreactor.

Nitrogen and Phosphorus: Two major nutrients needed by microalgae to conduct photosynthesis.

N:P Ratio: Ratio of nitrogen to phosphorus present in a medium.

Photobioreactor: Bioreactor technology that is used to cultivate microalgae through photosynthesis using a light source, nutrients, and CO_2.

Appendix

SYMBOLS

a: Water activity
a_1: Water activity of feed solution
a_2: Water activity of stripping solution
a_i : Activity of component i
A : Permeability constant containing structural factors
A_w : Water permeability
A_m : Membrane surface area
A_S : Internal surface area of packed bed porous media
atm: Atmospheric pressure
B : Pore size morphology constant
c_F : Concentration factor
C_{mlss} : Concentration of mixed liquor suspended solids
C_p : Heat capacity
cm: Centimeter
C : Concentration
C_0: Initial concentration of the target component in the retentate
C_f : Final concentration of the target component in the retentate
$C_{alum,m}$: Concentration of alum 'mother solution'
$C_{alum,opt}$: Concentration of optimum alum dosage
$C_{i_{downstream}}$: Concentration of the component i on the downstream side of membrane
$C_{i_{feed}}$: Concentration of the component i in feed stream
$C_{i_{permeate}}$: Concentration of the component i in permeate stream
$C_{i_{retentate}}$: Concentration of the component i in retentate stream
$C_{i_{upstream}}$: Concentration of the component i on the upstream side of membrane
C_b : Concentration of solute in bulk
C_m : Concentration of solute near the membrane/at the membrane interphase

$C_{m,f}$: Concentration of feed at the membrane interface
$C_{m,p}$: Concentration of permeate at the membrane interface
C_A : Concentration of component A
C_B : Concentration of component B
C_F : Concentration of feed
C_P : Concentration of permeate
C_R : Concentration of retentate
$C_{A_{feed}}$: Concentration of compound A in the feed
$C_{A_{permeate}}$: Concentration of compound A in the permeate
$C_{B_{feed}}$: Concentration of compound B in the feed
$C_{B_{permeate}}$: Concentration of compound B in the permeate
C_{NaCl} : Concentration of sodium chloride (salt), NaCl
C_W : Concentration of water
CC : Annual capita cost
CCI : Construction cost index
d: Day
d_b : Diameter of liquid bead
d_d : Diameter of liquid drop
d_h : Hydraulic diameter of membrane pores
d_p : Mean pore size, diameter
$d_{p,max}$: Maximum pore size, diameter
d_T : Outer diameter tip
D : Diffusivity/Diffusion coefficient
D_i : Diffusivity of component *i*
D_{max} : Maximum collecting distance
D_{min} : Minimum collecting distance
$D_{v/a}$: Diffusion coefficient of water vapour in air
DHR : Direct hourly wage rate
DSL : Diesel fuel price
ft: Feet
F_c : Tube-row correction factor
F_{dry} : Gas permeation velocity through a dried membrane
F_{wet} : Gas permeation velocity through a wetted membrane
FAU: Forced air unit
g: Gram
g : Gravitational force (9.81 m/s^2)
G: Energy gradient per second

G : Gibb's free energy
Gz : Graetz number
h: Hour
ha: Hectare
h : Heat transfer coefficient
h_c : Heat transfer coefficient of condensate
h_{cf} : Heat transfer coefficient of cooling fluid
h_f : Heat transfer coefficient of feed
h_m : Heat transfer coefficient of membrane
h_p : Heat transfer coefficient of permeate
H: Height of rectangular pipe
$I(q)$: Scattering intensity
J : Flux
J_1 : Flux of non-solvent
J_2 : Flux of solvent
J_{20} : Flux at 20 °C
J_i : Flux of component i
J_{Kn} : Flux due to Knudsen diffusion
J_{mol} : Flux due to molecular diffusion
J_{vis} : Flux due to viscous flow
J_{NaCl} : Flux of sodium chloride (solute)
J_W : Flux of water (solvent)
J_T : Flux at operating temperature, T
J_T : Total flux
k : Thermal conductivity
k_B : Boltzmann constant
k_g : Thermal conductivity of air/gas
k_m : Thermal conductivity of membrane
k_{mm} : Thermal conductivity of the membrane material
k_p : Thermal conductivity of cooling plate
k_i : Incident beam
k_f : Scattered beam
k_{LF} : Liquid lost factor
k_{SF} : Shrinkage factor
kDa: Kilo Dalton
kg: Kilogram
km: Kilometer
kmol: Kilo mole

kPa: Kilo Pascal
kV: Kilo Volt
kWh: Kilo Watt-hour
K : Carman-Kozeny constant
K : Mass transfer coefficient
K_{Kn} : Mass transfer coefficient for Knudsen diffusion
K_{mol} : Mass transfer coefficient for molecular diffusion
K_{vis} : Mass transfer coefficient in viscous flow
K_{Kn-mol} : Mass transfer coefficient in Knudsen-molecular transition mechanism
K_{Kn-vis} : Mass transfer coefficient in Knudsen-viscous transition mechanism
$K_{Kn-mol-vis}$: Mass transfer coefficient in Knudsen-viscous-molecular mechanism
Kn : Knudsen number
L: Litre
L/c/d: Litres per capita per day
L : Length of channel
L_i : Proportionality coefficient
LMH: Litre per square meter per hour
m: Position coordinate in the polymer-fixed frame of reference
m (unit): Meter
m_d : Weight of the dried membrane
m_i : Mass of component *i*
m_w : Weight of the wetted membrane
mg: Milligram
min: Minute
mL: Millilitre
mm: Millimeter
mN: Milli Newton
mol: Mole
mPA: Milli Pascal
M: Support position
M : Molecular concentration
MI : Miles
MLD: Million litres per day
MPa: Mega Pascal
MW : Molecular weight
MW_i : Molecular weight of component *i*
nm: Nanometer
N: Newton
NGS : Natural gas cost
NTU: Nephelometric turbidity units
n : Number of ions dissociation

n_A : Number of mole of component A
n_B : Number of mole of component B
$n_{i_{in}}$: Amount of substance of the component *i* in the entering stream
$n_{i_{out}}$: Amount of substance of the component *i* of the useful product
n_s : Number of mole of solvent
n_{sol} : Number of mole of solute
n_v : Vapour viscosity of the permeation component
N_i : Number of molecules of component i
N_j : Number of molecules of component j
N_E : Number of membrane element
$N_{E/V}$: Number of element per pressure vessel
N_V : Number of pressure vessel
$N_V(i)$: Number of pressure vessel in *i*th stage
$N_V(i+1)$: Number of pressure vessel in (*i*+1)th stage
$N_V(1)$: Number of pressure vessels in stage 1
Nu : Nusselt number
OC : Operating and maintenance cost
Oh: Ohnesorge number
pe: Population equivalent
ppm: Parts per million (fractional units of concentration, equal to mg/L)
ppt: Parts per trillion
Pa: Pascal
PPI : Producers price index
PR : Power cost
p_i : Vapour pressure of solution i
p_i° : Vapour pressure of pure solvent i
$p_{m,f}$: Vapour pressure at the membrane surface on feed side
$p_{m,p}$: Vapour pressure at the membrane surface on the permeate side
P : Pressure
$\underline{P}_a$: Average air pressure within the membrane pores
P_F : Pressure of feed
P_m : Average pressure within the membrane pores
P_P : Pressure of permeate
P_R : Pressure of retentate
P_T : Total pressure inside the membrane pores
Pr : Prandtl number
Pr_w : Prandtl number evaluated at tube-wall temperature

P_i: Permeability of component i
P_i: Permeance of component i
P_{NaCl}: Permeance of sodium chloride (solute)
P_W: Permeance of water (solvent)
$P(q)$: Form factor
q: Scattering vector
Q: Heat flux
Q_{cf}: Heat transfer at the cooling fluid side
Q_f: Heat flux through the feed boundary layer
Q_m: Heat flux through the membrane
Q_p: Heat flux through the permeate boundary layer
$Q_{p,ag}$: Heat transfer from membrane to condensate layer across the air gap
$Q_{p,c}$: Heat transfer from condensate layer to the cooling plate
$Q_{p,p}$: Heat transfer from cooling plate to the cooling fluid
Q_{alum}: Flow rate of optimum alum dosage
$Q_{w,raw}$: Flow rate of raw water
Q_{inf}: Influent flow rate
Q_F: Flow rate of feed
Q_P: Flow rate of permeate
Q_R: Flow rate of retentate
$Q_{W,permeate}$: Flow rate of water (solvent) permeate
rpm: Rotation per minute
r: Pore radius
r_{max}: Maximum pore size, radius
r_i: Retention factor
R: Universal gas constant
R_i: Rejection factor of component i
R_1: Resistance in phase 1
R_2: Resistance in phase 2
R_M: Membrane resistance
$R_{M,T}$: Total membrane resistance
R_s: Staging ratio
R_v: Removal efficiency
Re: Reynolds number
s: Second
S_i: Solubility of component i
S_c: Separation coefficient

S_F : Separation factor
$S(q)$: Structure factor
Sc : Schmidt number
Sh : Sherwood number
TCU: True colour units
t : Time
T : Temperature
$T_{b,f}$: Bulk temperature of the feed
$T_{b,p}$: Bulk temperature of the permeate
$T_{c,p}$: Temperature of condensate at the permeate side
T_f : Temperature of feed
T_m : Temperature of membrane
$T_{m,f}$: Temperature of membrane surface on the feed boundary layer
$T_{m,p}$: Temperature of membrane surface on the permeate boundary layer
T_p : Temperature of permeate
$T_{p,cf}$: Temperature of cooling plate at the cooling fluid side
$T_{p,p}$: Temperature of cooling plate at the permeate side
TCF : Temperature correction factor
TDH : Total dynamic head
$UNTS$: Number of process units
$USRT$: Design parameter
$V_{t,AS}$: Volume of activated sludge tank
v : Velocity
V : Volume
$\underline{V}_i$: Molar volume of component i
$\underline{V}_s$: Molar volume of solvent
W: Width of rectangular pipe
x: Spatial position coordinate normal to the membrane surface
x_i : Mole fraction of component i
x_j : Mole fraction of component j
x_s : Mole fraction of solvent
x_{sol} : Mole fraction of solute
x_{NaCl} : Mole fraction of sodium chloride (salt), NaCl
x_W : Mole fraction of water
X: Position of the interface between the film and the coagulation bath
X_{BOD_5} : Concentration of particulate BOD_5
y: Position coordinate that moves with the interface
Y : Fraction of system recovery

Z : Space coordinate measured normal to the membrane
Å: Angstrom
α : Membrane material constant for Equation 2-45
β : Membrane material constant for Equation 2-45
ΔH_v : Latent heat of vaporization of the feed
ΔM : Molecular concentration difference between feed and permeate
Δp^2 : Contrast factor
ΔP : Pressure drop
ΔP : Transmembrane pressure
ΔP_{T1} : Transmembrane pressure at temperature 1, $T1$
ΔP_{T2} : Transmembrane pressure at temperature 2, $T2$
$\Delta \pi$: Osmotic pressure difference
δ : Membrane thickness
δ_g : Thickness of air gap
δ_p : Thickness of cooling plate
ε: Membrane porosity
γ : Surface tension
γ_i : Activity coefficient
λ : Mean free path of water vapour
µm: Micrometer
µg: Micro gram
µS: Micro Siemens
μ : Viscosity
μ_T : Viscosity at operating temperature, T
μ_{T1} : Viscosity at temperature 1, $T1$
μ_{T2} : Viscosity at temperature 2, $T2$
μ_{20} : Viscosity at 20 °C
μ_i : Chemical potential of component i
μ_s : Chemical potential of solvent
μ_i° : Constant for the chemical potential of the pure compound i
ν : Kinematic viscosity
π : Osmotic pressure
ρ : Density
ρ_m : Density of membrane
ρ_p : Density of polymer
ρ_w : Density of the wetted membrane
σ_i : Collision diameter
τ : Membrane tortuosity
φ : Knudsen diffusion factor

η : Relative recovery
η_i : Relative recovery of component i
Λ : Wavelength of incident beam
$\varnothing$: Density of particles in volume
θ : Contact angle
2θ : Scattering angle
$|q|$: Magnitude of scattering vector
wt%: Weight percent
% (v/v): Percentage volume per volume
°C: Degree Celsius

ABBREVIATIONS

$Al_2(SO_4)_3$: Aluminium sulfate (alum)
CO_2: Carbon dioxide
$Fe(OH)_3$: Ferric hydroxide
H_2O: Water
O_2: Oxygen
2,4,5-T: 2,4,5-trichlorophenoxyacetic acid
2,4-D: 2,4-dichlorophenoxyacetic acid
2,4-DB: 2,4-dichlorophenoxybutyric acid
ABS: Acrylonitrile butadiene styrene
AC: Asbestos cement
AD: Anaerobic digestion
AFM: Atomic force microscopy
AJPH: American Journal of Public Health
Al: Aluminium
$Al(OH)_3$: Aluminium hydroxide
$Al_2(SO_4)_3$: Aluminium sulfate (Alum)
Al_2O_3: Aluminium oxide
AOX: Adsorbable organic halides/ organically bound halogens:
As: Arsenic
ASP: Alkaline-surfactant-polymer
ATS: Algal turf scrubber
Au: Gold
BAT: Best available technology
BOD: Biological Oxygen Demand
BOD/N: Biological Oxygen Demand to Nitrogen ratio
BOD_3: 3-day Biological Oxygen Demand
BOD_5: 5-day Biological Oxygen Demand
BTX: Benzene, toluene, xylene
C/N: Carbon-to-nitrogen ratio

CA: Cellulose acetate
$Ca(OH)_2$: Calcium hydroxide (Lime)
Ca^{2+}: Calcium ion
$CaSO_4$: Calcium sulfate
CFU: Colony-forming unit
CH_4: Methane gas
CI: Cast iron
Cl^-: Chloride ion
Cl_2: Chlorine gas
ClO^-: Hypochlorite ion
CN: Cyanide
CO_3^{2-}: Carbonate ion
COD: Chemical Oxygen Demand
COD_{Cr}: Chemical Oxygen Demand (determined by the dichromate method)
CPO: Crude palm oil
CR: Dewatering facilities
CSTR: Continuous flow stirred tank reactor
Cu: Copper
DBKK: *Dewan Bandaraya Kota Kinabalu*
DDT: Dichlorodiphenyltrichloroethane
DI: Ductile iron
DO: Dissolved oxygen
DoE: Department of Environment
EA: Extended aeration
EGSB: Expanded granular sludge bed
EIA: Environmental Impact Assessment
EOR: Enhanced oil recovery
EPS: Extracellular polymeric substances
ESEM: Environmental scanning electron microscopy
F: Fluoride
F/M: Food to microorganism ratio
Fe: Iron
$Fe(HCO_3)_2$: Ferrous bicarbonate
FESEM: Field emission scanning electron microscopy
FFB: Fresh fruit bunch
FTIR: Fourier transform infrared
H^+: Hydrogen ion
H_2: Hydrogen gas
H_2S: Hydrogen sulfide gas
HCl: Hydrochloric acid
HClO: Hypochlorous acid
HCO_3^-: Bicarbonate ion
HDPE: High-density Polyethylene

HPLC: High-performance liquid chromatography
HRAP: High-rate algal pond
HRT: Hydraulic retention time
HTI: Hydration Technologies Inc.
ID/OD: Inner diameter over outer diameter
IDEA: Intermittently decanted extended aeration
IMWQSM: Interim Marine Water Quality Standard for Malaysia:
INWQS: Interim National Water Quality Standard
IPA: Isopropylalcohol
IR: Infrared
IST: Individual septic tank
IUPAC: International Union of Pure and Applied Chemistry
JKR: *Jabatan Kerja Raya*
K: Potassium
K^+: Potassium ion
$K_2Cr_2O_7$: Potassium dichromate
$KMnO_4$: Potassium permanganate
KMPT: Knudsen-molecular diffusion-Poiseuille transition
KMT: Knudsen flow molecular transition
KPP: Kimanis Power Plant
LCHA: Latex Concentrate High Ammonia
LEP: Liquid entry pressure
LEP_w: Liquid entry pressure of water
LNG: Liquefied natural gas
LPG: Liquefied petroleum gas
Ltd.: limited
MBR: Membrane bioreactor
MCPA: 2-methyl-4-chlorophenoxyacetic acid
Mg: Magnesium
Mg^{2+}: Magnesium ion
MLSS: Mixed liquor suspended solids
mMR: Microalgae membrane reactor
Mn: Manganese
$Mn(HCO_3)_2$: Manganese bicarbonate
MnO_2: Manganese dioxide
MPN: Most probable number
MPOB: Malaysian Palm Oil Board
MS: Mild steel
MSS: Municipal sewage sludge
MSW: Municipal solid waste
MWCO: Molecular weight cut-off
MWCO: Molecular weight cut-off
N: Nitrogen

Na^+: Sodium ion
Na_2HPO_4: Disodium phosphate
NaCl: Sodium chloride (salt)
NAG: Non-Assosiated Gas
NaOCl: Sodium hypochlorite
NaOH: Sodium hydroxide
NH_3: Ammonia
NH_3-N: Ammonia nitrogen
NH_4: Ammonium
NO_2: Nitrogen dioxide
NO_3: Nitrate
NO_3^-: Nitrate ion
P: Phosphorus
PAN: Polyacrilonitrile
Pb: Lead
Pb^{2+}: Lead ion
PBT: *Pihak Berkuasa Tempatan*
Pd: Palladium
PVDF: Polyvinylidene difluoride
PE: Polyethylene
PEG: Polyethylene glycol
PES: Polyether sulfone
PET: Polyethylene terephthalate
PETRONAS: Petroliam Nasional Berhad
PI: Polyimide
PM: Particulate matter
PM0.5: Particulate matter 0.5 μm
PM2.5: Particulate matter 2.5 μm
PO_4: Phosphate
PO_4^{3-}: Phosphate ion
POME: Palm oil mill effluent
POMS: Palm oil mill sludge
PP: Polypropylene
PS: Primary settling tank
PSf: Polysulfone
Pt-Co: Platinum-Cobalt
PTFE: Polytetrafluoroethylene
PVC: Polyvinyl chloride
PVDF: Polyvinylidiene fluoride
PVP: Polyvinylpyrrolidone
QCM: Quartz crystal microbalance
RAB: Rotating algal biofilm
RAS: Recirculating aquaculture system

RBC: Rotating biological contactor
RBS: Rutherford backscattering spectrometry
RDR: Rotating drum reactor
REM: Reflection electron microscopy
RSTP: Regional sewage treatment plant
RT: Retention time
RWH: Rainwater harvesting
RWS: Raw Water Standard
SAMUR: Sabah Ammonia-Urea Plant Project
SANS: Small angle neutron scattering
SAS: Small angle scattering
SAXS: Small angle x-ray scattering
SBGAST: Sabah Gas Terminal
SBR: Sequencing Batch Reactor
SBR: Sequencing batch reactor
SEM: Scanning electron microscopy
SFI: Sabah Forest Industries
SI: Système International
Si: Silicon
SiO_2: Silicon dioxide
SKTM: School of Engineering and Information Technology
SMP: Soluble microbial products
SO_4^{2-}: Sulfate ion
SOGIP: Sipitang Oil & Gas Industrail Park
SOGT: Sabah Oil and Gas Terminal
SPAN: National Water Services Commission (*Suruhanjaya Perkhidmatan Air Negara*)
SR: Specialty Rubber
SS: Stainless steel
SS (Biology): Suspended solids
SSGP: Sabah-Sarawak Gas Pipeline
STEM: Scanning transmission electron microscopy
S-TiO_2: Sulfated-titanium dioxide
TDS: Total Dissolved Solids
TEM: Transmission electron microscopy
TFC: Thin-film composite
Ti: Titanium
TiO_2: Titanium dioxide
TIPS: Thermally-induced phase separation
TMP: Transmembrane pressure
TOC: Total organic carbon
TOCl: Total organochlorine
TSS: Total suspended solids
UASB: Upflow anaerobic sludge blanket

UF: Ultrafiltration
UMS: University Malaysia Sabah
uPVC: Unplasticized polyvinyl chloride
UV: Ultraviolet
VFA: Volatile fatty acids
VOC: Volatile organic compounds
WHO: World Health Organization
WWTP: Wastewater Treatment Plant
XRD: X-ray diffraction
Zn^{2+}: Zinc ion
ZnO: Zinc oxide
ZrO_2: Zirconium dioxide

About the Author

Rosalam Sarbatly is a senior lecturer at the Universiti Malaysia Sabah (UMS) has experienced for more than 22 years as an academician. He is currently the Chairman of Sabah's Education and Innovation Minister Advisory Council. Born in a rural area in Beaufort District of Sabah, he is the youngest child of Hj. Sarbatly bin Taib and Sa'adah Mohd Tahir. Dr Rosalam obtained his PhD in the field of Chemical Engineering specializing in Membrane Technology from the University of Bath UK, Master in Environmental Engineering and Bachelor in Chemical and Process Engineering from Universiti Kebangsaan Malaysia, and Diploma in Science from ITM Perlis. His early education was at SK. Klias Kecil Beaufort and secondary school at SMK Majakir Papar. He married to Puan Maslipah Idris a quantity survey degree holder from UTM, and was blessed with three children named Fatin Arisa studying at the Yonsei University Seoul, Mohd Fadlin Zaydan and Fahim Wafi both currently studying at the Science's Boarding School in Tuaran Sabah. Before joining as a lecturer at UMS, he worked as a mill engineer at the palm oil processing mill for five years. Dr Rosalam is a professional engineer with a practicing certification registered with the Board of Engineer Malaysia in the field of Chemical Engineering. He is an expert in membrane technology and nano-fibers specifically for the synthesis and applications of the membrane in water and wastewater treatment processes, nanofibre membrane for the air filter and landfill waste storage liner, fibers for sorption of the oil spill and the sustainability processes. He is an elected member of Malaysian Membrane Society, a group scientific adviser for the Petcorp Group of Companies formerly known as Palladium Enterprise Sdn Bhd, the former Dean of Faculty Engineering UMS, and the former Executive Director of UMS Oil & Gas Sdn Bhd. During his career as a lecturer, Dr Rosalam has published over 120 journals and technical research papers and is the author of the book "Inovasi Kejuruteraan Pertanian Pintar Desa (Smart Village Agriculture Engineering Innovation)" by UMS's publisher and co-editor of Advances in Biofuels published by Springer. He actively involves in social engineering and entrepreneurship innovation related to rural economic and educational development.

Index

G

H

I

M

N

O

P

R

S

T

U

V

W

Z

www.ingramcontent.com/pod-product-compliance
Lightning Source LLC
LaVergne TN
LVHW060413300926
849162LV00006B/1418